Latent Variable Models and Factor Analysis

WILEY SERIES IN PROBABILITY AND STATISTICS

A complete list of the titles in this series can be found on http://www.wiley.com/WileyCDA/Section/id-300611.html.

Latent Variable Models and Factor Analysis

A Unified Approach

3rd Edition

David Bartholomew • **Martin Knott** • **Irini Moustaki**
London School of Economics and Political Science, UK

A John Wiley & Sons, Ltd., Publication

This edition first published 2011

Registered office
John Wiley & Sons Ltd, The Atrium, Southern Gate, Chichester, West Sussex, PO19 8SQ, United Kingdom

For details of our global editorial offices, for customer services and for information about how to apply for permission to reuse the copyright material in this book please see our website at www.wiley.com.

Library of Congress Cataloging-in-Publication Data

Bartholomew, David J.
Latent variable models and factor analysis : a unified approach. – 3rd ed. / David Bartholomew, Martin Knott, Irini Moustaki.
p. cm.
Includes bibliographical references and index.
ISBN 978-0-470-97192-5 (cloth)
1. Latent variables. 2. Latent structure analysis. 3. Factor analysis. I. Knott, M. (Martin) II. Moustaki, Irini. III. Title.
QA278.6.B37 2011
519.5′35–dc22

2011007711

A catalogue record for this book is available from the British Library.

Print ISBN: 978-0-470-97192-5
ePDF ISBN: 978-1-119-97059-0
oBook ISBN: 978-1-119-97058-3
ePub ISBN: 978-1-119-97370-6
Mobi ISBN: 978-1-119-97371-3

Set in 10/12pt Times by Aptara Inc., New Delhi, India.

Contents

	Preface	**xi**
	Acknowledgements	**xv**
1	**Basic ideas and examples**	**1**
1.1	The statistical problem	1
1.2	The basic idea	3
1.3	Two examples	4
	1.3.1 Binary manifest variables and a single binary latent variable	4
	1.3.2 A model based on normal distributions	6
1.4	A broader theoretical view	6
1.5	Illustration of an alternative approach	8
1.6	An overview of special cases	10
1.7	Principal components	11
1.8	The historical context	12
1.9	Closely related fields in statistics	17
2	**The general linear latent variable model**	**19**
2.1	Introduction	19
2.2	The model	19
2.3	Some properties of the model	20
2.4	A special case	21
2.5	The sufficiency principle	22
2.6	Principal special cases	24
2.7	Latent variable models with non-linear terms	25
2.8	Fitting the models	27
2.9	Fitting by maximum likelihood	29
2.10	Fitting by Bayesian methods	30
2.11	Rotation	33
2.12	Interpretation	35
2.13	Sampling error of parameter estimates	38
2.14	The prior distribution	39
2.15	Posterior analysis	41
2.16	A further note on the prior	43
2.17	Psychometric inference	44

3 The normal linear factor model 47
3.1 The model 47
3.2 Some distributional properties 48
3.3 Constraints on the model 50
3.4 Maximum likelihood estimation 50
3.5 Maximum likelihood estimation by the E-M algorithm 53
3.6 Sampling variation of estimators 55
3.7 Goodness of fit and choice of q 58
3.7.1 Model selection criteria 58
3.8 Fitting without normality assumptions: least squares methods 59
3.9 Other methods of fitting 61
3.10 Approximate methods for estimating Ψ 62
3.11 Goodness of fit and choice of q for least squares methods 63
3.12 Further estimation issues 64
3.12.1 Consistency 64
3.12.2 Scale-invariant estimation 65
3.12.3 Heywood cases 67
3.13 Rotation and related matters 69
3.13.1 Orthogonal rotation 69
3.13.2 Oblique rotation 70
3.13.3 Related matters 70
3.14 Posterior analysis: the normal case 71
3.15 Posterior analysis: least squares 72
3.16 Posterior analysis: a reliability approach 74
3.17 Examples 74

4 Binary data: latent trait models 83
4.1 Preliminaries 83
4.2 The logit/normal model 84
4.3 The probit/normal model 86
4.4 The equivalence of the response function and underlying variable approaches 88
4.5 Fitting the logit/normal model: the E-M algorithm 90
4.5.1 Fitting the probit/normal model 93
4.5.2 Other methods for approximating the integral 93
4.6 Sampling properties of the maximum likelihood estimators 94
4.7 Approximate maximum likelihood estimators 95
4.8 Generalised least squares methods 96
4.9 Goodness of fit 97
4.10 Posterior analysis 100
4.11 Fitting the logit/normal and probit/normal models: Markov chain Monte Carlo 102
4.11.1 Gibbs sampling 102
4.11.2 Metropolis–Hastings 105

4.11.3 Choosing prior distributions 108
4.11.4 Convergence diagnostics in MCMC 108
4.12 Divergence of the estimation algorithm 109
4.13 Examples 109

5 Polytomous data: latent trait models 119
5.1 Introduction 119
5.2 A response function model based on the sufficiency principle 120
5.3 Parameter interpretation 124
5.4 Rotation 124
5.5 Maximum likelihood estimation of the polytomous logit model 125
5.6 An approximation to the likelihood 126
5.6.1 One factor 127
5.6.2 More than one factor 130
5.7 Binary data as a special case 134
5.8 Ordering of categories 136
5.8.1 A response function model for ordinal variables 136
5.8.2 Maximum likelihood estimation of the model with ordinal variables 138
5.8.3 The partial credit model 140
5.8.4 An underlying variable model 140
5.9 An alternative underlying variable model 144
5.10 Posterior analysis 147
5.11 Further observations 148
5.12 Examples of the analysis of polytomous data using the logit model 149

6 Latent class models 157
6.1 Introduction 157
6.2 The latent class model with binary manifest variables 158
6.3 The latent class model for binary data as a latent trait model 159
6.4 *K* latent classes within the GLLVM 161
6.5 Maximum likelihood estimation 162
6.6 Standard errors 164
6.7 Posterior analysis of the latent class model with binary manifest variables 166
6.8 Goodness of fit 167
6.9 Examples for binary data 167
6.10 Latent class models with unordered polytomous manifest variables 170
6.11 Latent class models with ordered polytomous manifest variables 171
6.12 Maximum likelihood estimation 172
6.12.1 Allocation of individuals to latent classes 174
6.13 Examples for unordered polytomous data 174
6.14 Identifiability 178
6.15 Starting values 180

6.16 Latent class models with metrical manifest variables 180
6.16.1 Maximum likelihood estimation 181
6.16.2 Other methods 182
6.16.3 Allocation to categories 185
6.17 Models with ordered latent classes 185
6.18 Hybrid models 186
6.18.1 Hybrid model with binary manifest variables 186
6.18.2 Maximum likelihood estimation 187

7 Models and methods for manifest variables of mixed type 191
7.1 Introduction 191
7.2 Principal results 192
7.3 Other members of the exponential family 193
7.3.1 The binomial distribution 193
7.3.2 The Poisson distribution 194
7.3.3 The gamma distribution 194
7.4 Maximum likelihood estimation 195
7.4.1 Bernoulli manifest variables 196
7.4.2 Normal manifest variables 197
7.4.3 A general E-M approach to solving the likelihood equations 199
7.4.4 Interpretation of latent variables 200
7.5 Sampling properties and goodness of fit 201
7.6 Mixed latent class models 202
7.7 Posterior analysis 203
7.8 Examples 204
7.9 Ordered categorical variables and other generalisations 208

8 Relationships between latent variables 213
8.1 Scope 213
8.2 Correlated latent variables 213
8.3 Procrustes methods 215
8.4 Sources of prior knowledge 215
8.5 Linear structural relations models 216
8.6 The LISREL model 218
8.6.1 The structural model 218
8.6.2 The measurement model 219
8.6.3 The model as a whole 219
8.7 Adequacy of a structural equation model 221
8.8 Structural relationships in a general setting 222
8.9 Generalisations of the LISREL model 223
8.10 Examples of models which are indistinguishable 224
8.11 Implications for analysis 227

9 Related techniques for investigating dependency 229

9.1 Introduction 229

9.2 Principal components analysis 229

9.2.1 A distributional treatment 229

9.2.2 A sample-based treatment 233

9.2.3 Unordered categorical data 235

9.2.4 Ordered categorical data 236

9.3 An alternative to the normal factor model 236

9.4 Replacing latent variables by linear functions of the manifest variables 238

9.5 Estimation of correlations and regressions between latent variables 240

9.6 Q-Methodology 242

9.7 Concluding reflections of the role of latent variables in statistical modelling 244

Software appendix 247

References 249

Author index 265

Subject index 271

Preface

It is more than 20 years since the first edition of this book appeared in 1987, and its subject, like statistics as a whole, has changed radically in that period. By far the greatest impact has been made by advances in computing. In 1987 adequate implementation of most latent variable methods, even the well-established factor analysis, was guided more by computational feasibility than by theoretical optimality. What was true of factor analysis was even more true of the assortment of other latent variable techniques, which were then seen as unconnected and very specific to different applications. The development of new models was seriously inhibited by the insuperable computational problems which they would have posed. This new edition aims to take full account of these changes.

The Griffin series of monographs, then edited by Alan Stuart, was designed to consolidate the literature of promising new developments into short books. Knowing that one of us (DJB) was attempting to develop and unify latent variable modelling from a statistical point of view, he proposed what appeared in 1987 as Volume 40 in the Griffin series. Ten years later the series had been absorbed into the Kendall Library of Statistics monographs designed to complement the evergreen volumes of Kendall and Stuart's *Advanced Theory of Statistics. Latent Variable Models and Factor Analysis* took its place as Volume 7 in that series in 1999. This second edition was somewhat different in character from its predecessor, and a second author (MK) brought his particular expertise into the project. After a further decade that book was in urgent need of revision, and this could only be done adequately by recruiting a third author (IM) who is actively involved at the frontiers of contemporary research. Throughout its long history the principal aim has remained unchanged and it is worth quoting at some length from the Preface of the second edition:

> the prime object of the book remains the same – that is, to provide a unified and coherent treatment of the field from a statistical perspective. This is achieved by setting up a sufficiently general framework to enable us to derive the commonly used models, and many more as special cases. The starting point is that all variables, manifest and latent, continuous or categorical, are treated as random variables. The subsequent analysis is then done wholly within the realm of the probability calculus and the theory of statistical inference.

The subtitle, added in this edition, merely serves to emphasise, rather than modify its original purpose.

Chapter 1 covers the same ground as before, but the order of the material has been changed. The aim of the revision is to provide a more natural progression of ideas from the most elementary to the more advanced.

Chapters 2 and 3, as before, are the heart of the book. Chapter 2 provides an overall treatment of the basic model together with an account of general questions of inference relating to it. It introduces what we call the general linear latent variable model (GLLVM) from which almost all of the models considered later in the book are derived as special cases. An important new feature is an introductory account of Markov chain Monte Carlo (MCMC) methods for parameter estimation. These are a good example of the computer-intensive methods which the growth in the power of computers has made possible. In principle, these methods are now capable of handling any of the models in this book and a general introduction is given in this chapter, leaving more detailed treatment until later.

In Chapter 3 the general model is specialised to the normal linear factor model. This includes traditional factor analysis, which is probably the most thoroughly studied and widely applied latent variable model. Little directly relevant research has appeared since the second edition, but our treatment has been revised and this chapter will serve as a source for the basic theory, much of which is now embodied in computer software.

Latent trait models are widely used, especially in educational testing, but they have a far wider field of application, as the examples in Chapter 4 show. The chapter begins with two versions of the model and then discusses the statistical methods available for their implementation. Although the traditional estimation methods, based on likelihood, are efficient and are present in the standard software, we have also taken the opportunity to demonstrate the MCMC method in some detail in a situation where it can easily be compared with established methods. There is no intention here to suggest that its use is limited to such relatively simple examples. On the contrary, this example is designed to illustrate the potential of the MCMC method in a broader context.

Chapters 5 and 7 extend the ideas into newer areas, particularly where ordered categorical variables are involved. A number of the models appeared for the first time in earlier editions. This work has been consolidated here and, now that computing is no longer a barrier, they should find wider application. Latent class models are often seen as among the simpler latent variable models, and in the first edition they appeared much earlier in the book. Here they appear in Chapter 6 where it can be seen more easily, perhaps, how they fit in to the broader scheme.

Chapter 8, on relationships between latent variables, has been supplemented by an account of methods of estimation and goodness-of-fit in the LISREL model, but otherwise is unchanged, apart from the transfer to Chapter 9 of some material noted below.

Chapter 9 is entirely new except for the inclusion of a little material from the old Chapter 8 which now fits more naturally in its new setting. It draws attention to a number of methods, especially principal components analysis, which serve much the same purpose as latent variable models but come from a different statistical tradition.

The examples are an important part of the text. They are intended not only to illustrate the mechanics of putting the theory into practice but they also bring to light many subtleties which are not immediately evident from the formal derivations. This is especially important in latent variable modelling where questions of interpretation need to be explored in numerical terms for their full implications to be appreciated. Many of the original examples have been retained because, although the data on which they are based are now necessarily older, it is the point that the examples make which is important. Where we felt that these could not be bettered, they have been retained. But, in some cases, we have replaced original examples and added new ones where we felt that an improvement could be made. However, all the original examples have been recalculated using the newer software described in the Appendix.

There was a website linked to the second edition which has been discontinued. There are two reasons for this. First, we have provided an appendix to this book which gives details of the more comprehensive software that is currently available: the new appendix has removed the need for the individual programs provided on the original website. Secondly, it is now much easier to find numerical examples on which the methods can be tried out. One convenient source is in Bartholomew *et al.* (2008) and its associated website, where there are extensive data sets and some of the methods are described in a form more suitable for users.

Acknowledgements

Alan Stuart died in 1998, but his encouragement and support in getting the first edition off the ground, when latent variable models were often viewed by statisticians with suspicion, if not hostility, still leave the statistical community in his debt.

Much of the earlier editions remains, as does our debt to those who contributed to them: Lilian de Menezes, Panagiota Tzamourani, Stephen Wood, Teresa Albanese and Brian Shea, all once at the London School of Economics. Fiona Steele read a draft of the new Chapter 9 and her comments have materially helped the exposition.

The anonymous advice garnered by our publisher, John Wiley, for this edition was invaluable both in encouraging us to proceed and in defining the changes and additions we have made.

We extensively used the IRTPRO software for producing output for the factor analysis model for categorical variables. The authors of the software, Li Cai, Stephen du Toit and David Thissen, have kindly provided us with a free version of the software, and Li Cai in particular helped us resolve any software-related questions. We would also like to thank Jay Magidson and Jeroen Vermunt for their help with Latent Gold and Albert Maydeu-Olivares for sharing with us the UK data on Eysenck's Personality Questionnaire–Revised.

The material relating to Sir Godfrey Thomson's work in Chapter 9 was covered in much greater detail in a research project at the University of Edinburgh in which one of us (DJB) was a principal investigator. References to relevant publications arising from the project are included here. This project was financed as part of research supported by the Economic and Social Research Council, grant no. RES-000-23-1246.

David J. Bartholomew
Martin Knott
Irini Moustaki
London School of Economics and Political Science
January 2011

1

Basic ideas and examples

1.1 The statistical problem

Latent variable models provide an important tool for the analysis of multivariate data. They offer a conceptual framework within which many disparate methods can be unified and a base from which new methods can be developed. A statistical model specifies the joint distribution of a set of random variables and it becomes a latent variable model when some of these variables – the latent variables – are unobservable. In a formal sense, therefore, there is nothing special about a latent variable model. The usual apparatus of model-based inference applies, in principle, to all models regardless of their type. The interesting questions concern why latent variables should be introduced into a model in the first place and how their presence contributes to scientific investigation.

One reason, common to many techniques of multivariate analysis, is to reduce dimensionality. If, in some sense, the information contained in the interrelationships of many variables can be conveyed, to a good approximation, in a much smaller set, our ability to 'see' the structure in the data will be much improved. This is the idea which lies behind much of factor analysis and the newer applications of linear structural models. Large-scale statistical enquiries, such as social surveys, generate much more information than can be easily absorbed without drastic summarisation. For example, the questionnaire used in a sample survey may have 50 or 100 questions and replies may be received from 1000 respondents. Elementary statistical methods help to summarise the data by looking at the frequency distributions of responses to individual questions or pairs of questions and by providing summary measures such as percentages and correlation coefficients. However, with so many variables it may still be difficult to see any pattern in their interrelationships. The fact that our ability to visualise relationships is limited to two or three dimensions places us under strong pressure to reduce the dimensionality of the data in a manner which preserves as much of the structure as possible. The reasonableness of such a course is often

---

Latent Variable Models and Factor Analysis: A Unified Approach, Third Edition.
David Bartholomew, Martin Knott and Irini Moustaki.

evident from the fact that many questions overlap in the sense that they seem to be getting at the same thing. For example, one's views about the desirability of private health care and of tax levels for high earners might both be regarded as a reflection of a basic political position. Indeed, many enquiries are designed to probe such basic attitudes from a variety of angles. The question is then one of how to condense the many variables with which we start into a much smaller number of indices with as little loss of information as possible. Latent variable models provide one way of doing this.

A second reason is that latent quantities figure prominently in many fields to which statistical methods are applied. This is especially true of the social sciences. A cursory inspection of the literature of social research or of public discussion in newspapers or on television will show that much of it centres on entities which are handled as if they were measurable quantities but for which no measuring instrument exists. Business confidence, for example, is spoken of as though it were a real variable, changes in which affect share prices or the value of the currency. Yet business confidence is an ill-defined concept which may be regarded as a convenient shorthand for a whole complex of beliefs and attitudes. The same is true of quality of life, conservatism, and general intelligence. It is virtually impossible to theorise about social phenomena without invoking such hypothetical variables. If such reasoning is to be expressed in the language of mathematics and thus made rigorous, some way must be found of representing such 'quantities' by numbers. The statistician's problem is to establish a theoretical framework within which this can be done. In practice one chooses a variety of indicators which can be measured, such as answers to a set of yes/no questions, and then attempts to extract what is common to them.

In both approaches we arrive at the point where a number of variables have to be summarised. The theoretical approach differs from the pragmatic in that in the former a pre-existing theory directs the search and provides some means of judging the plausibility of any measures which result. We have already spoken of these measures as indices or hypothetical variables. The usual terminology is *latent variables* or *factors*. The term *factor* is so vague as to be almost meaningless, but it is so firmly entrenched in this context that it would be fruitless to try to dislodge it now. We prefer to speak of latent variables since this accurately conveys the idea of something underlying what is observed. However, there is an important distinction to be drawn. In some applications, especially in economics, a latent variable may be real in the sense that it could, in principle at least, be measured. For example, personal wealth is a reasonably well-defined concept which could be expressed in monetary terms, but in practice we may not be able or willing to measure it. Nevertheless we may wish to include it as an explanatory variable in economic models and therefore there is a need to construct some proxy for it from more accessible variables. There will be room for argument about how best to do this, but wide agreement on the existence of the latent variable. In most social applications the latent variables do not have this status. Business confidence is not something which exists in the sense that personal wealth does. It is a summarising concept which comes prior to the indicators of it which we measure. Much of the philosophical debate which takes place on latent variable models centres on *reification*; that is, on speaking as though such things as quality of life and business confidence were real entities in the sense that length and weight are. However, the usefulness and validity of the methods to be described in this book do not depend primarily on whether one adopts a realist or an instrumentalist view of latent variables. Whether one regards the latent variables as existing in some real world or merely as a means of thinking economically about complex relationships, it is possible to use the methods for prediction or establishing relationships *as if* the theory were dealing with real entities. In fact, as we shall see, some methods, which appear to be purely empirical, lead their users to behave as if they had adopted a latent variable model. We shall return to the question of interpreting latent variables at the end of Chapter 9. In the meantime we note that an interesting discussion of the meaning of a latent variable can be found in Sobel (1994).

1.2 The basic idea

We begin with a very simple example which will be familiar to anyone who has met the notion of spurious correlation in elementary statistics. It concerns the interpretation of a 2×2 contingency table. Suppose that we are presented with Table 1.1. Leaving aside questions of statistical significance, the table exhibits an association between the two variables. If A was being a heavy smoker and B was having lung cancer someone might object that the association was spurious and that it was attributable to some third factor C with which A and B were both associated – such as living in an urban environment. If we go on to look at the association between A and B in the presence and absence of C we might obtain data as set out in Table 1.2. The original association has now vanished and we therefore conclude that the underlying variable C was wholly responsible for it. Although the correlation between the manifest variables might be described as spurious, it is here seen as pointing to an underlying latent variable whose influence we wish to determine.

Even in the absence of any suggestion about C it would still be pertinent to ask whether the original table could be decomposed into two tables exhibiting independence. If so, we might then look at the members of each subgroup to see if they had anything in common, such as most of one group living in an urban environment. The idea can be extended to a p-way table and again we can enquire whether it can be decomposed into sub-tables in which the variables are independent. If this were possible there would be grounds for supposing that there was some latent categorisation which fully explained the original association. The discovery of such a

Table 1.1 A familiar example.

	A	$\bar{A}$	Total
B	350	200	550
$\bar{B}$	150	300	450
	500	500	1000

Table 1.2 Effect of a hidden factor.

	C			$\bar{C}$		
	A	$\bar{A}$	Total	A	$\bar{A}$	Total
B	320	80	400	30	120	150
$\bar{B}$	80	20	100	70	280	350
	400	100	500	100	400	500

decomposition would amount to having found a latent categorical variable for which conditional independence held. The validity of the search does not require the assumption that the goal will be reached. In a similar fashion we can see how two categorical variables might be rendered independent by conditioning on a third continuous latent variable. We now illustrate these rather abstract ideas by showing how they arise with two of the best-known latent variable models.

1.3 Two examples

1.3.1 Binary manifest variables and a single binary latent variable

We now take the argument one step further by introducing a probability model for binary data. In order to do this we shall need to anticipate some of the notation required for the more general treatment given below. Thus suppose there are p binary variables, rather than two as in the last example. Let these be denoted by $x_1, x_2, \ldots, x_p$ with $x_i = 0$ or 1 for all i. Let us consider whether the mutual association of these variables could be accounted for by a single binary variable y. In other words, is it possible to divide the population into two parts so that the xs are mutually independent in each group? It is convenient to label the two hypothetical groups 1 and 0 (as with the xs, any other labelling would serve equally well). The prior distribution of y will be denoted $h(y)$, and this may be written

$$h(1) = P\{y = 1\} = \eta \quad \text{and} \quad h(0) = 1 - h(1). \tag{1.1}$$

The conditional distribution of x_i given y will be that of a Bernoulli random variable written

$$P\{x_i \mid y\} = \pi_{iy}^{x_i}(1 - \pi_{iy})^{1-x_i} \quad (x_i, y = 0, 1), \tag{1.2}$$

where π_{iy} is the probability that $x_i = 1$ when the latent class is y. Notice that in this simple case the form of the distributions h and $P\{x_i \mid y\}$ is not in question; it is only their parameters, η, $\{\pi_{i0}\}$ and $\{\pi_{i1}\}$ which are unspecified by the model.

For this model

$$f(\mathbf{x}) = \eta \prod_{i=1}^{p} \pi_{i1}^{x_i}(1-\pi_{i1})^{1-x_i} + (1-\eta)\prod_{i=1}^{p} \pi_{i0}^{x_i}(1-\pi_{i0})^{1-x_i}. \tag{1.3}$$

To test whether such a decomposition is adequate we would fit the probability distribution of (1.3) to the observed frequency distribution of **x**-vectors and apply a goodness-of-fit test. As we shall see later, the parameters of (1.3) can be estimated by maximum likelihood. If the fit were not adequate we might go on to consider three or more latent classes or, perhaps, to allow y to vary continuously.

If the fit were satisfactory we might wish to have a rule for allocating individuals to one or other of the latent classes on the basis of their **x**-vector. For this we need the posterior distribution

$$\begin{aligned} h(1 \mid \mathbf{x}) &= P\{y = 1 \mid \mathbf{x}\} \\ &= \eta \left(\prod_{i=1}^{p} \pi_{i1}^{x_i}(1-\pi_{i1})^{1-x_i} \right) \Big/ f(\mathbf{x}) \\ &= 1 \Big/ \left[1 + \left(\frac{1-\eta}{\eta}\right) \exp \sum_{i=1}^{p} \left\{ x_i \ln \frac{\pi_{i0}}{\pi_{i1}} + (1-x_i) \ln \frac{1-\pi_{i0}}{1-\pi_{i1}} \right\} \right]. \end{aligned} \tag{1.4}$$

Clearly individuals cannot be allocated with certainty, but if estimates of the parameters are available an allocation can be made on the basis of which group is more probable. Thus we could allocate to group 1 if

$$h(1 \mid \mathbf{x}) > h(0 \mid \mathbf{x}),$$

that is, if

$$\begin{aligned} X &= \sum_{i=1}^{p} x_i \{\operatorname{logit} \pi_{i1} - \operatorname{logit} \pi_{i0}\} \\ &> \sum_{i=1}^{p} \ln\{(1-\pi_{i0})/(1-\pi_{i1})\} - \operatorname{logit} \eta. \end{aligned} \tag{1.5}$$

where $\operatorname{logit} u = \ln\{u/(1-u)\}$. An interesting feature of this result is that the rule for discrimination depends on the xs in a linear fashion. Here, this is a direct consequence of the fact that the posterior distribution of (1.4) depends on **x** only through the linear combination which we may denote by X. In that sense X contains all the relevant information in the data about the latent variable. This is not peculiar to this example but will turn out to be a key idea which is at the heart of the theoretical treatment of Chapter 2.

It is worth emphasising again that much of the arbitrariness in the general approach with which we started has been avoided by fixing the number of latent classes and hence the form of the distribution h. There might, of course, be some prior grounds for expecting two latent groups, but nothing is lost by the assumption because, if it fails, we can go on to try more.

1.3.2 A model based on normal distributions

When $\mathbf{x}$ consists of metrical variables the writing down of a model is a little less straightforward. As before, we might postulate two latent classes and then we should have

$$f(\mathbf{x}) = \eta \prod_{i=1}^{p} g_i(x_i \mid y = 1) + (1 - \eta) \prod_{i=1}^{p} g_i(x_i \mid y = 0), \tag{1.6}$$

where $g_i(x_i \mid .)$ denotes the conditional density of x_i given the particular value of y. However, we are now faced with the choice of conditional distributions for x_i. There is now no natural choice as there was when the xs were binary. We could, of course, make a plausible guess, fit the resulting model and try to justify our choice retrospectively by a goodness-of-fit test. Thus if a normal conditional distribution seemed reasonable we could proceed along the same lines as in Section 1.3.1. Models constructed in this way will be discussed in Chapter 6.

1.4 A broader theoretical view

Having introduced the basic idea of a latent variable model, we are now ready to move on to the general case where the latent variables may not be determined in either number or form. As our primary concern is with the logic of the general model we shall treat all variables as continuous, but this is purely for notational simplicity and does not affect the key idea.

There are two sorts of variables to be considered and they will be distinguished as follows. Variables which can be directly observed, also known as *manifest* variables, will be denoted by x. A collection of p manifest variables will be distinguished by subscripts and written as a column vector $\mathbf{x} = (x_1, x_2, \ldots, x_p)'$. In the interests of notational economy we shall not usually distinguish between random variables and the values which they take. When necessary, the distinction will be effected by the addition of a second subscript, thus x_{ih} will be the observed value of random variable x_i for the hth sample member and $\mathbf{x}_h$ will be that member's $\mathbf{x}$-vector. The corresponding notation for latent variables will be y and q, and such variables will form the column vector $\mathbf{y}$. In practice we shall be concerned with the case where q is much smaller than p. Since both manifest and latent variables, by definition, vary from one individual to another they are represented in the theory by random variables. The relationships between them must therefore be expressed in terms of probability distributions, so, for example, after the xs have been observed the information we have about **y** is contained in its conditional distribution given **x**. Although we are expressing the theory in terms of continuous variables, the modifications required for the other combinations of Table 1.3 on page 11 are straightforward and do not bear upon the main points to be made.

As only **x** can be observed, any inference must be based on their joint distribution whose density may be expressed as

$$f(\mathbf{x}) = \int_{R_y} h(\mathbf{y})g(\mathbf{x} \mid \mathbf{y})d\mathbf{y}, \tag{1.7}$$

where $h(\mathbf{y})$ is the prior distribution of **y**, $g(\mathbf{x} \mid \mathbf{y})$ is the conditional distribution of **x** given **y** and R_y is the range space of **y** (this will usually be omitted). Our main interest is in what can be known about **y** after **x** has been observed. This information is wholly conveyed by the conditional density $h(\mathbf{y} \mid \mathbf{x})$, deduced from Bayes' theorem,

$$h(\mathbf{y} \mid \mathbf{x}) = h(\mathbf{y})g(\mathbf{x} \mid \mathbf{y})/f(\mathbf{x}). \tag{1.8}$$

We use h for both the prior distribution and the conditional distribution, but which one is meant is always clear from the notation. The nature of the problem we face is now clear. In order to find $h(\mathbf{y} \mid \mathbf{x})$ we need to know both h and g, but all that we can estimate is f. It is obvious that h and g are not uniquely determined by (1.7) and thus, at this level of generality, we cannot obtain a complete specification of $h(\mathbf{y} \mid \mathbf{x})$. For further progress to be made we must place some further restriction on the classes of functions to be considered. In fact (1.7) and (1.8) do not specify a model, they are merely different ways of expressing the fact that **x** and **y** are random variables that are mutually dependent on one another. No other assumption is involved. However, rather more is implied in our discussion than we have yet brought out. If the xs are each related to one or more of the ys then there will be correlations among the xs. Thus if x_1 and x_2 both depend on y_1 we may expect the common influence of y_1 to induce a correlation between x_1 and x_2. Conversely if x_1 and x_2 were uncorrelated there would be no grounds for supposing that they had anything in common. Taking this one step further, if x_1 and x_2 are uncorrelated *when* y_1 *is held fixed* we may infer that no other y is needed to account for their relationship since the existence of such a y would induce a correlation even if y_1 were fixed.

In general we are saying that if the dependencies among the xs are induced by a set of latent variables **y** then when all ys are accounted for, the xs will be independent if all the ys are held fixed. If this were not so the set of ys would not be *complete* and we should have to add at least one more. Thus q must be chosen so that

$$g(\mathbf{x} \mid \mathbf{y}) = \prod_{i=1}^{p} g_i(x_i \mid \mathbf{y}). \tag{1.9}$$

This is often spoken of as the assumption (or axiom) of conditional (or local) independence. But it is misleading to think of it as an assumption of the kind that could be tested empirically because there is no way in which **y** can be fixed and therefore no way in which the independence can be tested. It is better regarded as a definition of what we mean when we say that the set of latent variables $\mathbf{y}$ is complete. In other words, that $\mathbf{y}$ is sufficient to explain the dependencies among the xs. We are asking whether $f(\mathbf{x})$ admits the representation

$$f(\mathbf{x}) = \int h(\mathbf{y}) \prod_{i=1}^{p} g_i(x_i \mid \mathbf{y}) d\mathbf{y}, \tag{1.10}$$

for some q, h and $\{g_i\}$. In practice we are interested in whether (1.10) is an adequate representation for some small value of q. The dependence of (1.10) on q is concealed by the notation and is thus easily overlooked. We do not *assume* that (1.10) holds; a key part of our analysis is directed to discovering the smallest q for which such a representation is adequate.

The treatment we have just given is both general and abstract. In the following chapter we shall propose a family of conditional distributions to take the place of $g_i(x_i \mid \mathbf{y})$ which will meet most practical needs. However, there are several points which the foregoing treatment makes very clear which have often been overlooked when we come down to particulars. For example, once $f(\mathbf{x})$ is known, or estimated, we are not free to choose h and g independently. Our choice is constrained by the need for (1.7) to be satisfied. Thus if we want to talk of 'estimating' the prior distribution $h(\mathbf{y})$, as is sometimes done, such an estimate will be constrained by the choice already made for g. Similarly, any attempt to locate individuals in the latent space using the conditional distribution $h(\mathbf{y} \mid \mathbf{x})$ must recognise the essential arbitrariness of the prior distribution $h(\mathbf{y})$. As we shall see, this indeterminacy is central to understanding what is often called the *factor scores* problem.

A more subtle point concerns the interpretation of the prior distribution, $h(y)$ itself (we temporarily restrict the discussion to the one-dimensional case). The latent variable is essentially a construct and therefore there is no need for it to exist in the ordinary sense of that word. Consequently, its distribution does not exist either and it is therefore meaningless to speak of estimating it!

1.5 Illustration of an alternative approach

In the foregoing development we constructed the model from its basic ingredients, finally arriving at the joint distribution which we can actually observe. We might try to go in the other direction starting from what we observe, namely $f(\mathbf{x})$, and deducing what the ingredients would have to be if $f(\mathbf{x})$ is to be the end-product. Suppose, for example, our sample of $\mathbf{x}$s could be regarded as coming from a multivariate normal distribution with mean vector $\boldsymbol{\mu}$ and non-singular covariance matrix $\boldsymbol{\Sigma}$. We might then ask whether the multivariate normal distribution admits a representation of the form (1.10) and, if so, whether it is unique. It is easy to find one such representation using standard results of distribution theory. Suppose, for example, that

$$\mathbf{y} \sim N_q(\mathbf{0}, \mathbf{I})$$

and

$$\mathbf{x} \mid \mathbf{y} \sim N_p(\boldsymbol{\mu} + \boldsymbol{\Lambda}\mathbf{y}, \boldsymbol{\Psi}), \tag{1.11}$$

where $\boldsymbol{\Lambda}$ is a $p \times q$ matrix of coefficients and $\boldsymbol{\Psi}$ is a diagonal matrix of variances. It then follows that

$$\mathbf{x} \sim N_p(\boldsymbol{\mu}, \boldsymbol{\Lambda}\boldsymbol{\Lambda}' + \boldsymbol{\Psi}), \tag{1.12}$$

which is of the required form. Note that although this representation works for all $q \leq p$ there is no implication in general that a $\boldsymbol{\Lambda}$ and $\boldsymbol{\Psi}$ can be found such that $\boldsymbol{\Lambda}\boldsymbol{\Lambda}' + \boldsymbol{\Psi}$ is equal to the given $\boldsymbol{\Sigma}$. Every model specified by (1.11) leads to a multivariate normal $\mathbf{x}$, but if $q < p$ the converse is not true. The point of the argument is to show that the model (1.11) is worth entertaining if the xs have a multivariate normal distribution.

The posterior distribution of $\mathbf{y}$ is easily obtained by standard methods and it, too, turns out to be normal. Thus

$$\mathbf{y} \mid \mathbf{x} \sim N_q\left(\boldsymbol{\Lambda}'\boldsymbol{\Sigma}^{-1}(\mathbf{x} - \boldsymbol{\mu}), (\boldsymbol{\Lambda}'\boldsymbol{\Psi}^{-1}\boldsymbol{\Lambda} + \mathbf{I})^{-1}\right), \tag{1.13}$$

where $\boldsymbol{\Sigma} = \boldsymbol{\Lambda}\boldsymbol{\Lambda}' + \boldsymbol{\Psi}$ and $q < p$. The mean of this distribution might then be used to predict $\mathbf{y}$ for a given $\mathbf{x}$ and the precision of the predictions would be given by the elements of the covariance matrix.

Unfortunately the decomposition of (1.11) is not unique as we now show. In the traditional approach to factor analysis this feature is met in connection with *rotation*, but the point is a general one which applies whether or not $\mathbf{x}$ is normal.

Suppose that $\mathbf{y}$ is continuous and that we make a one-to-one transformation of the factor space from $\mathbf{y}$ to $\mathbf{v}$. This will have no effect on $f(\mathbf{x})$ since it is merely a change of variable in the integral (1.7), but both of the functions h and g will be changed. In the case of h the *form* of the prior distribution will, in general, be different, and in the case of g there will be a change, for example, in the regression of $\mathbf{x}$ on the latent variables. It is thus clear that there is no unique way of expressing $f(\mathbf{x})$ as in (1.10) and therefore no empirical means of distinguishing among the possibilities. We are thus not entitled to draw conclusions from any analysis which would be vitiated by a transformation of the latent space. However, there may be some representations which are easier to interpret than others. We note in the present case, from (1.11), that the regression of x_i on $\mathbf{y}$ is linear and this enables us to interpret the elements of $\boldsymbol{\Lambda}$ as weights determining the effect of each $\mathbf{y}$ on a particular x_i. Any non-linear transformation of $\mathbf{y}$ would destroy this relationship.

Another way of looking at the matter is to argue that the indeterminacy of h leaves us free to adopt a metric for $\mathbf{y}$ such that h has some convenient form. A normal scale is familiar so we might require each y_j to have a standard normal distribution. If, as a further convenience, we make the ys independent as in (1.11) then the form of g_i is uniquely determined and we would then note that we had the additional benefit of linear regressions. This choice is essentially a matter of *calibration*; we are deciding on the properties we wish our scale to have.

In general, if we find the representation (1.10) is possible, we may fix either h or $\{g_i\}$; in the normal case either approach leads us to (1.11).

If $\mathbf{x}$ is normal there is an important transformation which leaves the form of h unchanged and which thus still leaves a degree of arbitrariness about $\{g_i\}$. This is the rotation which we referred to above. Suppose $q \geq 2$; then the orthogonal transformation $\mathbf{v} = \mathbf{My}$ ($\mathbf{M}'\mathbf{M} = \mathbf{I}$) gives

$$\mathbf{v} \sim N_q(\mathbf{0}, \mathbf{I}),$$

which is the same distribution as $\mathbf{y}$ had. The conditional distribution is now

$$\mathbf{x} \mid \mathbf{v} \sim N_p(\boldsymbol{\mu} + \boldsymbol{\Lambda}\mathbf{M}'\mathbf{v}, \boldsymbol{\Psi}), \tag{1.14}$$

so that a model with weights $\boldsymbol{\Lambda}$ is indistinguishable from one with weights $\boldsymbol{\Lambda}\mathbf{M}'$. The effect of orthogonally transforming the latent space is thus exactly the same as transforming the weight matrix. The joint distribution of $\mathbf{x}$ is, of course, unaffected by this. In the one case the covariance matrix is $\boldsymbol{\Lambda}\boldsymbol{\Lambda}' + \boldsymbol{\Psi}$ and in the other it is $\boldsymbol{\Lambda}\mathbf{M}'\mathbf{M}\boldsymbol{\Lambda}' + \boldsymbol{\Psi}$, and these are equal because $\mathbf{M}'\mathbf{M} = \mathbf{I}$.

The indeterminacy of the factor model revealed by this analysis has advantages and disadvantages. So far as determining q, the dimensionality of the latent space, is concerned, there is no problem. But from a purely statistical point of view the arbitrariness is unsatisfactory and in Chapter 3 we shall consider how it might be removed. However, there may be practical advantages in allowing the analyst freedom to choose from among a set of transformations that which has most substantive meaning. This too is a matter to which we shall return.

The reader already familiar with factor analysis will have recognised many of the formulae in this section, even though the setting may be unfamiliar. The usual treatment, to which we shall come later, starts with the linear regression implicit in (1.11) and then adds the distributional assumptions in a convenient but more or less arbitrary way. In particular, the essential role of the conditional independence postulate is thereby obscured. The advantage of starting with the distribution of $\mathbf{x}$ is that it leads to the usual model in a more compelling way but, at the same time, makes the essential arbitrariness of some of the usual assumptions clearer. We shall return to these points in Chapter 2 where we shall see that the present approach lends itself more readily to generalisation when the rather special properties of normal distributions which make the usual linear model the natural one are no longer available.

1.6 An overview of special cases

One of the main purposes of this book is to present a unified account of latent variable models in which existing methods take their place within a single broad framework.

Table 1.3 Classification of latent variable methods.

		Manifest variables	
		Metrical	Categorical
Latent variables	Metrical	Factor analysis	Latent trait analysis
	Categorical	Latent profile analysis	Latent class analysis

This framework can be conveniently set out in tabular form as in Table 1.3. The techniques mentioned there will be defined in later chapters.

It is common to classify the level of measurement of variables as nominal, ordinal, interval or ratio. For our purposes it is convenient to adopt a twofold classification: *metrical* and *categorical*. Metrical variables have realised values in the set of real numbers and may be discrete or continuous. Categorical variables assign individuals to one of a set of categories. They may be unordered or ordered; ordering commonly arises when the categories have been formed by grouping metrical variables. The two-way classification in Table 1.3 shows how the commonly used techniques are related.

It is perfectly feasible to mix types of variables – both manifest and latent. A model including both continuous and categorical *x*s has been given by Moustaki (1996) for continuous *y*s and by Moustaki and Papageorgiou (2004) for categorical *y*s , and these are described in Chapters 6 and 7. When latent variables are of mixed type we obtain what we shall later call *hybrid* models.

1.7 Principal components

We remarked above that the representation

$$\mathbf{\Sigma} = \mathbf{\Lambda}\mathbf{\Lambda}' + \mathbf{\Psi} \tag{1.15}$$

is always possible when $q = p$. This follows from the fact that $\mathbf{\Sigma}$ is a symmetric matrix and so can be expressed as

$$\mathbf{\Sigma} = \mathbf{M}'\mathbf{\Delta}\mathbf{M},$$

where $\mathbf{\Delta}$ is a diagonal matrix whose elements are the eigenvalues, $\{\delta_i\}$, of $\mathbf{\Sigma}$ and $\mathbf{M}'$ is an orthogonal matrix whose columns are the corresponding eigenvectors of $\mathbf{\Sigma}$. Consequently, if we choose

$$\mathbf{\Lambda} = \mathbf{M}'\mathbf{\Delta}^{1/2} \quad \text{and} \quad \mathbf{\Psi} = \mathbf{0},$$

then (1.15) follows. The conditional distribution of $\mathbf{x}$ given $\mathbf{y}$ in (1.11) is now degenerate with all the probability concentrated at the mean given by

$$\mathbf{x} = \boldsymbol{\mu} + \boldsymbol{\Lambda}\mathbf{y} = \boldsymbol{\mu} + \mathbf{M}'\boldsymbol{\Delta}^{1/2}\mathbf{y}, \tag{1.16}$$

with $\mathbf{x}$ having been expressed exactly as a linear combination of independent standard normal variables. The variables

$$\mathbf{y}^* = \boldsymbol{\Delta}^{1/2}\mathbf{y} = \mathbf{M}(\mathbf{x} - \boldsymbol{\mu}) \tag{1.17}$$

are known as the *principal components*. We shall return to the subject of principal components in Chapter 9 where other approaches will also be presented.

1.8 The historical context

Latent variable models and factor analysis are among the oldest multivariate methods, but their origin and development lie almost entirely outside of mainstream statistics. For this reason topics such as latent class analysis and factor analysis had separate origins and have remained almost totally isolated from one another. This means that they have been regarded as separate fields with virtually no cross-fertilisation. The distinctive feature of the treatment of latent variable models given in this book is that all are presented as special cases of the family described in Section 1.4 and Chapter 2. The statistical approach to such matters, which we have followed, is to begin with a probability model and then use standard statistical ideas to develop the theory. This approach was foreshadowed in Anderson (1959) and taken up again in Fielding (1977), but the first explicit statement, of which we are aware, of the general approach of Section 1.4 was given in Bartholomew (1980) where it was applied to categorical data. The general class of models derived from the exponential family, which underlies the general linear latent variable model of Chapter 2, was proposed in Bartholomew (1984). This, in turn, led to the first edition of the present book (Bartholomew 1987) which demonstrated that most existing latent variable models, and some new ones, could be subsumed under the general framework. The practical potential of this idea has been worked out for an even wider field of applications in Skrondal and Rabe-Hesketh (2004).

All of this means that the extensive literature, extending over almost a century, presents a disjointed picture both notationally and conceptually. It may, therefore, help the reader if we identify some of the key elements in their historical context. This task is made the more necessary by the fact that much of the development has been the work of psychologists and sociologists whose technical language often appears unfamiliar or even meaningless to statisticians. (A good example of this is provided by Maraun (1996) and the ensuing discussion.)

The origins of latent variable modelling can be traced back to the early part of the last century in psychology, notably in the study of human abilities. The origins of factor analysis, in particular, are usually attributed to Spearman (1904). However, the key idea was already present in Galton (1888) as the following quotation, drawn to our attention by John Aldrich of the University of Southampton, shows:

> Two variable organs are said to be co-related when the variation of the one is accompanied on the average by more or less variation of the other, and in the same direction.... It is easy to see that co-relation must be the consequence of the variations of the two organs being partly due to common causes. If they were wholly due to common causes, the co-relation would be perfect, as is approximately the case with the symmetrically disposed parts of the body.

Whether or not Spearman knew of this must remain a matter for speculation, but an account of Spearman's innovative role in the development of the subject can be found in Bartholomew (1995b). Spearman was concerned with the fact that people, especially children, who performed well in one test of mental ability also tended to do well in others. This led to the idea that all individuals' scores were manifestations of some underlying general ability which might be called general ability or g. However, the scores on different items were certainly not perfectly correlated, and this was explained by invoking factors specific to each item to account for the variation in performance from one item to another. The common and specific factors were thus supposed to combine in some way to produce the actual performance. This idea was expressed in what was originally known as Spearman's two-factor model. He supposed that any test score could be expressed as the sum of two parts – or factors as he would have called them – thus:

$$\text{Score} = \text{common factor} + \text{specific factor}. \tag{1.18}$$

The common factor was named g by Spearman and it appears that he regarded it as a characteristic of the brain, though he deliberately did not call it intelligence in order to avoid the contemporary connotations of that term. If (1.18) were true it could be deduced that the correlation matrix would have a particularly simple form which can be characterised in a variety of ways. One of the most useful was based on the fact that the *tetrad differences* were all zero. If a correlation matrix has typical element ρ_{ij}, the tetrad differences are given by

$$\rho_{ij}\rho_{i+h,j+k} - \rho_{i,j+k}\rho_{i+h,j} = 0,$$

for all integer values of i, j, h, k within the bounds of the matrix.

It was noted quite early by Thomson (1916) that although the two-factor theory was a sufficient condition for the tetrad differences to be zero, it was not necessary. Thomson proposed what has become known as his *bonds model* as an alternative, and we shall return to this in Chapter 9.

At the time the two-factor theory was proposed the idea of a probability model was not current among either statisticians or psychologists. However, to clothe Spearman's idea in modern dress we must translate (1.18) into a statement about the distribution of random variables. It was probably Bartlett who first recognised that the 'specific factor' of (1.18) had all the characteristics of an error term as used in statistics. From there it is a short step to write (1.18) in the form

$$x_i = \mu_i + \lambda_i y + e_i, \tag{1.19}$$

where x_i is a random variable representing the score on test i, y the common factor and e_i the factor specific to item i. (The notation has been chosen to conform to that used in the early part of this chapter.) If y and the es are mutually independent the correlation coefficient of x_i and x_j will have the form

$$\rho_{ij} = a_i a_j, \tag{1.20}$$

which is the zero tetrad condition in another guise. A a rough test of the plausibility of the model can be made by inspection of the correlation matrix. It follows from (1.20) that

$$\rho_{ih}/\rho_{ij} = a_h/a_j. \tag{1.21}$$

If we form the ratios on the left-hand side for any pair of columns of the matrix they should be constant across rows. Another way of exhibiting the pattern in the matrix is by observing that if $\rho_{ij} \neq 0$ then

$$\rho_{ih}\rho_{hj}/\rho_{ij} = a_h^2, \tag{1.22}$$

for all pairs i and j. Since

$$a_h^2 = \lambda_h^2 \operatorname{var}(y)/\{\lambda_h^2 \operatorname{var}(y) + \operatorname{var}(e_h)\},$$

it is thus possible to estimate the ratio $\lambda_h^2 \operatorname{var}(y)/\operatorname{var}(e_h)$ and so determine the relative contributions of the general and specific factors to the observed score.

If the correlation matrix fails to exhibit the simple pattern required by (1.20) it is natural to add further general factors to the model in the hope of reproducing the observed pattern more exactly.

Within psychology there was a new impetus in the 1930s from Thurstone and his associates in Chicago. He advocated the desirability of obtaining solutions possessing what he called 'simple structure', which meant replacing Spearman's single general factor by a cluster of distinct but possibly correlated factors representing different abilities. The common or general factor of earlier work was thus seen as a sort of average of the 'real' factors and thus of little substantive meaning. Much of the work in this tradition was published in the then new journal *Psychometrika*, the fiftieth anniversary of which was the occasion of a historical review by Mulaik (1986). Although this was primarily concerned with the subject as it has developed in the pages of that journal there are indications of what was happening on a broader front.

Thurstone's writings are now mainly of historical interest, but his book *Multiple Factor Analysis* (Thurstone 1947) is still worth reading for its insight into the essential character and purpose of factor analysis – something which easily becomes overlaid by technical details. The same is true of Thomson's *The Factorial Analysis of Human Ability*. The first edition of this book appeared in 1939 and the fifth and final edition in 1951 with a reprint in 1956; this anticipates many topics which have occupied post-war theorists.

A second fiftieth anniversary, this time of the *British Journal of Mathematical and Statistical Psychology* in 1997, was also marked by a special issue which contains much relevant historical material. Within the psychological tradition there has been a steady stream of texts at all levels, including Harman (1976), Mulaik (1972) and McDonald (1985). Mulaik gives a comprehensive statement of the theory as it then existed, interspersed with illuminating historical material. It is still a useful source of many of the basic algebraic derivations. The book by Harman, which ran to three editions, lays considerable emphasis on statistical methods and computational matters but with less mathematical apparatus than Mulaik or McDonald. However, Mulaik (2009a) is a second edition of the earlier book and brings the subject up to date from the psychological perspective.

A third anniversary occurred in 2004, this time marking 100 years since the publication of Spearman's path-breaking paper of 1904. This was marked by a conference at the Thurstone Laboratory at the University of North Carolina at Chapel Hill in 2004, the 16 contributions to which were published in Cudeck and MacCallum (2007). These represent a diverse set of perspectives looking forward as well as back.

A major treatment of an entirely different kind can be found in the much older book by Cattell (1978) who is at pains to emphasise the limitations of a purely statistical approach to factor analysis as against its use as part of scientific method. His exposition is made without any explicit use of models. It is worth adding that factor analysis has been widely used, mainly by psychologists, for purposes for which it was not intended. We shall note in Chapter 9 its use instead of cluster analysis and multidimensional scaling in Q-analysis. Hotelling (1957) also drew attention to other problems for which factor analysis had been used inappropriately.

The present book, like the first and second editions (Bartholomew 1987; Bartholomew and Knott 1999), lies at the opposite pole to Cattell in that it gives priority to modelling. A model serves to clarify the conceptual basis of the subject and provides a framework for analysis and interpretation. In a subject which has been criticised as arbitrary and hence too subjective, it is especially necessary to clarify in as rigourous a way as possible what is being assumed and what can be legitimately inferred. As already noted, the general approach used here goes back to Anderson (1959), but his work pre-dated the computer era and the time was not ripe for its exploitation.

The first major book-length treatment of factor analysis within the statistical tradition was that of Lawley and Maxwell (1971). The first edition appeared in 1963 and the second, larger, edition followed in 1971; this remains a valuable source of results on the normal theory factor model – especially in the area of estimation and hypothesis testing. Basilevsky (1994) is a more recent comprehensive theoretical treatment, though *Factor Analysis* in its title is used with a broader connotation than usual. Factor analysis in the sense used here is largely confined to two or three chapters.

More recently the focus of research has shifted to linear structural relations models. These originated with Jöreskog (1970) and may be regarded as a generalisation of the factor model. They incorporate not only the basic factor model but also linear relationships among the *y*s (or factors). Our general framework can be extended to include such models and an introductory account is given in Chapter 8. A key reference is Bollen (1989), but work in this area is focused on three major software packages, LISREL (Jöreskog and Sörbom 2006), Mplus (Muthén and Muthén 2010) and EQS (Bentler 2008). All these authors have made many fundamental contributions to the whole field of latent variable modelling and a selection of their relevant publications is included in the References.

Statisticians have often preferred principal components analysis to factor analysis and some have seen little value in the latter (see, for example, Chatfield and Collins (1980), Hills (1977) and Seber (1984)). The idea of this technique goes back to Pearson (1901) but it was developed as a multivariate technique by Hotelling (1933). Although, as we shall see, it is quite distinct from factor analysis in that it does not depend on any probability model, the close affinities between the two techniques outlined above mean that they are often confused.

Latent structure analysis has its origins in sociology and concerns models in which the latent variables are categorical. Its development has, until recently, been entirely separate from factor analysis. It originated with Lazarsfeld as a tool for sociological analysis and was expounded in the book by Lazarsfeld and Henry (1968). Here again modern computing facilities have greatly extended the applicability of the methods. A more up-to-date account is in Everitt (1984) and Heinen (1996). Langeheine and Rost (1988) and Rost and Langeheine (1997) contain many examples of more recent work.

A third, distinct, strand which has contributed to the current range of models comes from educational testing. The manifest variables in this case are usually categorical, often indicating whether an individual got an item right or wrong in a test. It is often assumed that there is a single continuous latent variable corresponding to an ability of some kind, and the practical objective is to locate individuals on some suitable scale. The contribution of Birnbaum (1968) was a major step forward in this field, which is now the scene of substantial research activity and some controversy. A good introductory account is given by Hambleton *et al.* (1991). Much of the controversy has centred upon the so-called Rasch model (see, for example, Rasch (1960), Andersen (1980b) or Bartholomew (1996)) which has many appealing statistical properties. However, a feature of much of this work on latent trait models, as they are called, which tends to obscure their connection with the general family set out in Section 1.4, is that the latent traits are not treated as random variables. Instead a parameter is introduced to represent each individual's position on the latent scale. Such a model may be termed a *fixed effects* model by analogy with usage in other branches of statistics. Its use would be appropriate if we were interested only in the abilities of the particular individual in the sample and not in the distribution of ability in any population from which they might have been drawn. It is not easy to find practical examples where this would be the case, and for this reason such models will receive only passing mention in the rest of the book. However, some of the later work on the Rasch model (for example, Andersen (1973, 1977) and Andersen and Madsen (1977)) treats 'ability' as a latent variable and thus falls within our territory. In Chapters 2 and 4, in particular, we shall have occasion to note the simplifications that result from specialisation to the Rasch model.

Fixed effects models have also been considered in factor analysis by Whittle (1953) and Anderson and Rubin (1956), for example, but have not attracted much practical interest. One of their chief disadvantages is that the number of parameters goes up in proportion to the sample size and this creates problems with the behaviour of maximum likelihood estimators. However, there are circumstances in which such methods are relatively simple and can be made to yield estimates of the item parameters which are virtually the same as those derived from a random effects model. They thus have a certain practical interest but in spite of a voluminous and often polemical literature they are, from our standpoint, outside the mainstream of theoretical development.

There is clearly a need for this whole area to be part of current statistical theory and practice. The combination of a unified theoretical treatment of the whole field with the flexibility and power of current computing facilities offers ample scope for new developments in many directions. Some of the gaps in the theory, which were particularly obvious when the first edition of this book (Bartholomew 1987) appeared, have been filled largely by the advent of more adequate computing facilities. But much more remains to be done, especially in introducing to new audiences the unifying ideas on which this book is based.

1.9 Closely related fields in statistics

Latent variables have a long history in statistics, but their ubiquity has been partly obscured by variations in terminology. Lee and Nelder (2009), for example, mention *random effects*, *latent processes*, *factor*, *missing data*, *unobserved future observations* and *potential outcomes*, among others. Even where there is no specific name, the idea is present. For instance, the latent class model is an example of a finite mixture model. There are many such statistical problems where an observed frequency distribution is thought to be composed of a known or unknown number of samples from populations with different distributions. A general treatment of such problems is given in Titterington *et al.* (1985). Mixture models have also been used as a basis for cluster analysis. In fact any analysis based on a latent variable model which assumes that the latent space is categorical can be regarded as a method of cluster analysis.

The idea may be illustrated further by the example of accident proneness which has received much attention in the statistical literature. If all members of a large population have the same small risk of an accident, which is constant over time, then the number of accidents per unit time will have a Poisson distribution. Often it is found that the actual frequency distribution is over-dispersed. This can be explained by supposing that the Poisson rate parameter is a random variable. Proneness is a construct which cannot be directly observed and so, in our terminology, is a latent variable. It is usual to assume that the rate has a gamma distribution and this leads to a negative binomial distribution for the number of accidents.

More recently, latent variables have been used under the name *hidden variables* especially in the context of discrete time series where Markov chains are used. This term may have been borrowed from quantum mechanics where it is used to refer to unobserved variables which can be invoked, by some, to explain the otherwise random character of quantum behaviour. If the movement of a system among a finite set of states is described by a Markov chain it may happen that the states of the chain cannot be observed directly, but only via manifest indicators. The aim will be to make inferences about the transition matrix of the underlying chain. An introduction to such models is given in Zucchini and MacDonald (2009). This work extends the scope of latent variable modelling into the domain of time series which should prove a fertile field for new developments.

Another long-standing use of latent variables, though under another name, arises in many applications in econometrics and related fields. Econometric models often allow for what is termed unobserved heterogeneity. Discussion of this will be found in the contributions of Chamberlain, Hoem, Trydman and Singer and Tuma to Heckman and Singer (1985) where further references will be found. Another source is Tuma and Hannan (1984).

It is clear that not all the types of latent variable mentioned have the same scientific status. Missing values, for example, are observable *in principle*. They may even have been observed and then lost or mislaid. With finite mixtures the position is more equivocal. We may know that we have a mixture but do not have the means of identifying the underlying populations. On the other hand, we may be exploring the data to see whether the sample could have arisen from a mixture. The latent variables with which we are concerned in this book might be better described as *hypothetical* because, like intelligence, they are constructed for a particular purpose, which is to provide a parsimonious and meaningful description of the data. However, we shall follow well-established custom by using the term *latent*,

2

The general linear latent variable model

2.1 Introduction

The principal aim of this book is to give a unified treatment of latent variable models. This is to be achieved by deriving them as special cases of what we shall call the *general linear latent variable model* (GLLVM). In a later section we shall show that there are rather compelling reasons for choosing this general model, but first we introduce it informally and use it to motivate the general treatment which follows. Having introduced the model and justified our choice we shall then use it to establish a number of properties shared by all members of the family. This is not only more economical but it gives a greater insight into the strengths and weaknesses of all latent variable models.

2.2 The model

We saw in Chapter 1 that a latent variable model consists of two parts. First there is the *prior distribution* of the latent variables represented in (1.10) by the density function $h(\mathbf{y})$. This was seen to be essentially arbitrary and, as we shall argue later, its choice is largely a matter of convention. In any event, most of the results of this chapter do not depend on the choice of $h(\mathbf{y})$ which, for most purposes, is therefore irrelevant. The second element in the model is the set of conditional distributions of the manifest variables $\{x_i\}$ given the latent variables. In (1.10) these were denoted by $g_i(x_i \mid \mathbf{y})$ $(i = 1, 2, \ldots, p)$ where the subscript i on g reminds us that the form of the distribution can vary with i.

A convenient family of distributions which turns out to have many useful properties in other branches of statistics is the *one-parameter exponential family*. If we

Latent Variable Models and Factor Analysis: A Unified Approach, Third Edition.
David Bartholomew, Martin Knott and Irini Moustaki.

suppose that the latent variables combine to determine the value of the parameter then we may write

$$g_i(x_i \mid \theta_i) = F_i(x_i)G_i(\theta_i)\exp(\theta_i u_i(x_i)) \quad (i = 1, 2, \ldots, p), \tag{2.1}$$

where θ_i is some function of $\mathbf{y}$. The simplest assumption about the form of this function is to suppose that it is a linear function, in which case we have

$$\theta_i = \alpha_{i0} + \alpha_{i1}y_1 + \alpha_{i2}y_2 + \cdots + \alpha_{iq}y_q \quad (i = 1, 2, \ldots, p). \tag{2.2}$$

This is the general linear latent variable model. The term *linear* refers to its linearity in the αs for reasons which will appear shortly.

The similarity between this model and the so-called generalised linear model (GLIM) used in statistics will be immediately apparent. The chief difference is that here we have a set of xs rather than a single dependent variable and here the ys are unobservable. In view of this connection it may seem somewhat perverse of us to have used x to denote the dependent variable and y the independent (or regressor) variables when the usual convention is to reverse their roles. The reason for this will become apparent when we come to discuss the posterior analysis of the model. In that context we shall be predicting the ys given the xs, in which case our notation conforms to the standard usage.

The exponential family of (2.1) includes the normal, gamma and Bernoulli distributions as special cases. If we allow x_i and θ to be vector-valued it also includes the multinomial distribution which we shall need for our treatment of categorical variables. However, we continue for the moment with the scalar version in order not to obscure the essential simplicity of the model by mathematical complications.

2.3 Some properties of the model

We noted in Chapter 1 that all the information about the latent variables in the vector $\mathbf{x}$ is conveyed by the posterior distribution with density

$$h(\mathbf{y} \mid \mathbf{x}) = h(\mathbf{y}) \prod_{i=1}^{p} g_i(x_i \mid \mathbf{y}) \Big/ f(\mathbf{x}). \tag{2.3}$$

Substituting for $g_i(x_i \mid \mathbf{y})$ as given by (2.1), we find

$$\begin{aligned} h(\mathbf{y} \mid \mathbf{x}) &= \frac{h(\mathbf{y}) \left[\prod_{i=1}^{p} F_i(x_i)G_i(\theta_i)\right] \exp \sum_{i=1}^{p} \theta_i u_i(x_i)}{\prod_{i=1}^{p} \int h(\mathbf{y}) \left[F_i(x_i)G_i(\theta_i)\right] \exp \sum_{i=1}^{p} \theta_i u_i(x_i) d\mathbf{y}} \\ &\propto h(\mathbf{y}) \left[\prod_{i=1}^{p} G_i(\theta_i)\right] \exp \sum_{j=1}^{q} y_j X_j, \end{aligned} \tag{2.4}$$

where $X_j = \sum_{i=1}^{p} \alpha_{ij} u_i(x_i)$ $(j = 1, 2, \ldots, q)$. The first important thing to notice here is that the posterior distribution of $\mathbf{y}$ depends on $\mathbf{x}$ only through the q-dimensional vector $\mathbf{X}' = (X_1, X_2, \ldots, X_q)$. This has dimension precisely equal to that of $\mathbf{y}$. In the Bayesian sense $\mathbf{X}$ is a *minimal sufficient statistic* for $\mathbf{y}$. It represents a reduction in dimensionality from the p-vector $\mathbf{x}$ to the q-vector $\mathbf{X}$. We shall call these sufficient statistics *components*. They are not *principal components* but they will play a similar role in our treatment to principal components which they include as a special (degenerate) case of one of our models. The second important thing to notice is that the reduction effected by (2.4) does not depend on the prior distribution of the latent variables. This is reassuring in view of the arbitrariness in the choice of $h(\mathbf{y})$ and it demonstrates that the reduction of dimensionality is achieved, without loss of information, and without any reference to the distribution of the latent variables.

2.4 A special case

Although we shall consider various special cases in detail later in this chapter and in other chapters, it may help to fix ideas if we give a simple example now. Suppose that x_i is a Bernoulli random variable arising from putting a proposition to people and inviting them to agree or disagree. If the answers are coded 1 (agree) or 0 (disagree) the conditional distribution of x_i may be written

$$\begin{aligned} g(x_i \mid y) &= \pi_i^{x_i}(1 - \pi_i)^{1-x_i} \quad (x_i = 0 \text{ or } 1) \\ &= (1 - \pi_i)\exp\{x_i \operatorname{logit}(\pi_i)\}, \end{aligned}$$

where π_i is the probability of agreeing. Comparing this with (2.1), we see that $\theta_i = \operatorname{logit} \pi_i = \ln\{\pi_i/(1 - \pi_i)\}$, $G_i(\theta_i) = 1 - \pi_i$ and $u_i(x_i) = x_i$. The GLLVM is thus

$$\operatorname{logit}(\pi_i) = \alpha_{i0} + \sum_{j=1}^{q} \alpha_{ij} y_j, \tag{2.5}$$

and the components are given by $X_j = \sum_{i=1}^{p} \alpha_{ij} x_i$. The logit here is known as the *link function*. In the special case $q = 1$ this reduces to the familiar logistic latent trait model with *response function*

$$\pi_i(y) = \frac{1}{1 + e^{-\alpha_{i0} - \alpha_{i1} y}}.$$

The model tells us that if we have a collection of scores $\{x_i\}$ all conforming to this model then, if there is a single latent dimension underlying them, all that we can learn about that latent variable is summed up in the linear combination of the responses given by $X = \sum_{i=1}^{p} \alpha_{i1} x_i$. This is a remarkable result which is not true for an alternative model which is often used in this context and which we discuss in Section 4.3.

This example also serves to highlight other questions which still have to be answered. The parameters $\{\alpha_{ij}\}$ are, of course, unknown, and will have to be estimated from the data. Any method, and maximum likelihood in particular, will depend on the joint distribution of the xs. Equation (1.10) shows that this does depend on the prior distribution. Similarly, q is unknown and will have to be estimated. Although X is sufficient for y, the theory so far tells us nothing about its relationship with y. If, for example, one individual scores higher than another on X, what can we say about their ranking on the y scale?

A final point to note is that there is nothing in the theory which requires y to be a continuous variable. To take an extreme case, suppose $q = 1$ and that y is binary, taking one of the two values (0, 1). We would then have a simple example of a *latent class* model with two latent classes. Having constrained y to just two values in this way the prior distribution is no longer completely arbitrary but must consist of two probability masses located at $y = 0$ and $y = 1$. All that remains arbitrary in this case is the location and scale of y. This illustrates in a simple manner the intimate relationship which exists between the prior distribution and the linear model.

2.5 The sufficiency principle

The attractive feature of the GLLVM is that it leads to a q-dimensional sufficient statistic which, in a precise sense, can be used as a proxy for the unobservable $\mathbf{y}$. Our choice of this model is thus justified retrospectively, but it is not clear from our analysis whether this is the only family of models which admits this reduction. We now show that there is a larger class of models which lead to a minimal sufficient statistic. In determining this class we shall provide a more satisfactory basis for using the GLLVM.

Let us begin with the idea of dimension reduction. If we wish to construct a model with q latent variables the best we can hope for is to find q functions of $\mathbf{x}$, $X_1, X_2, \ldots, X_q$, say, such that the conditional distribution given $\mathbf{X}$ does not depend on $\mathbf{y}$. Then $\mathbf{X}$ is a minimal sufficient statistic. The first question to ask, therefore, is whether there is a class of conditional distributions $g_i(x_i \mid \mathbf{y})$ $(i = 1, 2, \ldots, p)$ for which such a minimal sufficient set exists. The answer has been provided by Barankin and Maitra (1963, Theorems 5.1, 5.2 and 5.3). They give necessary and sufficient conditions which amount to the requirement that, subject to weak regularity conditions, at least $p - q$ of the g_i shall be of exponential type with

$$g_i(x_i \mid \mathbf{y}) = F_i(x_i)G_i(\mathbf{y}) \exp \sum_{j=1}^{q} u_{ij}(x_i)\phi_j(\mathbf{y}). \tag{2.6}$$

This is more general than (2.1). For instance, taking all g_i to be of the form (2.6), the posterior distribution is easily found to be

$$h(\mathbf{y} \mid \mathbf{x}) = \frac{h(\mathbf{y}) \prod_{i=1}^{p} [F_i(x_i)G_i(\mathbf{y})] \exp \sum_{j=1}^{q} X_j \phi_j(\mathbf{y})}{\int h(\mathbf{y}) \prod_{i=1}^{p} [F_i(x_i)G_i(\mathbf{y})] \exp \sum_{j=1}^{q} X_j \phi_j(\mathbf{y}) d\mathbf{y}}, \tag{2.7}$$

where $X_j = \sum_{i=1}^{p} u_{ij}(x_i)$ $(j = 1, 2, \ldots, q)$, yielding $\mathbf{X}$ as the minimal sufficient statistic. It should be noted that the functions $\{u_{ij}(x)\}$ may involve unknown parameters and then, in speaking of X_j as a statistic, we are extending the conventional usage according to which sufficient statistics are functions of random variables only. In effect, we are treating any such parameters as known. The term 'sufficient' has exactly the meaning we wish to convey and so we will use it in this extended sense.

Special cases of our general result when the xs are binary go back at least to Birnbaum (1968). From our perspective his object was to determine the class of models for which the minimal sufficient statistic was a linear combination of the manifest variables. Andersen (1973) similarly asked what class of model yields a statistic which is parameter-free. In our notation this means that the functions $\{u_{ij}(x_i)\}$ of (2.7) involve no unknown parameters. The answer to this question led to the Rasch model for which the sufficient statistic is the total score. Later Andersen (1977) extended the result to polytomous xs.

The GLLVM of (2.2) arises as a special case when $u_{ij}(x_i) = \alpha_{ij} u_i(x_i)$, in which case what is often called the natural parameter of the exponential family is a linear function of the latent variables. In order to see what further scope the model offers, consider the following simple example. Let $q = 2$ and suppose that

$$x_i \mid \mathbf{y} \sim N(y_1, y_2).$$

This may be written in the form (2.6) by setting

$$F_i(x_i) = 1, \quad G_i(\mathbf{y}) = y_2^{-1/2} \exp\left\{-\frac{1}{2}\frac{y_1^2}{y_2}\right\},$$

$$u_{i1}(x_i) = x_i, \quad u_{2i}(x_i) = x_i^2, \quad \phi_1(\mathbf{y}) = y_1/y_2, \quad \phi_2(\mathbf{y}) = -1/(2y_2).$$

The two sufficient statistics are

$$X_1 = \sum_{i=1}^{p} x_i, \quad X_2 = \sum_{i=1}^{p} x_i^2.$$

The more general model therefore allows us to have the two parameters of the normal distribution depending simultaneously on different latent variables. Given the difficulties of handling one-parameter models which we shall encounter, there seem likely to be severe practical problems in implementing the many-parameter generalisations permitted by the more general result of (2.6). For the remainder of the book we shall therefore confine the discussion to the general linear latent variable model of (2.2).

2.6 Principal special cases

The bulk of the work on latent variable models is confined to a few special cases of the GLLVM. It is convenient to introduce them at this point in order to make the link between the general theory of this chapter and the detailed treatment to be given in subsequent chapters. The classification of models in Table 1.3 provides the starting point.

The latent trait model has already been illustrated in Section 2.4. It was obtained by supposing that each of the xs was a Bernoulli random variable for which the probability of a positive response depended on a single latent variable. In practice this latent variable is often supposed to represent a human ability or attitude. The sufficiency principle shows that to achieve the full dimensional reduction, the logit of the response probability must be a linear function of the latent variable. The traditional model generalises immediately to one with q latent variables when we have

$$\operatorname{logit} \pi_i(\mathbf{y}) = \alpha_{i0} + \sum_{j=1}^{q} \alpha_{ij}\, y_j. \tag{2.8}$$

This model is the subject of Chapter 4 where it will be compared with other models which do not achieve the reduction in dimensionality without loss of information.

If x_i is a categorical variable with more than two categories a further generalisation is required. This will be given in Chapter 5 and is achieved by allowing x_i to be a vector-valued indicator variable. Thus, if there are c_i categories, x_i would consist of $c_i - 1$ zeros and a single 1 occurring in the position corresponding to the category in which the individual falls.

The latent class model can be thought of as a special kind of latent trait model in which the latent variable is categorical. As shown in Section 2.4, if we have a single binary latent variable, we arrive at a two-class latent class model. By allowing $\mathbf{y}$ to be a vector-valued indicator variable we can introduce more latent classes into the model. Latent class models will be treated in Chapter 6.

The remaining two cells of the Table 1.3 relate to models in which the xs are continuous. The distinction between a latent profile and a factor model lies in the form of the latent variable. The latent profile model can be thought of as a factor model with a discrete (multivariate) prior distribution. (As we shall see later in Section 6.16.2, this fact leads to serious problems of identifiability.) For this reason we shall not give an extended treatment to latent profile models.

Many variants of the factor model are possible. The one in common use is discussed in Chapter 3 and is known as the *normal linear factor model*. We need to take some care in deriving it from the GLLVM because the normal distribution which is used for the conditional distribution is a two-parameter family and it can be parameterised to give two distinct models.

We suppose first that

$$x_i \mid \mathbf{y} \sim N(\mu_i, \sigma_i^2)$$

or

$$g_i(x_i \mid \mathbf{y}) = \frac{1}{\sqrt{2\pi}\sigma_i} \exp\left\{-\frac{1}{2}\left(\frac{x_i^2}{\sigma_i^2} - \frac{2x_i\mu_i}{\sigma_i^2} + \frac{\mu_i^2}{\sigma_i^2}\right)\right\}. \tag{2.9}$$

If σ_i is known this may be written in the required form by setting

$$\theta_i = \mu_i/\sigma_i, \quad u_i(x_i) = x_i/\sigma_i, \quad G_i(\theta_i) = e^{-\frac{1}{2}\frac{\mu_i^2}{\sigma_i^2}} \quad F_i(x_i) = \frac{1}{\sqrt{2\pi}\sigma_i} \exp\left\{-\frac{1}{2}\frac{x_i^2}{\sigma_i^2}\right\}.$$

The GLLVM is then

$$\frac{\mu_i}{\sigma_i} = \alpha_{i0} + \sum_{j=1}^{p} \alpha_{ij} y_j \quad (i = 1, 2, \ldots, p),$$

according to which the latent variables are supposed to influence the mean only and in an additive fashion. If the ys are assumed to be normal this is the standard linear normal model referred to above. The sufficient statistics are

$$X_j = \sum_{i=1}^{p} \alpha_{ij} \frac{x_i}{\sigma_i} \quad (j = 1, 2, \ldots, q).$$

An alternative model arises if we treat μ_i as known. The density of (2.9) may then be written in the exponential form with

$$\theta_i = \frac{1}{\sigma_i^2}, \quad u_i(x_i) = -\frac{1}{2}(x_i - \mu_i)^2, \quad G_i(\theta_i) = \frac{\sqrt{\theta_i}}{\sqrt{2\pi}}, \quad F_i(x_i) = \text{constant}.$$

The GLLVM is then a linear model for the inverse of the variance (the precision) and the sufficient statistics are

$$X_j = -\frac{1}{2}\sum_{i=1}^{p} \alpha_{ij}(x_i - \mu_i)^2 \quad (j = 1, 2, \ldots, q). \tag{2.10}$$

This model would enable us to investigate the effect of latent variables on the variance rather than the mean. The link function would have to be chosen so that θ_i was always positive. No study of models of this kind appears to have been made but it serves as a further example to show the range of models encompassed by our general approach.

2.7 Latent variable models with non-linear terms

It is worth pointing out that since nothing has been assumed so far about the prior distribution there is nothing to prevent some of the latent variables being functionally related (polynomial terms of latent variables or interactions among latent variables). This fact enables us to include a variety of models which are non-linear in the *y*s, in particular polynomial or interaction models of the kind discussed by McDonald (1962, 1967a,b,c, 1985) and Etezadi-Amoli and McDonald (1983). The examples given are the typical middle item of an attitude scale and variables that measure the intensity of attitude that are known to have a U-shape relation to anxiety. They were viewed from the present angle in Bartholomew (1985) and Bartholomew and McDonald (1986) but without taking full advantage of the flexibility which the present approach offers, as the work by Rizopoulos and Moustaki (2008) shows. Kenny and Judd (1984) also modelled latent variable interaction terms under the structural equation modelling (SEM) framework. In their formulation, products of the observed variables are used as indicators for the non-linear terms of the model. Their paper led to many methodological papers discussing several aspects of non-linear latent variable modelling mainly under the SEM approach – in particular, Jöreskog and Yang (1996), Yang Jonsson (1997), Arminger and Muthén (1998), Bollen (1995), Bollen and Paxton (1998) and Wall and Amemiya (2000, 2001). Bayesian estimation methods have also been proposed by Zhu and Lee (1999), Arminger and Muthén (1998), Lee and Zhu (2002), Lee and Song (2004a,b) and Song and Lee (2004, 2006). Applications of latent variable models with non-linear terms in different disciplines such as marketing, social psychology and political theory can be found in Schumacker and Marcoulides (1998) and references therein.

Bartlett (1953), McDonald (1983), Etezadi-Amoli and McDonald (1983) and Bartholomew (1985) have suggested graphical ways (bivariate plots of factor scores and components) of investigating the necessity of non-linear terms. In addition, Bartlett (1953) pointed out that the inclusion of the interaction between two latent variables in the linear factor analysis model will produce the same correlation properties of the manifest variables as if a third genuine factor had been included. We illustrate the point by considering a one-factor model with σ_i^2 known in which

$$x_i \mid y \sim N(\mu_i + \alpha_i y + \beta_i y^2, \sigma_i^2) \quad (i = 1, 2, \ldots, p).$$

So far as finding the sufficient statistics is concerned, this is equivalent to a two-factor model with latent variables $y_1 = y$ and $y_2 = y^2$. The components are then

$$X_1 = \sum_{i=1}^{p} \alpha_i x_i / \sigma_i$$

and

$$X_2 = \sum_{i=1}^{p} \beta_i x_i / \sigma_i.$$

A two-dimensional statistic is needed to fix the path of the quadratic mean in the one-dimensional latent space.

The GLLVM can be written to include non-linear terms in its parameter θ_i:

$$\theta_i = \alpha_{i0} + \boldsymbol{\alpha}_i' G(\mathbf{y}) \quad (i = 1, \ldots, p), \tag{2.11}$$

where $G(\mathbf{y})$ is a vector function of $\mathbf{y}$ that can allow for any relationship between the manifest and latent variables with an associated item-specific parameter vector $\boldsymbol{\alpha}_i$. For example, for the model with two latent variables and an interaction term, $G'(\mathbf{y}) = (y_1, y_2, y_1 y_2)$. Although non-linear terms increase the complexity of the exponential parameter θ, their inclusion might be necessary to describe more complex relationships between the latent and observed variables. The current formulation has certain numerical advantages. First, the model is based on conditional distributions of observed variables given the latent variables that keep their form within the known distributions of the exponential family of (2.1). Second, in the case of a two-factor model with non-linear terms, the marginal distribution of $\mathbf{x}$ is computed using a double integral instead of a higher-order integration, while allowing for more complex latent structures compared with a two-factor model. The model with non-linear terms can be fitted with maximum likelihood where the likelihood function is given in (2.19).

Regarding the estimation of standard errors, it has been recognised (Busemeyer and Jones 1983) that the inclusion of non-linear terms could lead to potential underestimation. Rizopoulos and Moustaki (2008) propose a robust estimation for the standard errors of $\widehat{\boldsymbol{\alpha}}$ based on the sandwich estimator (see also White (1982) for more details):

$$\widehat{\mathrm{var}(\widehat{\boldsymbol{\alpha}})} = C(\widehat{\boldsymbol{\alpha}}) = H^{-1}(\widehat{\boldsymbol{\alpha}}) K(\widehat{\boldsymbol{\alpha}}) H^{-1}(\widehat{\boldsymbol{\alpha}}), \tag{2.12}$$

where $H(\widehat{\boldsymbol{\alpha}})$ is the observed information matrix evaluated at the maximum likelihood estimates, and

$$K(\widehat{\boldsymbol{\alpha}}) = D(\widehat{\boldsymbol{\alpha}}) D(\widehat{\boldsymbol{\alpha}})^T,$$

with $D(\widehat{\boldsymbol{\alpha}})$ denoting the score vector evaluated at the maximum likelihood estimates.

At this point the reader who is interested in a particular model can proceed to the appropriate chapter. However, since many of the questions which will arise to do with fitting, rotation and so on are common to most models they will be investigated in the remainder of this chapter, which will thus serve also as a point of reference for later use.

2.8 Fitting the models

Fitting the models includes estimating the parameters and testing goodness of fit. For this purpose some assumptions must be made about the prior distribution $h(\mathbf{y})$. However, considerable progress can be made with some of the models using methods which may be described as distribution-free.

We note first that, without loss of generality, the ys may be assumed to have zero means and unit standard deviations. Any shift in location can be absorbed into the constant, α_{i0}, and any change of scale into the α_{ij} $(j > 0)$. For the present we assume the ys to be independent because this has advantages when we come to the matter of interpretation.

The distribution-free method depends on the following result. Suppose that we can find a transformation $\xi(x_i)$ of x_i such that

$$E\{\xi(x_i) \mid \mathbf{y}\} = \theta_i = \alpha_{i0} + \sum_{j=1}^{q} \alpha_{ij} y_j \quad (i = 1, 2, \ldots, p).$$

Then

$$\begin{aligned}\operatorname{cov}\{\xi(x_u), \xi(x_v)\} &= E\left[E\left\{\xi(x_u)\xi(x_v) \mid \mathbf{y}\right\}\right] - E\xi(x_u)E\xi(x_v) \quad (u \neq v)\\ &= \sum_{j=1}^{q} \alpha_{uj}\alpha_{vj}\end{aligned}$$

and

$$\operatorname{var}(\xi(x_i)) = \sum_{j=1}^{q} \alpha_{ij}^2 + E \operatorname{var}\{\xi(x_i) \mid \mathbf{y}\}. \tag{2.13}$$

In matrix notation we may thus express the covariance matrix as

$$D(\boldsymbol{\xi}) = \mathbf{A}\mathbf{A}' + \mathbf{V}, \tag{2.14}$$

where $\mathbf{A} = \{\alpha_{ij}\}$ and $\mathbf{V}$ is a diagonal matrix of variances given by (2.13).

If this is possible, we can estimate the parameters by minimising the distance between the theoretical covariance matrix, given by (2.14), and the sample matrix estimated from the data matrix. Methods for doing this are available and will be reviewed in the discussion of the normal linear model in Chapters 3 and 8.

The generality of this method depends, of course, on whether a suitable transformation can be found. In the case of the linear factor model where $x_i \sim N(\mu_i, \sigma_i^2)$ it follows at once that $E(x_i \mid \mathbf{y}) = \mu_i = \theta_i$, so that no transformation is needed.

If the conditional distribution of x_i has the gamma form with

$$g_i(x_i \mid \theta_i) = \theta_i^{\nu_i} x_i^{\nu_i - 1} \exp\{-x_i\theta_i\}/\Gamma(\nu_i), \tag{2.15}$$

then, provided that $\nu_i > 1$, $E(1/x_i) = \theta_i$ and so we proceed to use the reciprocals of the xs. (See Section 7.3.3 for further treatment of the gamma distribution.)

Even if there is no such transformation, as when $\nu_i = 1$ in the foregoing example, it may still be possible to find an approximation. One method of doing so utilises the fact that for any distribution of exponential type

$$E\{u_i(x_i) \mid \theta_i\} = -\frac{d}{d\theta_i} \ln G_i(\theta_i) = T(\theta_i), \tag{2.16}$$

say. This suggests trying the transformation

$$\xi(x_i) = T^{-1}\{u_i(x_i)\},$$

or some modification of it.

As an illustration, consider the case when x_i is a Poisson variable with mean m_i (see also Section 7.3.2). The natural parameter is $\theta_i = \ln m_i$, $u_i(x_i) = x_i$ and $T(\theta_i) = \exp \theta_i$, which suggests the transformation $\xi(x_i) = \ln x_i$. However, $\ln x_i$ has infinite expectation, but if we take

$$\xi(x_i) = \ln(x_i + c)$$

to avoid this, we find that with $c = \frac{1}{2}$,

$$E\left\{\ln\left(x_i + \frac{1}{2}\right)\middle| y\right\} = \theta_i + O(m_i^{-2}). \tag{2.17}$$

Thus, provided that the means are large enough, we could fit the model by least squares using the covariances of the transformed variables.

A similar argument for binomial random variables when x_i has parameters n_i and p_i yields the transformation

$$\xi(x_i) = \log\left\{\frac{x_i + \frac{1}{2}}{n_i - x_i - \frac{1}{2}}\right\}, \tag{2.18}$$

which requires n_i to be reasonably large. A very important special case which cannot be handled by this means occurs when $n_i = 1$ for all i. This requires the special treatment given in Chapter 4.

2.9 Fitting by maximum likelihood

All models can, in principle, be fitted by maximum likelihood. This method also provides asymptotic standard errors of the parameter estimates and a global test of goodness of fit. The likelihood function is formed from the joint distribution of the xs which, for the one-parameter exponential family given in (2.1), is

$$f(\mathbf{x}) = \int h(\mathbf{y}) \left[\prod_{i=1}^{p} F_i(x_i) G_i(\theta_i)\right] \exp\left(\sum_{i=1}^{p} u_i(x_i)\theta_i\right) d\mathbf{y}, \tag{2.19}$$

where $\theta_i = \alpha_{i0} + \sum_j^q \alpha_{ij} y_j$ $(i = 1, 2, \ldots, p)$. When dealing with individual special cases we shall discover that the choice of $h(\mathbf{y})$ seems to have very little effect on the parameter estimates. This suggests that the choice is not critical. The following heuristic argument provides some further support for this.

The marginal distribution of any particular x_i may be written

$$f(x_i) = F_i(x_i)E\{G_i(\theta_i)\exp(u_i(x_i)\theta_i)\},$$

where the expectation is taken with respect to the distribution of θ_i which, of course, is a linear function of $\mathbf{y}$. Hence if q is not too small θ_i will be approximately normal, by the central limit theorem, whatever the distribution of $\mathbf{y}$. Thus the marginal distributions will be close to what they would be if the ys were normal. The near normality of the univariate margins does not, of course, guarantee the approximate multivariate normality of $\mathbf{x}$ but one would expect them to be close.

2.10 Fitting by Bayesian methods

In a similar manner, all models can be fitted using Bayesian estimation methods. Bayesian inference is based on the posterior distribution of the vector of all random parameters conditionally on the data. First and second moments or percentiles of the posterior distribution are used for computing point estimates and their corresponding standard deviations. Bayesian methods typically require computer-intensive techniques that can transform a mathematically intractable problem of finding the posterior distribution into a computationally feasible one. For parameter spaces of high dimension, Bayesian inference uses methods such as Markov chain Monte Carlo (MCMC). Monte Carlo integration is a long-established technique for numerically approximating high-dimensional integrals by drawing values from a desired probability distribution. MCMC methods are more sophisticated methods for achieving the same goal by constructing chains of those sampled values. One can then obtain estimates as expectations, or modes, calculated from their empirical distribution. The basic idea depends upon our being able to find a Markov chain defined on the set of states, comprising all possible values of the unknown quantities, which has as its steady-state distribution the desired posterior distribution. A single run gives one sample value; repeated simulations will give a sample for the desired distribution. The key question is how to choose transition probabilities from the chain (the kernel).

More specifically, MCMC methods are iterative algorithms that aim to construct a Markov chain that can be used as a sample from a desired probability distribution. The term 'Markov chain' is due to the fact that results from the current iteration are used for the next iteration, resulting in a chain. MCMC methods avoid the complexity of computing multiple integrals often required in the computation of posterior distributions by taking draws directly from that distribution function. They are approximate methods that might require a large number of steps before the values sampled are from the desired distribution. MCMC methods can be seen as an alternative to marginal maximum likelihood estimation shortly to be discussed in Chapters 3 to 7. Marginal maximum likelihood estimation and MCMC methods have been compared and found to give close if not identical parameter estimates (Kim 2001; Moustaki and Knott 2005; Wollack *et al.* 2002).

MCMC methods, such as the Metropolis–Hastings algorithm and its special case, Gibbs sampling, (Gelfand and Smith 1990) have become popular for estimating the parameters of latent variable models mainly because they simplify the estimation of complex models; see, for example Tanner and Wong (1987), Albert (1992), Albert and Chib (1993), Baker (1998), Johnson and Albert (1999), Patz and Junker (1999a,b), Dunson (2000, 2003), Béguin and Glas (2001), Fox and Glas (2001) and Shi and Lee (1998). Lee and Song (2003), Lee (2007) and Palomo *et al.* (2007) have also proposed Bayesian estimation methods for structural equation models for continuous and categorical responses. Lopez and West (2004) also looked at the problem of estimating the number of factors in the classical factor analysis model using reversible jump MCMC algorithms. Earlier, researchers such as Birnbaum (1969) and Owen (1975) proposed Bayes estimates of ability parameters in an item response model under the assumptions that the item parameters are known.

In a Bayesian estimation model, parameters are not considered fixed but are stochastic random variables and so probability statements are made about those parameters. The model is defined as the joint distribution over all unobserved (parameters) and observed quantities (data). The way information is obtained about the unobserved quantities is through the posterior distribution of the unobserved quantities conditional on the data. The marginalisation of the posterior distribution with respect to individual parameters of interest is done through Gibbs sampling (Gelfand and Smith 1990) or the Metropolis–Hastings algorithm. We will consider both those sampling techniques for factor analysis models. Albert (1992) applied Gibbs sampling for obtaining parameter estimates of the two-parameter probit/normal model, whereas Patz and Junker (1999b) used the Metropolis–Hastings algorithm for estimating a family of item response theory models. Gibbs sampling is easier to implement when one can obtain closed-form expressions for the conditional densities of the model parameters. This is feasible in the probit/normal model as shown in Albert (1992) but not in other item response theory models. To overcome the limitation of Gibbs sampling one can use the Metropolis–Hastings algorithm within Gibbs sampling. The Metropolis–Hastings algorithm introduces an efficient acceptance/rejection sampling at each Gibbs step that helps to sample from the 'complete' conditional densities.

Bayesian estimation has two main advantages. First, it provides information for the whole distribution of the model parameters and latent variables; in particular, the standard errors of the parameters. Secondly, it is appropriate when one wants to incorporate prior information about the parameters (e.g. imposing limits on the range of values for the factor loadings or for the variances of the error term in an attempt to avoid Heywood cases; see also Section 3.12.3). However, the implications of rotation (see Section 2.11) and label switching have not been thoroughly investigated under the Bayesian framework and workable solutions still need to be given. As with rotation, any random permutation of the set of factors leaves the joint distribution of the observed variables unchanged, which often leads to multimodal posterior distributions for the parameters of interest. To our knowledge no research has been published that addresses label switching in latent variable models. For the classical factor analysis model, restrictions similar to those applied for model identifiability have usually been adopted as a practice for avoiding label switching.

A sketch of estimation in the Bayesian framework

Let us denote by $\mathbf{v}' = (\mathbf{y}', \boldsymbol{\alpha}')$ the vector with all the unknown parameters including now the vector of latent variables, where the vector $\boldsymbol{\alpha}$ contains all the parameters that appear in the parameter θ_i of (2.2) for all i. The joint posterior distribution of the parameter vector $\mathbf{v}$ is

$$h(\mathbf{v} \mid \mathbf{x}) = \frac{g(\mathbf{x} \mid \mathbf{v})\psi(\mathbf{v})}{\int g(\mathbf{x} \mid \mathbf{v})\psi(\mathbf{v})d\mathbf{v}} \propto g(\mathbf{x} \mid \mathbf{v})\psi(\mathbf{v}), \tag{2.20}$$

which is the likelihood of the data multiplied by the prior distribution ($\psi(\mathbf{v})$) of all model parameters including the latent variables and divided by a normalising constant. Calculating the normalising constant requires multidimensional integration that can be very heavy computationally, if not infeasible. As will be explained in Chapter 4, MCMC techniques make use of the joint posterior distribution possible without having necessarily to compute the normalising constant. We shall take the form of $g(\mathbf{x} \mid \mathbf{v})$ from the exponential family with parameter $\theta_i = \alpha_{i0} + \sum_{j=1}^{q} \alpha_{ij} y_j$ already defined in (2.1) and (2.2) respectively. The main steps of the Bayesian approach are as follows:

1. Inference is based on the posterior distribution, $h(\mathbf{v} \mid \mathbf{x})$, of the unknown parameters, $\boldsymbol{\alpha}$ and $\mathbf{y}$, conditional on the observations $\mathbf{x}$. Depending on the model fitted, the form of the posterior distribution can be very complex.

2. The mean, mode or any other percentile vector, and the standard deviation of the posterior distribution and $h(\mathbf{v} \mid \mathbf{x})$ can be used as an estimator of $\mathbf{v}$ and its corresponding standard error.

3. In general, we may use the posterior mean $E(\psi(\mathbf{v}) \mid \mathbf{x})$ as a point estimate of a function of the parameters $\psi(\mathbf{v})$, where $E(\psi(\mathbf{v}) \mid \mathbf{x}) = \int \cdots \int \psi(\mathbf{v})h(\mathbf{v} \mid \mathbf{x})d\mathbf{v}$.

4. Analytic evaluation of the above expectation is impossible. Alternatives include numerical evaluation, analytic approximations and Monte Carlo integration.

Markov chain Monte Carlo

To avoid the integration required in the posterior expectation, Monte Carlo integration is used to approximate that integral by a sample average of quantities calculated by the method. R samples are drawn from the posterior distribution of all the unknown parameters, $h(\mathbf{v} \mid \mathbf{x})$. Then the expectation over the posterior is approximated by the average

$$\frac{1}{R}\sum_{r=1}^{R} \psi(\mathbf{v}^{(r)} \mid \mathbf{x}).$$

The samples drawn from the posterior distribution do not have to be independent. Samples are drawn from the posterior distribution through a Markov chain with $h(\mathbf{v} \mid \mathbf{x})$ as its stationary distribution. Algorithms such as the Gibbs sampler and the Metropolis–Hastings algorithm are used to get the unique stationary distribution.

We start by defining a Markov chain $M_0, M_1, \ldots$ with states $M_k = (\mathbf{v}^k) = (\mathbf{y}^k, \boldsymbol{\alpha}^k)$. We proceed by simulating states from the Markov chain, and those simulated states will be used to make inferences about the parameters $(\mathbf{y}, \boldsymbol{\alpha})$. Under some regularity conditions and as the number of draws k increases, the distribution of state M_k will converge to the chain's stationary distribution $\pi(\mathbf{y}, \boldsymbol{\alpha})$. The aim is to define the state M_k in such a way that $\pi(\mathbf{y}, \boldsymbol{\alpha})$ will be the desired distribution $h(\mathbf{y}, \boldsymbol{\alpha} \mid \mathbf{x})$. The stationary distribution $\pi(\mathbf{y}, \boldsymbol{\alpha})$ satisfies

$$\int t[(\mathbf{y}^0, \boldsymbol{\alpha}^0), (\mathbf{y}^1, \boldsymbol{\alpha}^1)]\pi(\mathbf{y}^0, \boldsymbol{\alpha}^0) d(\mathbf{y}^0, \boldsymbol{\alpha}^0) = \pi(\mathbf{y}^1, \boldsymbol{\alpha}^1), \qquad (2.21)$$

where $t[(\mathbf{y}^0, \boldsymbol{\alpha}^0), (\mathbf{y}^1, \boldsymbol{\alpha}^1)]$ is the transition kernel defined as

$$P(M_{k+1} = (\mathbf{y}^1, \boldsymbol{\alpha}^1) \mid M_k = (\mathbf{y}^0, \boldsymbol{\alpha}^0)).$$

The success of the whole process depends on the choice of the transition kernel. There is always a 'burn-in' period where the first k observations are thrown away and the remaining observations are used to make inference since they will be, in theory, draws from the desired posterior distribution $h(\mathbf{y}, \boldsymbol{\alpha} \mid \mathbf{x})$. The choice of the transition kernel depends on the sampling scheme used, and this will be illustrated in Chapter 4.

2.11 Rotation

If $\mathbf{y}$ is continuous, no latent variable model is unique. By this we mean that there are other models which lead to exactly the same joint distribution of the observable variables and hence are empirically indistinguishable from the one in question. This is easily seen by noting that any transformation of the latent variable in (1.7) leaves $f(\mathbf{x})$ unchanged. However, in general, such a transformation changes both the prior distribution and the conditional distribution. In the special case of the GLLVM the natural parameter of the conditional distribution continues to be linear in the parameters but not in the latent variables.

It is important to appreciate the full implications of this simple fact, so we pursue it with the logit model for binary data before proceeding with the main topic of this section. For the case of a single latent variable we have

$$\operatorname{logit} \pi_i(y) = \alpha_{i0} + \alpha_{i1} y \quad (i = 1, 2, \ldots, p). \tag{2.22}$$

Suppose we now transform to $z = H(y)$, where H is the cumulative distribution function of the prior distribution. Then instead of a model with prior $h(y)$ and response function given by (2.22) we have one with a uniform prior on (0, 1) and response function

$$\operatorname{logit} \pi_i(z) = \alpha_{i0} + \alpha_{i1} H^{-1}(z) \quad (i = 1, 2, \ldots, p). \tag{2.23}$$

Both models lead to exactly the same distribution of the xs. The second version is of some independent interest but is only one of infinitely many which can be generated in this manner. One other, very simple, transformation is to take $z = -y$. This has an effect equivalent to changing the sign of α_{i1}, thus drawing our attention to the general fact that the signs of the αs can be changed without affecting the joint distribution.

This example makes it clear that the prior and the response function have to be considered as a pair. For any change by transformation we make in one, there is a compensating change we can make in the other which leaves the joint distribution unchanged. It is in this sense that the prior is arbitrary. It means, for example, that it is meaningless to talk about estimating one member of the pair without first fixing the other. This argument applies equally to any member of the GLLVM where the natural parameter takes over the role of the response function. It also applies however many latent variables there may be.

In general, transformations may exist which leave the form of the prior distribution and of the natural parameter unchanged. In the example given above this would be the case for the transformation $z = -y$ if the prior were symmetrical. This example is, in fact, a very special case of an *orthogonal rotation* which we now consider.

If we start with the GLLVM given by

$$\boldsymbol{\theta} = \mathbf{Ay} \tag{2.24}$$

and make the non-singular transformation $\mathbf{z} = \mathbf{My}$ then

$$\boldsymbol{\theta} = \mathbf{AM}^{-1}\mathbf{z}. \tag{2.25}$$

If $\mathbf{y} \sim N(\mathbf{0}, \mathbf{I})$ then

$$\mathbf{z} \sim N(\mathbf{0}, \mathbf{MM}'). \tag{2.26}$$

If the transformation is orthogonal $\mathbf{MM}' = \mathbf{I}$ and hence $\mathbf{y}$ and $\mathbf{z}$ have the same joint distribution, as we noted in Section 1.5. We thus have two indistinguishable models with the same prior but different θs, both linear in the latent variables. This means that, with a multivariate normal prior, we cannot distinguish either conceptually or empirically between a GLLVM with loading matrix $\mathbf{A}$ and one with $\mathbf{AM}'$ (noting that $\mathbf{M}' = \mathbf{M}^{-1}$ for an orthogonal transformation). This result depends upon the assumption of a normal prior. Any orthogonal transformation takes the standard normal prior in which the ys are independent into a transformed prior of exactly the same form. The normal prior is the only form for which this is true. This is a consequence of what is sometimes called Maxwell's theorem (see, for example, (Feller 1966, corollary on p. 77)). In the bivariate case this states that if the unrotated and rotated ys are both to be independent they must also be normal. Normality and independence thus go together. If this is to be true for all ys in the multivariate case it must be true for every pair. Thus the normal prior is special in the sense that it is necessary to guarantee invariance under rotation.

Viewed geometrically, the transformation is an orthogonal rotation in the latent variable space and so the generation of alternative solutions by this means is known as *rotation*. Rotation assumes a very important role when we come to the interpretation of latent variables and it has played a central part in the classical treatment of the normal factor model. The foregoing discussion shows that rotation is a much more general concept, and we shall see it at work in relation to other models in later chapters.

There are two further points to be added. If, instead of assuming $\mathbf{y}$ to be standard normal, we supposed $\mathbf{y} \sim N(\mathbf{0}, \mathbf{\Phi})$ then any non-singular transformation, not necessarily orthogonal, would have yielded

$$\mathbf{z} \sim N(\mathbf{0}, \mathbf{M\Phi M}'); \tag{2.27}$$

that is, the transformed prior would have belonged to the same family as the initial prior. This is an example of an *oblique rotation* which, as we shall see later in Chapter 3, leads to a wider range of options for interpretation.

Secondly, we note that the equivalence of the initial and rotated priors depends on the assumed normality of $\mathbf{y}$. If that assumption is not made the two models will be distinguishable conceptually if not empirically. Even so, important features of the distribution may be unaffected – in particular, the covariance structure of the xs.

2.12 Interpretation

Interpretation is concerned with naming the latent variables – that is, with linking them with theoretical constructs in the substantive field to which the data relate. There are two, closely related, approaches to this question which can be used with any member of the GLLVM family. They can be elaborated in various ways to exploit the characteristics of particular models, but here we shall confine ourselves to those aspects which are common to all.

The first approach is along the lines of traditional factor analysis in terms of the α_{ij} which in this context are known as the *factor loadings*. A large value of α_{ij} means that the jth factor exerts a big influence on the ith manifest variable. By identifying the set of manifest variables for which y_j is an important determinant we might hope to find an interpretation for y_j. In essence we ask what it is that these variables have in common. It often happens that the dominant latent variable has a large influence on all manifest variables, in which case it is sometimes called a general factor. In tests of ability, for example, general ability is indicated by a tendency to score similarly on all items.

The idea of 'looking for what is common' to a set of manifest variables is central to interpretation. We begin with defining the notion of *simple structure* for a loading matrix because this is particularly easy to interpret. Then we consider how such a pattern might be achieved by rotation.

As an example, consider the following matrix of loadings:

$$\mathbf{A} = \begin{bmatrix} + & \cdot & \cdot \\ + & \cdot & \cdot \\ + & \cdot & \cdot \\ \cdot & + & \cdot \\ \cdot & + & \cdot \\ \cdot & + & \cdot \\ \cdot & + & \cdot \\ \cdot & \cdot & + \\ \cdot & \cdot & + \\ \cdot & \cdot & + \end{bmatrix} \tag{2.28}$$

where $+$ denotes a large positive loading and $\cdot$ a small loading. Presented with such a pattern, we would seek to identify y_1 with what is the common influence on x_1, x_2 and x_3; y_2 with the common influence on x_4, x_5, x_6, x_7; and y_3 with the common influence on x_8, x_9 and x_{10}. If such a pattern occurred we would have partitioned the manifest variables into three non-overlapping groups each of which was indicative of a different latent variable. The matrix is then said to have *simple* structure.

In practice, of course, it would be rare to find simple structure in such a clear-cut form, but given that there is an unlimited number of rotations to choose from it might be possible to find one which provides a reasonable approximation. A number of algorithms for this purpose are in common use in factor analysis, and we shall meet some of them in Chapter 3. For the moment we shall give a simple method which will often serve the purpose.

If we had a matrix of the form $\mathbf{A}$ in which the small loadings were all precisely zero it would follow that $\mathbf{A}'\mathbf{A}$ was diagonal. Given an arbitrary $\mathbf{A}$, we now show how to find a rotation such that the transformed matrix satisfies this diagonality condition.

Any $p \times q$ loading matrix can be expressed in the form

$$\mathbf{A} = \mathbf{MDN}', \tag{2.29}$$

where $\mathbf{D}$ is a $q \times q$ diagonal matrix of singular values, $\mathbf{M}$ is a $p \times q$ matrix with orthogonal columns and $\mathbf{N}$ is a $q \times q$ orthogonal matrix. This is known as the singular value decomposition of **A** (we shall meet this again in Chapter 9). The matrix $\tilde{\mathbf{A}} = \mathbf{MD}$ is an orthogonal rotation of **A** (since it is obtained by post-multiplying $\tilde{\mathbf{A}}$ by the orthogonal matrix $\mathbf{N}'$). It also has the diagonality property because

$$\tilde{\mathbf{A}}'\tilde{\mathbf{A}} = \mathbf{DM}'\mathbf{MD} = \mathbf{D}^2,$$

since $\mathbf{M}'\mathbf{M} = \mathbf{I}$. An alternative means of finding $\tilde{\mathbf{A}}$ from **A** is provided by the fact that the elements of $\mathbf{D}^2$ are the non-zero eigenvalues values of $\mathbf{A}'\mathbf{A}$ and **M** consists of the associated eigenvectors. Facilities for making this calculation are readily available in many software packages, for instance R. It should be noted that what we have shown is that we can always find a rotation which makes $\tilde{\mathbf{A}}'\tilde{\mathbf{A}}$ diagonal and that this is also a property of any loading matrix which has simple structure. It does not follow that the transformed matrix will necessarily have simple structure. However, if it does our problem is solved.

The second approach to interpretation starts with the components and is similar in spirit to the usual method of interpreting principal components. It was presented in the first edition (Bartholomew 1987) as a way of interpreting factors in the normal linear factor model, but it is quite general. We have shown that the GLLVM leads to sufficient statistics (components) which contain all the information about the latent variables. When interpreting the components we are then, in a sense, interpreting the corresponding latent variables. For the GLLVM family of models the *j*th component is

$$X_j = \sum_{i=1}^{p} \alpha_{ij} u_i(x_i). \tag{2.30}$$

The question we now address is what X_j measures. If $u_i(x_i) \propto x_i$, as for the logit model for binary data and the normal factor model, the component is simply a linear combination of the *x*s. The interpretation of X_j – and hence y_j – then depends on the signs and magnitudes of the αs. If α_{ij} is large and positive then the larger x_i, the larger X_j – and conversely if α_{ij} is large and negative. If **A** had simple structure X_j would be essentially a weighted average of a subset of the variables. Such an average is a natural way of measuring what they have in common.

The idea of rotation carries over to the interpretation of the *X*s. The set of sufficient statistics **X** is not unique. Any one-to-one transformation is also sufficient and this includes, of course, orthogonal linear transformations. We could therefore search for simple structure via this route using the same method as before.

In essence then, we see that the two approaches are equivalent, though the connection is a little less direct if $u_i(x_i)$ is not proportional to x_i. The advantage of the approach based on **X** is that it can legitimately be regarded as a sensible measure in its own right. Sums or weighted averages of manifest variables have often been used as measures without any attempt to validate them by reference to a probability model. Users who might be cautious about adopting the full latent variable framework might nevertheless be prepared to adopt component-based measures on purely pragmatic grounds.

2.13 Sampling error of parameter estimates

Since the interpretation of the latent variables depends on the estimated parameter values it is important to know how precisely they have been estimated. For most practical purposes it will suffice to have estimates of their standard errors. The traditional approach, at least for maximum likelihood estimates, is to obtain asymptotic standard errors from the second derivatives evaluated at the maximum of the likelihood. An alternative is to use repeated sampling methods such as the bootstrap. Until recently, the latter method has been prohibitively expensive in computing time but this is changing rapidly and we anticipate that this will become the preferred method. We shall deal with the technical details when we come to discuss individual models in later chapters. However, there is one conceptual issue which is common to all models of the family we are considering.

If there are two or more latent variables we have seen that there is no unique set of estimates. Rather, there is a set of values which may be obtained from one another by orthogonal rotation as described above. This requires us to think carefully about what we mean when we ask for an estimate of sampling variability. Suppose we were to fix the parameter values in the model and to generate a random sample of data. This data set would then yield a set of estimates in the parameter space all having the same likelihood. Repeated sampling in this manner would therefore produce a collection of such sets and our problem would be to describe their collective variation. Variation of points in a parameter space can be summarised in terms of variances and covariances but it is not clear how best to describe the variation of sets.

Two avenues appear to be open. One is to choose some typical member of the set and to describe the variation of the set in terms of the variation of that member. This can conveniently be done by specifying constraints satisfied by only one member of the set. In the GLLVM this can be achieved by requiring the matrix $\mathbf{A}'\mathbf{A}$ to be diagonal, and fixing the sign for one loading for each factor. As we have seen, such solutions may facilitate the interpretation. An alternative is to specify a maximisation procedure and regard the member of the optimal set to which it leads as the representative member. In either case we can estimate the standard errors of other members of the set using the linear relationships which link all members of the optimal set.

The second avenue is to describe sampling variation in terms of parameters which are invariant over the optimal set. For example, in the general model the elements of the matrix $\mathbf{A}\mathbf{A}'$ are invariant under orthogonal rotation. Any functions of these elements will share the same property. Although standard errors of such parameters will tell us something about how well the model has been estimated, they may not be very helpful in the matter of interpretation. This is not necessarily so, as we shall see when we consider the normal linear factor model in more detail in Chapter 3. In that case the parameters, known as communalities, are invariant under rotation and, as we shall see, they give us useful information about the importance of individual variables.

2.14 The prior distribution

We have already noted that for many purposes the form of the prior distribution is irrelevant. Nevertheless there are two purposes for which it is necessary to make some assumption about the prior. The first is for estimating the parameters of the model by a method such as maximum likelihood which depends on $f(\mathbf{x})$. The second is for finding the posterior distribution $h(\mathbf{y} \mid \mathbf{x})$ and quantities derived from it such as $E(\mathbf{y} \mid \mathbf{x})$ and $\text{var}(\mathbf{y} \mid \mathbf{x})$.

We noted in Section 2.8 that, without loss of generality, we may fix the mean and variance of the prior. Thus if in the model

$$\theta_i = \alpha_{i0} + \alpha_{i1}y_1 + \alpha_{i2}y_2 + \cdots \quad (i = 1, 2, \ldots, p), \tag{2.31}$$

we make the transformation

$$z_j = a_j y_j + b_j,$$

the model becomes

$$\theta_i = \left(\alpha_{i0} - \alpha_{i1}\frac{b_1}{a_1} - \alpha_{i2}\frac{b_2}{a_2}\cdots\right) + \frac{\alpha_{i1}z_1}{a_1} + \frac{\alpha_{i2}z_1}{a_2} + \cdots. \tag{2.32}$$

This is statistically indistinguishable from (2.31) and shows that changes in the location or scale of $\mathbf{y}$ can be absorbed into the αs. As a matter of convention we shall usually suppose the ys to have zero mean and unit variance. It is, therefore, only through the effect of its shape that the prior can influence the parameter estimates. We have already seen in Section 2.8 that there are methods of fitting which depend only on the second moment properties and these are therefore independent of the prior distribution. This suggests that the effect of the shape of the prior on the estimates might be slight, and we shall meet empirical evidence for this later.

The most obvious way to deal with the problem of the prior is to treat it as unknown and then to estimate it. This can only be attempted if we first fix the form of θ, otherwise we run into the inherent indeterminacy discussed in Section 2.11. A natural choice is to suppose that θ is linear in the ys, as well as the αs. In that case we can, in principle, estimate the prior either parametrically, by choosing a family of distributions and estimating their shape parameters, or non-parametrically. In the latter case – see, for instance, Tzamourani and Knott (2002) – we would approximate $h(\mathbf{y})$ by a discrete distribution defined on a grid of points. Experience with estimating mixing distributions in general statistical practice suggests that it may be difficult to do this with adequate precision. Difficulty in estimating the prior is a consequence of its weak influence on the joint distribution of $\mathbf{x}$ noted above. For estimation purposes at least, there appears to be little at stake in the precise choice of $h(\mathbf{y})$ and the conventional choice of a standard normal prior is both convenient and innocuous.

We shall consider analysis based on the posterior distribution in the following section where it will appear that the effect of the prior is again slight. However, it is worth asking first whether there is anything implicit in our specification of the model which might constrain our choice. There are two points to be made. First, the range of $\mathbf{y}$ is not, in general, arbitrary. For example, in the logit model with a single y,

$$\operatorname{logit} \pi_i(y) = \alpha_{i0} + \alpha_{i1} y, \tag{2.33}$$

y must range over the interval $(-\infty, +\infty)$ if $\pi_i(y)$ is to span the interval $(0, 1)$. If we wish all possible values of $\pi_i(y)$ to be admissible, the range of y must be unrestricted. If the left-hand side of (2.33) were to be a quantity like a rate or a variance, as in the case of the gamma distribution, the range of y would have to be constrained to make the right-hand side non-negative.

A second, more powerful, constraint arises from our discussion of rotation. For it to be invariant under rotation we saw that the prior distribution had to be standard normal. Thus if we wish to have the rotational properties that are linked to the normal prior we are constrained in our choice to that prior.

Finally, there are a number of loosely related considerations of a more subjective kind which bear upon the question. In some fields, notably ability and attitude testing, the intuitive idea of the underlying scale seems to carry with it some sense of how individuals should be spaced along it. Heinen (1996, p. 105), for example, remarks that the normal prior in attitudinal data 'is often not very appropriate with regard to the population distribution'. Goldstein (1980) has made a similar point in relation to ability measurement in proposing the complementary log function in place of the logit in (2.33). Since such judgements cannot be substantiated empirically they must be understood as something implicit in the meaning of the terms 'attitude' or 'ability' to which the theory is required to give expression. We note that any monotonic stretching or shrinking of the scale is allowable within the framework we have adopted. In fact, as we shall see in the next section, only an ordinal scaling can be justified empirically.

A related point concerns the question of what it is that the prior distribution physically represents. Hitherto we have implicitly regarded it as the frequency distribution of the latent variable in some population from which the individuals under consideration have been sampled. The 'ability' or 'attitude' is thus regarded as a fixed 'property' of the individual: $h(\mathbf{y})$ describes how it varies in the population. Heinen (1996, p. 167) calls this the *random score* or *structural* model. Adopting a specific form for $h(\mathbf{y})$, such as the standard normal, then amounts to an agreement to measure y on a scale which renders its distribution normal. Assuming that $\mathbf{y}$ is everywhere continuous, this can always be done.

The whole of the argument thus far is predicated on the assumption that we are dealing with a single population. All inferences we make are relative to that population. If we wish to make comparisons between populations a new set of questions arise. Suppose, for example, that we had two populations with normal priors but widely different locations. We could estimate αs for samples from each population separately and scale individuals but each scaling will be with reference to its own population: it will tell us nothing about differences between populations. We take up this point in Section 2.16.

An alternative way of regarding $h(\mathbf{y})$ is to use what Holland (1990b) has called the *stochastic subject rationale*. This approach treats all individuals as inherently the same and regards y as a random response to the stimulus of the test item or question. It is latent because it cannot be observed directly but only indirectly through an associated set of variables, $\mathbf{x}$. Interest is focused on y, rather than $\mathbf{x}$, because it is, in some sense, considered to be more fundamental. On this interpretation we would be likely to have little interest in knowing what y was on any particular occasion but more in the form of its distribution.

2.15 Posterior analysis

We use this term to refer to any analysis relating to the distribution of $\mathbf{y}$ after $\mathbf{x}$ has been observed. It is concerned with making statements about an individual's latent location given their $\mathbf{x}$. All of the relevant information needed for this is contained in the posterior distribution $h(\mathbf{y} \mid \mathbf{x})$. In factor analysis what we call posterior analysis has traditionally included the 'problem of factor scores' which is concerned with how to find an 'estimate' of $\mathbf{y}$ for a given $\mathbf{x}$. The literature on this topic is extensive and controversial as the recent debate on the subject (Maraun 1996) bears witness. In our view, a great deal of needless confusion has arisen which is avoided by adopting the general approach to latent variable modelling used in this book.

For the latent class model where $\mathbf{y}$ is a finite set of points representing the c latent classes, inference about $\mathbf{y}$ given $\mathbf{x}$ is a straightforward application of Bayes' theorem. If the prior probabilities are $\{\eta_j\}$ then the posterior probability that an individual with observed $\mathbf{x}$ belongs to class j is

$$h(j \mid \mathbf{x}) = \eta_j \prod_{i=1}^{p} g_i(x_i \mid j) \Big/ f(\mathbf{x}) \quad (j = 0, 1, 2, \ldots, c-1). \tag{2.34}$$

A special case was given in (1.4). Individuals can be allocated to the class for which the estimate of this probability is greatest. We have seen that if $\{g_i(x_i \mid j)\}$ is a member of the exponential family, the posterior probability will depend on $\mathbf{x}$ only through a linear function which therefore acts as a linear discriminant function.

If the latent space is continuous we face the problem that the posterior distribution is proportional to the unknown $h(\mathbf{y})$. We can, of course, circumvent the difficulty by making an assumption about θ, but this merely shifts the arbitrariness from one function to another. However, it turns out that a surprising amount can be said without making any arbitrary assumption about $h(\mathbf{y})$. This is made possible by the following theorem, where we assume that the latent variable is arranged so that in (2.7) we have $\phi_j(\mathbf{y}) = y_j$.

Theorem 2.15.1 *The vector $E[\mathbf{y} \mid \mathbf{x}]$ is the gradient of a convex function of the component vector* $\mathbf{X}$.

Proof. From (2.7) it is easy to see that the moment generating function of $h(\mathbf{y} \mid \mathbf{x})$ is given by

$$M_{\mathbf{y}\mid\mathbf{x}}(\mathbf{t}) = M_{\mathbf{y}\mid\mathbf{0}}(\mathbf{X} - \mathbf{X}_0 + t)/M_{\mathbf{y}\mid\mathbf{0}}(\mathbf{X} - \mathbf{X}_0),$$

where $\mathbf{X}_0$ is the component vector for $\mathbf{X} = \mathbf{0}$. So, on taking logarithms and differentiating with respect to $\mathbf{t}$, we have

$$E(\mathbf{y} \mid \mathbf{x}) = \frac{\partial}{\partial \mathbf{t}} K_{\mathbf{y}\mid\mathbf{0}}(\mathbf{X} - \mathbf{X}_0 + \mathbf{t})\bigg|_{\mathbf{t}=\mathbf{0}} = \frac{\partial}{\partial \mathbf{X}} K_{\mathbf{y}\mid\mathbf{0}}(\mathbf{X} - \mathbf{X}_0),$$

$$\text{cov}(\mathbf{y} \mid \mathbf{x}) = \frac{\partial^2}{\partial \mathbf{t} \partial \mathbf{t}'} K_{\mathbf{y}\mid\mathbf{0}}(\mathbf{X} - \mathbf{X}_0 + \mathbf{t})\bigg|_{\mathbf{t}=\mathbf{0}} = \frac{\partial^2}{\partial \mathbf{X} \partial \mathbf{X}'} K_{\mathbf{y}\mid\mathbf{0}}(\mathbf{X} - \mathbf{X}_0),$$

where $K_{\mathbf{y}\mid\mathbf{x}}(t)$ is the cumulant generating function of the posterior distribution.

Since the covariance matrix is non-negative definite, it is clear that $E(\mathbf{y} \mid \mathbf{x})$ is the gradient of a convex function of the component vector $\mathbf{X}$. □

It follows immediately that $E(y_j \mid \mathbf{x})$ is non-decreasing in X_j, if all other components are held fixed. A slightly stronger conclusion is that if two individuals have $\mathbf{x}$-vectors $\mathbf{x}_1$, and $\mathbf{x}_2$, and component vectors $\mathbf{X}_1$, $\mathbf{X}_2$, then the vectors $E(\mathbf{y} \mid \mathbf{x}_1) - E(\mathbf{y} \mid \mathbf{x}_2)$ and $\mathbf{X}_1 - \mathbf{X}_2$ make an acute angle with each other, that is,

$$[E(\mathbf{y} \mid \mathbf{x}_1) - E(\mathbf{y} \mid \mathbf{x}_2)]'[E(\mathbf{y} \mid \mathbf{x}_1) - E(\mathbf{y} \mid \mathbf{x}_2)] \geq 0.$$

These results linking the rankings of individuals on the $\mathbf{y}$ scale and the scale based on $\mathbf{X}$ are valid *whatever the distribution of* $\mathbf{y}$.

The results above are weaker than those given in the second edition (Bartholomew and Knott 1999), which Jouni Kuha (personal communication) pointed out were too strong.

Similar results in a special case were given by Knott and Albanese (1993) and slightly extended in Bartholomew (1996). They are related to the *Dutch identity* (Holland 1990a). Another slight extension of the work by Albanese and Knott gives information about the form of the relationship between $\mathbf{X}$ and $E(\mathbf{y} \mid \mathbf{x})$. Using the moment generating function, it can be shown that if the posterior distribution is normal when $\mathbf{X} = \mathbf{0}$ then it is normal for all $\mathbf{X}$. The relevance of this result for our present purposes is that posterior distributions often turn out to be close to normal. and we conjecture that this is generally true as $p \to \infty$. Given normality of the posterior distribution, it follows that

$$E(\mathbf{y} \mid \mathbf{x}) = E(\mathbf{y} \mid \mathbf{0}) + \text{cov}(\mathbf{y} \mid \mathbf{0})'(\mathbf{X} - \mathbf{X}_0). \tag{2.35}$$

This result helps to explain why the empirical relationship between $\mathbf{X}$ and $E(\mathbf{y} \mid \mathbf{X})$ often turns out to be nearly linear – a point first noted by Upton (1980).

These results underpin the use of $\mathbf{X}$ for factor scores. Not only does it arise naturally from our general approach but it has a strong intuitive appeal in its own right.

The weak dependence of the posterior distribution on the prior has close affinities with Savage's *principle of precise measurement* in Bayesian inference. Roughly speaking, this says that as the sample size increases the dependence of the posterior distribution on the prior diminishes. The situation here is a little more complicated but what we would like to be true is that a similar result hold as p, the number of variables, increases. The variables are not identically distributed and the idea of sampling variables from an infinite sequence is not entirely straightforward. Nevertheless, it should be possible to establish an extended principle of precise measurement, and the notion of 'consistency at large' in Schneeweiss (1993) is a step in that direction. See also Chang and Stout (1993).

2.16 A further note on the prior

We have emphasised the unimportance of the prior distribution for most of the things we wish to do. In particular, we have argued that the choice of a standard normal prior is a matter of convention and is adopted largely as a matter of convenience. This gives rise to no problems as long as we are concerned with a single population. The αs are, in effect, calibrated with respect to a standardised prior for that population and the choice of the normal form is then tantamount to adopting a particular scaling.

The trouble with this approach arises when a second population comes into the picture. For definiteness let us discuss the problem as it would arise in the context of ability testing. Suppose that we estimate the item response curve of the logit model for a given population by drawing a sample from it. Each item will thus have its own difficulty and discrimination values. Scale values for any individual can then be predicted on the basis of their αs and this will place them at some point on a standard normal scale.

Suppose next that we draw a sample from a second population (in practice this might be the same population at a later time or a different population, country say, at the same time). Any scale values we calculate for this second sample will relate to this second population. Since our methods automatically transform the population mean to be zero (and the standard deviation 1) we are thus precluded from making any comparisons between populations. Yet this is often exactly what we wish to do. We may wish to know, for example, whether the general level of ability (the population mean) has increased over time. Is there any way we can get at this information?

One method is to argue as follows. The difficulty and discrimination of an item are intrinsic properties of that item (albeit calibrated with respect to a particular population). If these same items are administered to members of the second population their difficulty and discriminating ability should be the same. Hence any change observed must be attributable to population differences. The first population is assumed to have mean 0 and standard deviation 1: let the corresponding values for the second population be μ and σ. Let the model for the first population be

$$\operatorname{logit} \pi_i(y) = \alpha_{i0} + \alpha_{i1} y \tag{2.36}$$

and that for the second

$$\operatorname{logit} \pi_i(z) = \alpha_{i0} + \alpha_{i1} z \tag{2.37}$$

with $z = (y + \mu)\sigma$. The second model may be written in terms of y as

$$\operatorname{logit} \pi_i(y) = (\alpha_{i0} + \mu\sigma\alpha_{i1}) + \alpha_{i1}\sigma y. \tag{2.38}$$

If we assume that α_{i0} and α_{i1} have the same values in the second population as the first, the only unknowns in the second case would be μ and σ, which could then be estimated from the likelihood. This would give us estimates of the mean and standard deviation of the second population in units of the first. A rather more sophisticated estimation procedure would be to pool the likelihoods and estimate the αs, μ and σ simultaneously. However, here we are concerned with what is possible in principle and not with the technical details.

The problem simplifies considerably if the $\{\alpha_{i1}\}$s can be assumed to be equal, in which case we have the random effects Rasch model. In this case the common value of $\{\alpha_{i1}\}$ $(i = 1, \ldots, p)$ can be absorbed into σ in (2.37). This special case is discussed by Andersen (1980a) who gave a maximum likelihood method of estimating the means and variances of the group distributions. The method was illustrated on examples. The relevant likelihood is simplified by the fact that, in this case, the sufficient statistic is the sum of the x_i, and this takes the integer values $0, 1, \ldots, p$. Andersen's method requires the α_{i0}s to be known (or estimated independently with sufficient accuracy). Such known values must, of course be relative to some standard population.

A second approach would be to consider the mixture of the two populations as a single population. In fitting the logit model to the combined samples we would be adopting a different reference population and our estimates of $\{\alpha_{i0}\}$ and $\{\alpha_{i1}\}$ would not be the same as before, even though they related to the same items. Any difference between the two populations would now be evidenced by the way in which the predicted scale values for the individuals from the two samples were distributed with respect to one another. For example, if there had been an increase in the population mean from the first to the second occasion, we would expect the members of the second sample to tend to have higher factor scores than those of the first sample. This method is essentially a special case of what, in factor analysis, is called multi-group factor analysis; see Basilevsky (1994, Chapters 5 and 6) for further details and references. These methods could easily be extended to the GLLVM, as shown in Bock and Moustaki (2007), and this is a potentially fruitful area for future research. However, it is important not to lose sight of the implied assumption about the prior distribution.

2.17 Psychometric inference

Our approach to inference, here and in subsequent chapters, follows the standard lines of statistical inference, but we digress at this point to draw attention to another kind of inference which has sometimes been thought to be relevant in some psychological situations. This is known as psychometric inference. Psychometric inference is best explained by reference to the practical problem which gave rise to it.

It appears to have originated in educational testing where a sample of children take a series of tests in order to evaluate an ability of some kind. Frank Lord at the Educational Testing Centre in Princeton pioneered this work in the 1960s, and he referred to it as 'item selection'. This term indicates more precisely the problem that testers faced. In a typical testing situation n individuals take p test items. The n individuals will be a sample (usually random) of the population of individuals to which one wishes to generalise the result of the test. Standard statistical inference provides the tool for doing this. But if the items are also a selection from the set of all possible test items, we may well wish to generalise the result to apply also to the population of items from which they were selected. This requires psychometric inference.

Workers in educational testing wanted their conclusions to relate not only to all children but also to the whole population of test items which might have been used. The basic approach has been to ask whether the results from the sample of items actually used can be generalised to the whole population of items. (Hence what we are about to describe is part of what has been called generalisability theory.) The idea is as follows. If all the items are very similar (in difficulty or whatever distinguishes them from one another) then each will lead to very similar responses – two identical items would presumably lead to identical scores. If we can show that we would have got very similar results, *whatever sample of items had been selected*, then we can legitimately generalise the result to the whole population of test items. This can be judged by constructing some measure of the similarity of the sets of scores resulting from each item.

Such a measure is provided immediately by noting that the set-up is essentially the same as that encountered in the one-way analysis of a random effects model in the analysis of variance. The 'groups' of the analysis of variance are the test items, which we suppose have been selected from some population of items. Psychometric inference is based on a comparison of the 'between items' sum of squares and the 'between persons within items' sum of squares. The ratio of the expected mean squares under the standard model will be unity if the items show no variation and will increase as that variation increases. Cronbach (1951) proposed the complement of this ratio as an index of generalisability and called it coefficient alpha. Conventionally, it is sometimes suggested that alpha should be greater than 0.7, say, for generalisation to be permissible, but this value is not chosen with reference to any sampling criterion. It should be emphasised that the rationale for Cronbach's alpha involves the assumption that items have been selected at random, and in many applications this will not be valid. Educational testing is a partial exception, as a simple example will show. If one wished to test the ability of children to add up three 2-digit numbers the population size of such sums is finite but large. In such a case it would be possible to select items at random. Even if we excluded very simple cases there will be something approaching a million possible sums and no child could be expected to attempt that number in a single sitting.

In a factor analysis context, this approach requires us to estimate the factor structure in such a way that it is as close as possible to what we would have found if we had the full (possibly infinite) set of manifest variables available to us. For this purpose Kaiser and Caffrey (1965) used Cronbach's coefficient as a basis for what they called alpha factor analysis. Cronbach's alpha is essentially based on a ratio of variances and is very similar to newer measures of reliability; the link is discussed in Bartholomew and Schuessler (1991). Since it turns out that the generalisability coefficient is 1 for a one-factor model, the approach only comes into play if there are two or more factors.

The idea of sampling variables (items in the above context) has arisen several times in connection with latent variable modelling. Apart from the closely related topic of alpha factor analysis, which we briefly touch on again in Chapter 3, it has arisen in the debate on factor scores where Williams (1978), for example, has defined an infinite sequence of random variables, known as the domain, in order to study the behaviour of factor score 'estimates' as the number of elements in the domain increases without limit. Schneeweiss (1993) introduced the concept of 'consistency at large' to study the limiting behaviour of estimation procedures in infinite domains, and the idea is also implicit in Q-analysis to which we come in Chapter 9.

However, the great drawback of psychometric inference, as currently understood, is that it lacks an adequate theoretical probability framework. A promising new development, which is not open to this criticism, is foreshadowed in de Boeck (2008). Although his approach is directed primarily at the educational testing situation, it offers the prospect of wider applicability. The essential idea is to allow the parameters of the response function (see Chapter 4) to be random variables instead of being fixed. This new source of variation might arise because the items are selected from some population or because the items yield an outcome which is inherently uncertain. If this idea proves fruitful, the general approach adopted in this book could be generalized to accommodate it. For the present we emphasise that our general approach does not depend in any way on any manifest variables (xs) which have not been observed. The posterior distribution of $\mathbf{y}$, which encapsulates all that the available xs have to tell us about $\mathbf{y}$, depends only on the conditional distribution of $\mathbf{x}$ given $\mathbf{y}$ and the prior distribution of $\mathbf{y}$.

3

The normal linear factor model

3.1 The model

We have already introduced the normal linear factor (NLFM) model in Chapter 1 as an example of our general approach. Its main properties arise as special cases of the GLLVM given in Chapter 2. The emphasis in this chapter is on the implementation of the model and, especially, its interpretation.

The NLFM is the oldest and most widely used latent variable model. The analyses based on the model may be motivated in various ways and not all of them require the normality assumption. We shall therefore include sections at appropriate points which show what can be done if we do not assume normality. The distinctive feature of our presentation is that its development flows naturally from the framework provided in Chapter 2.

The NLFM assumes that

$$\mathbf{x} \mid \mathbf{y} \sim N_p(\boldsymbol{\mu} + \boldsymbol{\Lambda}\mathbf{y},\ \boldsymbol{\Psi}) \tag{3.1}$$

and

$$\mathbf{y} \sim N_q(\mathbf{0},\ \mathbf{I}), \tag{3.2}$$

where $\boldsymbol{\Psi}$ is a $p \times p$ diagonal matrix of *specific variances* and $\boldsymbol{\Lambda}$ is a $p \times q$ matrix of *factor loadings*. We have already noted that there is no loss of generality in assuming zero means and unit variances for the *y*s. For some purposes, to be considered later, we may wish to allow correlations among the *y*s, but for the moment this is excluded. Note that there are two assumptions of normality involved.

Latent Variable Models and Factor Analysis: A Unified Approach, Third Edition.
David Bartholomew, Martin Knott and Irini Moustaki.

An equivalent, and more common, way of writing the model is as a linear equation in normal random variables. Thus

$$\mathbf{x} = \boldsymbol{\mu} + \boldsymbol{\Lambda}\mathbf{y} + \mathbf{e}, \tag{3.3}$$

where $\mathbf{y} \sim N_q(\mathbf{0}, \mathbf{I})$, $\mathbf{e} \sim N_p(\mathbf{0}, \boldsymbol{\Psi})$ and $\mathbf{y}$ is independent of $\mathbf{e}$.

The difference between the two versions lies in the manner in which the manifest variables are supposed to have been generated. In the first version this takes place in two stages. First we select the individual from the population; the values of $\mathbf{x}$ for that individual are then sampled from the conditional distribution of $\mathbf{x}$. In the second version $\mathbf{y}$ and $\mathbf{e}$ are selected simultaneously and independently and are then combined according to (3.3) to give $\mathbf{x}$. Whether we write the model in terms of probability distributions, as in (3.1) and (3.2), or as an equation in random variables, as in (3.3), is perhaps more a matter of taste than of substance. However, the blurring of the distinction between random variables and their realised values which (3.3) encourages can lead to confusion when we come to the question of what can be said about $\mathbf{y}$ after $\mathbf{x}$ has been observed.

Since the conditional distribution of $\mathbf{x}$ in (3.1) belongs to the exponential family there is a sufficient statistic which we gave by way of example in Section 2.6. In matrix notation it is

$$\mathbf{X} = \boldsymbol{\Lambda}'\boldsymbol{\Psi}^{-1}\mathbf{x}, \tag{3.4}$$

in the parameterisation of the present section.

3.2 Some distributional properties

The two main properties have already been given in (1.12) and (1.13), namely

$$\mathbf{x} \sim N_p(\boldsymbol{\mu}, \boldsymbol{\Lambda}\boldsymbol{\Lambda}' + \boldsymbol{\Psi}) \tag{3.5}$$

and

$$\mathbf{y} \mid \mathbf{x} \sim N_q(\boldsymbol{\Lambda}'(\boldsymbol{\Lambda}\boldsymbol{\Lambda}' + \boldsymbol{\Psi})^{-1}(\mathbf{x} - \boldsymbol{\mu}), (\boldsymbol{\Lambda}'\boldsymbol{\Psi}^{-1}\boldsymbol{\Lambda} + \mathbf{I})^{-1}). \tag{3.6}$$

The first result is required for fitting the model by maximum likelihood; the second for making inferences about the latent variables on the basis of the observed xs.

The link between the conditional expectation of $\mathbf{y}$ and the components given in (3.4) is established by noting that

$$\boldsymbol{\Lambda}'\boldsymbol{\Psi}^{-1}(\boldsymbol{\Lambda}\boldsymbol{\Lambda}' + \boldsymbol{\Psi}) = (\mathbf{I} + \boldsymbol{\Gamma})\boldsymbol{\Lambda}', \tag{3.7}$$

where $\boldsymbol{\Gamma} = \boldsymbol{\Lambda}'\boldsymbol{\Psi}^{-1}\boldsymbol{\Lambda}$. From this it follows that

$$\boldsymbol{\Lambda}'(\boldsymbol{\Lambda}\boldsymbol{\Lambda}' + \boldsymbol{\Psi})^{-1} = (\mathbf{I} + \boldsymbol{\Gamma})^{-1}\boldsymbol{\Lambda}'\boldsymbol{\Psi}^{-1} \tag{3.8}$$

and hence that

$$E(\mathbf{y} \mid \mathbf{x}) = (\mathbf{I} + \mathbf{\Gamma})^{-1}(\mathbf{X} - E(\mathbf{X})). \tag{3.9}$$

The matrix $\mathbf{\Gamma}$ plays a key role in what follows. If it is diagonal we see that, given $\mathbf{x}$, the ys are independent and each posterior mean is proportional to the corresponding component. Other consequences will emerge below.

It follows from (3.5) that

$$\text{var}(x_i) = \sum_{j=1}^{q} \lambda_{ij}^2 + \psi_i \quad (i = 1, 2, \ldots, p). \tag{3.10}$$

The variance of x_i is thus composed of two parts. The first, $\sum \lambda_{ij}^2$, arises from what is common to all xs. For this reason it is known as the *communality*. The complementary part, ψ_i, is the variance specific to that particular x_i.

Other distributional properties, which are useful for interpreting the model, concern the relationship between the factors on the one hand and the observables, $\mathbf{x}$ and $\mathbf{X}$, on the other.

(a) The covariance between the manifest variables and the factor is given by

$$\begin{aligned} E(\mathbf{x} - \boldsymbol{\mu})\mathbf{y}' &= E\left[E(\mathbf{x} - \boldsymbol{\mu})\mathbf{y}' \mid \mathbf{y}\right] \\ &= E\left[E\{(\mathbf{x} - \boldsymbol{\mu}) \mid \mathbf{y}\}\mathbf{y}'\right] \\ &= E(\mathbf{\Lambda}\mathbf{y}\mathbf{y}') = \mathbf{\Lambda}. \end{aligned}$$

The factor loadings can therefore be interpreted as covariances between individual manifest variables and factors. The correlations are given by

$$\{\text{diag}\, \mathbf{\Sigma}\}^{-1/2}\, \mathbf{\Lambda}.$$

(b) The covariance between the components and the factors is

$$\begin{aligned} E(\mathbf{X} - E(\mathbf{X}))\mathbf{y}' &= \mathbf{\Lambda}'\mathbf{\Psi}^{-1} E[(\mathbf{x} - \boldsymbol{\mu})\mathbf{y}'] \\ &= \mathbf{\Lambda}'\mathbf{\Psi}^{-1}\mathbf{\Lambda} = \mathbf{\Gamma}. \end{aligned}$$

If $\mathbf{\Gamma}$ is diagonal there are no cross-correlations between components and factors. For the correlations we require the covariance matrix of $\mathbf{X}$ which is

$$\begin{aligned} \text{var}(\mathbf{X}) &= E\left[\mathbf{\Lambda}'\mathbf{\Psi}^{-1}(\mathbf{x} - \boldsymbol{\mu})(\mathbf{x} - \boldsymbol{\mu})'\mathbf{\Psi}^{-1}\mathbf{\Lambda}\right] \\ &= \mathbf{\Lambda}'\mathbf{\Psi}^{-1}(\mathbf{\Lambda}\mathbf{\Lambda}' + \mathbf{\Psi})\mathbf{\Psi}^{-1}\mathbf{\Lambda} \\ &= \mathbf{\Gamma}^2 + \mathbf{\Gamma}. \end{aligned}$$

The required correlation matrix is thus

$$\mathbf{\Gamma}\{\mathrm{diag}(\mathbf{\Gamma}^2 + \mathbf{\Gamma})\}^{-1/2}.$$

So if $\mathbf{\Gamma}$ is diagonal the correlation between X_j and y_j is $(1 + \Gamma_j^{-1})^{-1/2}$, where Γ_j is the jth diagonal element of $\mathbf{\Gamma}$. The larger Γ_j, the larger will the correlation be. Recalling the role which $\mathbf{\Gamma}$ plays in the posterior distribution of $\mathbf{y}$, we can say that the smaller the variance of y_j (i.e. the more precisely it is determined by the data) the larger will Γ_j be and hence the more closely related will X_j and y_j be.

3.3 Constraints on the model

There are circumstances in which we may wish to place constraints on the parameters. For example, we have seen above that if the matrix $\mathbf{\Gamma} = \mathbf{\Lambda}'\mathbf{\Psi}^{-1}\mathbf{\Lambda}$ is diagonal the ys will be independent *a posteriori* and the relationship between the ys and the components is particularly simple. This may help to facilitate the interpretation of the model. This particular constraint also removes the freedom to arbitrarily rotate $\mathbf{\Lambda}$ (unless some elements of $\mathbf{\Gamma}$ are equal).

In confirmatory factor analysis, to which we come in Chapter 8, there may be grounds for specifying the pattern of the elements in $\mathbf{\Lambda}$. This usually takes the form of setting the values of certain λs equal to zero. This too will usually remove the rotational freedom.

A third type of constraint arises from the arbitrariness of the scale of the manifest variables in many applications, especially in social sciences. Any change of scale in an x will be reflected in the covariances and hence in the parameter estimates and their interpretation. We therefore need a parameterisation which is unaffected by arbitrary changes of scale. This can be done by expressing the xs in units of their standard deviation. The effect of this is to make all the diagonal elements of $\mathbf{\Sigma}$ equal to 1, that is,

$$\sum_{j=1}^{q} \lambda_{ij}^2 + \Psi_i = 1 \quad (i = 1, 2, \ldots, p).$$

The implications of this constraint for maximum likelihood estimation will be explained in Section 3.12.2.

3.4 Maximum likelihood estimation

Estimation by maximum likelihood is not easy. The treatment below follows somewhat the same lines as Lawley and Maxwell (1971), with some help from matrix differentiation results in Magnus and Neudecker (1988). There is an attempt to fill some of the gaps in previously available treatments, including that of the second edition of the present volume (Bartholomew and Knott 1999), by including Heywood cases, and being more careful about the choice of stationary values. The derivation here is, as far as we know, different from any already published.

If $\mathbf{x} \sim N_p(\boldsymbol{\mu}, \boldsymbol{\Sigma})$ then the likelihood function for observations $\mathbf{x}_1, \ldots, \mathbf{x}_n$ may be written

$$l(\boldsymbol{\Sigma}) = \text{constant} + \frac{n}{2}[\ln |\boldsymbol{\Sigma}^{-1}| - \text{trace}[\boldsymbol{\Sigma}^{-1}\mathbf{S}]] \tag{3.11}$$

where $\mathbf{S} = \sum_{i=1}^{n}(\mathbf{x}_i - \boldsymbol{\mu})(\mathbf{x}_i - \boldsymbol{\mu})'/n$. The first step is to maximise with respect to $\boldsymbol{\mu}$. This is a standard problem and it is easily shown that $\hat{\boldsymbol{\mu}} = \bar{\mathbf{x}}$. Henceforth we shall suppose this estimate to have been substituted into $\mathbf{S}$. The novelty in the second stage is that we wish to maximise with respect to $\boldsymbol{\Lambda}$ and $\boldsymbol{\Psi}$, where

$$\boldsymbol{\Sigma} = \boldsymbol{\Lambda}\boldsymbol{\Lambda}' + \boldsymbol{\Psi} \tag{3.12}$$

and where $\boldsymbol{\Lambda}$ is a $p \times q$ matrix of rank q, and $\boldsymbol{\Psi}$ is diagonal.

It is convenient to use matrix differentiation results; see, for instance, Rao (1973), Magnus and Neudecker (1988). One can safely ignore the symmetry constraints. We have

$$\text{d}l(\boldsymbol{\Sigma}) = \frac{n}{2}[-\text{trace}(\boldsymbol{\Sigma}^{-1}\text{d}\boldsymbol{\Sigma} + \boldsymbol{\Sigma}^{-1}(\text{d}\boldsymbol{\Sigma})\boldsymbol{\Sigma}^{-1}\mathbf{S})] \tag{3.13}$$

$$= -\frac{n}{2}\text{trace}[(\boldsymbol{\Sigma}^{-1} - \boldsymbol{\Sigma}^{-1}\mathbf{S}\boldsymbol{\Sigma}^{-1})\text{d}\boldsymbol{\Sigma}]. \tag{3.14}$$

From (3.14)

$$\text{d}l(\boldsymbol{\Sigma}) = -\frac{n}{2}\text{trace}[(\boldsymbol{\Sigma}^{-1} - \boldsymbol{\Sigma}^{-1}\mathbf{S}\boldsymbol{\Sigma}^{-1})((\text{d}\boldsymbol{\Lambda})\boldsymbol{\Lambda}' + \boldsymbol{\Lambda}(\text{d}\boldsymbol{\Lambda})')] \tag{3.15}$$

$$= -n\ \text{trace}[\boldsymbol{\Lambda}'(\boldsymbol{\Sigma}^{-1} - \boldsymbol{\Sigma}^{-1}\mathbf{S}\boldsymbol{\Sigma}^{-1})\text{d}\boldsymbol{\Lambda}]. \tag{3.16}$$

So the stationarity conditions for $\boldsymbol{\Lambda}$ can be written

$$(\boldsymbol{\Sigma}^{-1} - \boldsymbol{\Sigma}^{-1}\mathbf{S}\boldsymbol{\Sigma}^{-1})\boldsymbol{\Lambda} = \mathbf{0}. \tag{3.17}$$

Similarly, one can see that

$$\text{d}l(\boldsymbol{\Sigma}) = -\frac{n}{2}\text{trace}[(\boldsymbol{\Sigma}^{-1} - \boldsymbol{\Sigma}^{-1}\mathbf{S}\boldsymbol{\Sigma}^{-1})\text{d}\boldsymbol{\Psi}], \tag{3.18}$$

so that stationarity conditions for $\boldsymbol{\Psi}$ are

$$\text{diag}(\boldsymbol{\Sigma}^{-1} - \boldsymbol{\Sigma}^{-1}\mathbf{S}\boldsymbol{\Sigma}^{-1}) = \mathbf{0}, \tag{3.19}$$

where diag $\mathbf{A}$ is the matrix with diagonal elements as for $\mathbf{A}$ and zeros elsewhere. Pre- and post-multiplying the matrices in (3.19) by the diagonal matrix $\boldsymbol{\Psi} = \boldsymbol{\Sigma} - \boldsymbol{\Lambda}\boldsymbol{\Lambda}'$ and using (3.17) gives the stationarity conditions for $\boldsymbol{\Psi}$ in the form

$$\text{diag}(\boldsymbol{\Sigma} - \mathbf{S}) = \mathbf{0}. \tag{3.20}$$

Now we follow Lawley and Maxwell in spirit, but unlike them allow $\boldsymbol{\Psi}$ to be singular to accommodate the Heywood cases that arise in practice (see Section 3.12.3). The form of the stationary values for $\mathbf{\Lambda}$ can be made more transparent. Pre-multiplying (3.17) by $\mathbf{S}^{-1/2}\mathbf{\Sigma}\mathbf{S}^{-1}\mathbf{\Sigma}$ leads to

$$[\mathbf{S}^{-1/2}\mathbf{\Sigma}\mathbf{S}^{-1/2} - \mathbf{I}]\mathbf{S}^{-1/2}\mathbf{\Lambda} = \mathbf{0}. \tag{3.21}$$

So at stationary values of $\mathbf{\Lambda}$, the columns of $\mathbf{S}^{-1/2}\mathbf{\Lambda}$ are eigenvectors of $\mathbf{S}^{-1/2}\mathbf{\Sigma}\mathbf{S}^{-1/2}$ with eigenvalues 1. As might be expected, one can rotate the columns of $\mathbf{\Lambda}$ by post-multiplying by an orthogonal matrix, and (3.12) is still satisfied. Replacing $\mathbf{\Sigma}$ by $\mathbf{\Psi} + \mathbf{\Lambda}\mathbf{\Lambda}'$ gives

$$[\mathbf{S}^{-1/2}\mathbf{\Psi}\mathbf{S}^{-1/2}]\mathbf{S}^{-1/2}\mathbf{\Lambda} = \mathbf{S}^{-1/2}\mathbf{\Lambda}[\mathbf{I} - \mathbf{\Lambda}'\mathbf{S}^{-1}\mathbf{\Lambda}]. \tag{3.22}$$

So, if the columns of $\mathbf{\Lambda}$ are scaled to make $\mathbf{\Lambda}'\mathbf{S}^{-1}\mathbf{\Lambda}$ a diagonal matrix, one can see that, at stationary values of $\mathbf{\Lambda}$, we then have $\mathbf{S}^{-1/2}\mathbf{\Lambda}$ with columns which are eigenvectors of $\mathbf{S}^{-1/2}\mathbf{\Psi}\mathbf{S}^{-1/2}$ with eigenvalues no greater than 1. It is always possible to choose $\mathbf{\Psi}$ so that there are q eigenvalues no greater than 1. The scaling is always possible. It is easy to verify directly that if $\mathbf{V}$ is a $p \times q$ set of orthogonal eigenvectors of $\mathbf{S}^{-1/2}\mathbf{\Psi}\mathbf{S}^{-1/2}$ with eigenvalues (all no greater than 1) on the diagonal of the $q \times q$ matrix $\mathbf{\Delta}$, one can take

$$\mathbf{\Lambda} = \mathbf{S}^{1/2}\mathbf{V}(\mathbf{I} - \mathbf{\Delta})^{1/2}. \tag{3.23}$$

It is interesting to note that $\mathbf{\Lambda}$ from (3.23) cannot be arbitrarily rotated while still keeping $\mathbf{\Lambda}'\mathbf{S}^{-1}\mathbf{\Lambda}$ diagonal. A rotation is chosen by the way in which we solve the likelihood equations.

The choice of a particular stationary value for $\mathbf{\Lambda}$ can be made looking to see which one maximises the likelihood. To maximise the log-likelihood is to minimise

$$-\ln|\mathbf{\Sigma}^{-1}\mathbf{S}| + \text{trace}[\mathbf{\Sigma}^{-1}\mathbf{S}] = -\ln|\mathbf{S}^{1/2}\mathbf{\Sigma}^{-1}\mathbf{S}^{1/2}| + \text{trace}[\mathbf{S}^{1/2}\mathbf{\Sigma}^{-1}\mathbf{S}^{1/2}]. \tag{3.24}$$

So, if θ_i are the eigenvalues of $\mathbf{S}^{-1/2}\mathbf{\Sigma}\mathbf{S}^{-1/2}$, one must minimise

$$\sum_{i=1}^{n}[-\ln\theta_i + 1/\theta_i]. \tag{3.25}$$

As θ varies, $-\ln\theta + 1/\theta$ can be seen to take a minimum value of 1 when $\theta = 1$. For any stationary value of $\mathbf{\Lambda}$, q of the $\{\theta_i\}$ are equal to 1, and the other $p - q$ of the $\{\theta_i\}$ are also eigenvalues of $\mathbf{S}^{-1/2}\mathbf{\Psi}\mathbf{S}^{-1/2}$, and are less than 1. So the log-likelihood is maximised by choosing the columns of $\mathbf{S}^{-1/2}\mathbf{\Lambda}$ to be the eigenvectors corresponding to the q smallest of the eigenvalues of $\mathbf{S}^{-1/2}\mathbf{\Psi}\mathbf{S}^{-1/2}$.

Having found $\mathbf{\Lambda}$, one can use (3.20) to find a new value for $\mathbf{\Psi}$ and then iterate. As Jöreskog (1967) first showed, the hard part of maximum likelihood estimation is the determination of $\mathbf{\Psi}$, while given $\mathbf{\Psi}$ the maximisation over $\mathbf{\Lambda}$ is easy. One can see from the presentation above that the maximum likelihood estimation of $\mathbf{\Psi}$ chooses it so that the largest $p - q$ eigenvalues $\{\theta_i\}$, all less than 1, of $\mathbf{S}^{-1/2}\mathbf{\Psi}\mathbf{S}^{-1/2}$ minimise $\sum(-\ln\theta_i + 1/\theta_i)$.

It is interesting to see what happens if there is a Heywood case (see also Section 3.12.3). The simplest way this arises is if the maximum likelihood estimate of one of the diagonal elements of $\mathbf{\Psi}$, say the first one, is zero. Then $\mathbf{S}^{-1/2}\mathbf{\Psi}\mathbf{S}^{-1/2}$ is singular, so one of the $\{\theta_i\}$ is zero. The eigenvector associated with that zero eigenvalue will therefore be one of the columns of the maximum likelihood estimate of $\mathbf{S}^{-1/2}\mathbf{\Lambda}$. It is easy to check directly that a suitable column of $\mathbf{\Lambda}$ is $\frac{1}{s_{11}}\mathbf{S}_1$, where $\mathbf{S}_1$ is the first column of $\mathbf{S}$, and s_{11} is its first element. The maximum likelihood estimates for the remaining factors, from (3.22), are found by using the sub-matrix of $\mathbf{S}^{-1/2}$ obtained by omitting its first row and first column. That is the same as using the conditional covariance matrix for variables $2, \ldots, p$ given variable 1. Jöreskog (1967) has this result, but no obvious way of integrating it into his analytic treatment of maximum likelihood estimation.

3.5 Maximum likelihood estimation by the E-M algorithm

Another way of carrying out the maximum likelihood estimation is to use the E-M method, which is an iterative technique well suited to maximum likelihood estimation for models where there is missing information. The missing pieces of information here are the values of the latent variables. The E-M approach was introduced in Dempster *et al.* (1977), and applied directly to the normal factor analysis model in Rubin and Thayer (1982).

Starting with some initial values for the parameters, the procedure is to write down the joint likelihood of $(\mathbf{x}_i, \mathbf{y}_i)$ for $i = 1, \ldots, n$. The log-likelihood is replaced by its expected value (E-step) conditional on the $\mathbf{x}_i$. The expected values are worked out with the current values of the parameters at that iteration.

This modified likelihood is maximised (M-step) to give new values for the parameters, and the whole procedure iterated until convergence. Although convergence to the global maximum is not guaranteed, the marginal likelihood of $\mathbf{x}_i$ for $i = 1, \ldots, n$ will never decrease with each change in the parameters. This follows from use of the maximising property of the M-step, and a well-known information theory inequality on the conditional distributions of $\mathbf{y}$ given $\mathbf{x}$; see Dempster *et al.* (1977, equation (3.10)).

In practice it is easier to set the conditional expected value of the score function from the joint likelihood of $(\mathbf{x}_i, \mathbf{y}_i)$ given $\mathbf{x}_i$ equal to zero. Notice that this is in fact the score function from the marginal likelihood of the $\mathbf{x}_i$ (McLachlan and Krishnan 1997, (3.42), for instance).

The log-likelihood of the $(\mathbf{x}_i, \mathbf{y}_i)$ for $i = 1, \ldots, n$ is

$$\text{constant} - \frac{n}{2}\log|\mathbf{\Psi}| - \frac{n}{2}\text{trace}\,\mathbf{\Psi}^{-1}\left[\frac{1}{n}\sum_{i=1}^{n}(\mathbf{x}_i - \boldsymbol{\mu} - \mathbf{\Lambda}\mathbf{y}_i)(\mathbf{x}_i - \boldsymbol{\mu} - \mathbf{\Lambda}\mathbf{y}_i)'\right]$$

$$-\frac{n}{2}\text{trace}\,\frac{1}{n}\sum_{i=1}^{n}(\mathbf{y}_i\mathbf{y}_i'),$$

which gives a score function for $\boldsymbol{\mu}$,

$$n\boldsymbol{\Psi}^{-1}(\bar{\mathbf{x}} - \boldsymbol{\mu} - \boldsymbol{\Lambda}\bar{\mathbf{y}}), \tag{3.26}$$

for $\boldsymbol{\Lambda}$,

$$n\boldsymbol{\Psi}^{-1}(\mathbf{S}'_{xy} - \boldsymbol{\mu}\bar{\mathbf{y}}' - \boldsymbol{\Lambda}\mathbf{S}'_{yy}), \tag{3.27}$$

and for $\boldsymbol{\Psi}$ the diagonal elements of

$$\begin{aligned}-\frac{n}{2}\boldsymbol{\Psi}^{-1} + \frac{n}{2}\boldsymbol{\Psi}^{-1}[\mathbf{S}'_{xx} - \boldsymbol{\mu}\bar{\mathbf{x}}' - \bar{\mathbf{x}}\boldsymbol{\mu}' - \mathbf{S}'_{xy}\boldsymbol{\Lambda}' - \boldsymbol{\Lambda}\mathbf{S}'_{yx} \\ + \boldsymbol{\mu}\bar{\mathbf{y}}'\boldsymbol{\Lambda}' + \boldsymbol{\Lambda}\bar{\mathbf{y}}\boldsymbol{\mu}' + \boldsymbol{\mu}\boldsymbol{\mu}' + \boldsymbol{\Lambda}\mathbf{S}'_{yy}\boldsymbol{\Lambda}']\boldsymbol{\Psi}^{-1},\end{aligned} \tag{3.28}$$

where

$$\mathbf{S}'_{xx} = \frac{1}{n}\sum_{i=1}^{n}\mathbf{x}_i\mathbf{x}'_i.$$

To find the conditional expected values of the score functions, it is enough here to do so for the sufficient statistics:

$$\begin{aligned}\bar{\mathbf{y}} &= \frac{1}{n}\sum \mathbf{y}_i, \\ \mathbf{S}'_{xy} &= \frac{1}{n}\sum \mathbf{x}_i\mathbf{y}'_i, \\ \mathbf{S}'_{yy} &= \frac{1}{n}\sum \mathbf{y}_i\mathbf{y}'_i.\end{aligned}$$

Now,

$$E[\bar{\mathbf{y}} \mid \mathbf{x}_i, i = 1, \ldots, n] = \boldsymbol{\Lambda}'\boldsymbol{\Sigma}^{-1}(\bar{\mathbf{x}} - \boldsymbol{\mu}) = \hat{\bar{\mathbf{y}}}, \tag{3.29}$$

$$\begin{aligned}E[\mathbf{S}'_{xy} \mid \mathbf{x}_i, i = 1, \ldots, n] &= \frac{1}{n}\sum_{i=1}^{n}\mathbf{x}_i E[\mathbf{y}'_i \mid \mathbf{x}_i] \\ &= \frac{1}{n}\sum_{i=1}^{n}\mathbf{x}_i(\boldsymbol{\Lambda}'\boldsymbol{\Sigma}^{-1}(\mathbf{x}_i - \boldsymbol{\mu}))' \\ &= (\mathbf{S}'_{xx} - \bar{\mathbf{x}}\boldsymbol{\mu}')\boldsymbol{\Sigma}^{-1}\boldsymbol{\Lambda} = \hat{\mathbf{S}}'_{xy},\end{aligned} \tag{3.30}$$

$$\begin{aligned}E[\mathbf{S}'_{yy} \mid \mathbf{x}_i, i = 1, \ldots, n] &= (\mathbf{I}_q + \boldsymbol{\Lambda}'\boldsymbol{\Psi}^{-1}\boldsymbol{\Lambda})^{-1} + \boldsymbol{\Lambda}'\boldsymbol{\Sigma}^{-1}(\mathbf{S}'_{xx} - \boldsymbol{\mu}\bar{\mathbf{x}}' - \bar{\mathbf{x}}\boldsymbol{\mu}' + \boldsymbol{\mu}\boldsymbol{\mu}')\boldsymbol{\Sigma}^{-1}\boldsymbol{\Lambda} \\ &= \hat{\mathbf{S}}'_{yy}.\end{aligned} \tag{3.31}$$

These are calculated replacing $\mathbf{\Sigma}$ and $\mathbf{\Lambda}$ by their current estimates and then substituted into the score functions of (3.26), (3.27), (3.28) above. Setting the score functions equal to zero and solving,

$$\hat{\boldsymbol{\mu}} = \bar{\mathbf{x}} - \mathbf{\Lambda}\hat{\bar{\mathbf{y}}}, \tag{3.32}$$

$$\hat{\mathbf{\Lambda}} = (\hat{\mathbf{S}}'_{xy} - \bar{\mathbf{x}}\hat{\bar{\mathbf{y}}}')(\hat{\mathbf{S}}'_{yy} - \hat{\bar{\mathbf{y}}}\hat{\bar{\mathbf{y}}}')^{-1}, \tag{3.33}$$

$$\begin{aligned}\hat{\mathbf{\Psi}} = \operatorname{diag}(&\mathbf{S}'_{xx} - \hat{\boldsymbol{\mu}}\bar{\mathbf{x}}' - \bar{\mathbf{x}}\hat{\boldsymbol{\mu}}' - \hat{\mathbf{S}}_{xy}\hat{\mathbf{\Lambda}}' - \hat{\mathbf{\Lambda}}\hat{\mathbf{S}}_{yx}\\ &+ \hat{\boldsymbol{\mu}}\hat{\bar{\mathbf{y}}}'\hat{\mathbf{\Lambda}}' + \hat{\mathbf{\Lambda}}\hat{\bar{\mathbf{y}}}\hat{\boldsymbol{\mu}}' + \hat{\boldsymbol{\mu}}\hat{\boldsymbol{\mu}}' + \hat{\mathbf{\Lambda}}\hat{\mathbf{S}}'_{yy}\hat{\mathbf{\Lambda}}').\end{aligned} \tag{3.34}$$

The procedure is iterated until it converges. It is easier, as Rubin and Thayer noticed, to leave out $\boldsymbol{\mu}$, which has maximum likelihood estimator $\bar{\mathbf{x}}$. The natural way to allow for this would seem to be to substitute $\boldsymbol{\mu} = \bar{\mathbf{x}}$ into (3.29), (3.30) and (3.31) to give

$$\begin{aligned}\hat{\bar{\mathbf{y}}} &= 0,\\ \hat{\mathbf{S}}'_{xy} &= \mathbf{S}_{xx}\mathbf{\Sigma}^{-1}\mathbf{\Lambda},\\ \hat{\mathbf{S}}'_{yy} &= (\mathbf{I}_q + \mathbf{\Lambda}'\mathbf{\Psi}^{-1}\mathbf{\Lambda})^{-1} + \mathbf{\Lambda}'\mathbf{\Sigma}^{-1}\mathbf{S}_{xx}\mathbf{\Sigma}^{-1}\mathbf{\Lambda}.\end{aligned}$$

These are then used in (3.33) and (3.34) to provide new updating equations. The updating equation for $\hat{\mathbf{\Psi}}$ that results from this approach is not that obtained by Rubin and Thayer who, in effect, substitute the result of updating $\hat{\mathbf{\Lambda}}$ into the updating equation for $\hat{\mathbf{\Psi}}$.

3.6 Sampling variation of estimators

The lack of information about the sampling behaviour of parameter estimates in the factor model when used in exploratory mode is a serious defect of the standard software packages. Standard errors, whether asymptotic or exact, are only crude indicators of uncertainty in problems with many parameters, especially if they turn out to be highly correlated. Nevertheless, without some indication of the precision of estimators the interpretation of factors, which depends critically on those values, must be suspect. The need for such information was recognised from an early stage, as the Cudeck and O'Dell (1994) account shows, and the lack of adequate theory has led to the adoption of a variety of rules of thumb which bear no relation to the sampling variability of the estimators themselves. For example, it is common to regard factor loadings greater than 0.3 as 'significant' and it is not unusual to find published tables of loadings with the smaller loadings deleted. This may make the interpretation easier but it owes nothing to the statistical significance of the loadings themselves.

In the absence of exact sampling theory we can have recourse to the asymptotic theory of maximum likelihood. For the normal factor model this was first provided by Lawley (1967) for the case of unstandardised variables and extended to standardised variables by Lawley and Maxwell (1971) who also gave a numerical illustration. Subsequently, Jennrich and Thayer (1973) provided a small correction to Lawley's results. As we pointed out in Section 2.11, the estimation process yields a set of estimates rather than a single point in the parameter space. What the Lawley (1967) method did was to give the sampling behaviour of the estimator given by the standard maximum likelihood routine, namely that for which $\mathbf{\Lambda}'\mathbf{\Psi}^{-1}\mathbf{\Lambda}$ is diagonal. Estimated standard errors for the loadings of rotated factors can be obtained by the usual 'delta' methods, and this was done for orthogonal rotation by Archer and Jennrich (1973) and for oblique rotations by Jennrich (1973). An account of the history of the subject is given in Cudeck and O'Dell (1994) who go on to extend the theory in various ways, including simultaneous inference for sets of parameters based on confidence intervals. Further results can be found in Ogasawara (1998).

The heavy computations required may have inhibited the implementation of these methods in a routine fashion but they are well within the scope of modern desktop computers. The Mplus software (Muthén and Muthén 2010) and the CEFA computer program (Browne *et al.* 1998) fit the NLFM and give standard deviations for many forms of rotated loadings, whether standardised or not.

However, one does not know how large the sample size need be before the approximations provided by the asymptotic results are adequate in practice. Results reported by de Menezes (1999), and discussed in Chapter 6 below for the latent class model, show that the asymptotic theory may provide a very poor approximation even if the sample size is of the order of 1000. Similar investigations for factor analysis seem to be lacking but, given the complexities of multi-parameter models, much more comparative information is needed before the asymptotic results can be used with confidence.

An alternative approach to studying sampling behaviour is through resampling or simulation methods of the bootstrap or jackknife variety. These too have a long history in factor analysis but have not been widely used because of the amount of computing needed. However, the situation is rapidly changing and it is now possible to contemplate using the approach with complex models and large sample sizes. Tucker *et al.* (1969) carried out a detailed simulation study and Seber (1984) reported a series of simulation studies on samples of size 50 by Francis (1974).

Pennell (1972) used the jackknife to find confidence intervals for factor loadings. In the first edition (Bartholomew 1987) we reported an application of the bootstrap by Chatterjee (1984). He used data taken from Johnson and Wichern (1982) concerning seven variables measured on 50 randomly chosen salesmen. Three variables were measures of sales performance and four were from tests of aptitude. The method of fitting was the principal factor method with the ψs set equal to zero (i.e. a principal components analysis). One would have preferred a method which allowed estimation of the ψs, but with only two or three factors turning out to be significant the results are not likely to be seriously affected and the computational ease of this method commended it for this particular study. Chatterjee settled the question of how many repeated samples to draw empirically, and it appeared that 300 gave reasonable stability. Some of the results are given in Table 3.1 for factor loadings (standardised λs) for the first three factors.

Table 3.1 Estimated loadings and their standard deviations for Johnson and Wichern's (1982) data.

	Original estimates			Bootstrap estimates on 300 samples					
Var.	Fac. 1	Fac. 2	Fac. 3	Fac. 1	s.d.	Fac. 2	s.d.	Fac. 3	s.d.
1	0.973	−0.110	0.054	0.972	0.006	−0.096	0.055	0.038	0.052
2	0.943	0.029	0.312	0.945	0.013	0.010	0.091	0.192	0.158
3	0.945	0.010	−0.144	0.943	0.020	0.014	0.079	0.076	0.168
4	0.660	0.646	−0.318	0.657	0.095	0.577	0.276	−0.179	0.254
5	0.783	0.286	−0.005	0.774	0.064	0.268	0.165	0.041	0.373
6	0.649	−0.620	−0.427	0.644	0.093	−0.516	0.358	−0.266	0.229
7	0.914	−0.193	0.306	0.916	0.017	−0.185	0.091	0.197	0.173

Table 3.1 shows that the first factor is well determined but that we should be wary of attributing much significance to the other two. None of the loadings for factor 3 differ from zero by much more than their standard deviation, and only two or three of those for factor 2 come anywhere near significance. An advantage of the bootstrap method is that we can also look at the frequency distributions of the estimators. Chatterjee (1984) gives a number of examples, of which that given in Table 3.2 is particularly instructive. We notice that the large standard deviation is due to the extreme skewness of the distribution. Six percent of the samples actually gave negative loadings (if *all* loadings on a factor were negative the signs should have been reversed) but even on the positive half of the scale the scatter is considerable.

Although this example is very limited in both scope and method, it provides a warning against taking estimated loadings at their face value for sample sizes as small as 50. The results reported by Seber (1984) and others support this. Much more work is needed to extend and consolidate our limited knowledge in this important area. Results reported in later chapters for categorical variables strongly suggest that much larger samples (say, 500 or more) are needed if parameters are to be estimated with precision. The lack of reproducibility of the results of factor analysis, which has somewhat tarnished its image among practitioners, doubtless owes much to the use of inadequate sample sizes.

A much simpler approach conceptually, which can be used if the sample is large enough, is known as cross-validation. By splitting the sample randomly into two (or more) equal parts and fitting the model to each part, some limited idea can be gained about the stability of the estimates. An illustration for the latent class model, due to Pickering and Forbes (1984), is reported in Chapter 6.

Table 3.2 Frequency distribution of factor 2 loading for variable 4 (300 samples).

Mid-point of interval	0.9	0.8	0.7	0.6	0.5	0.4	0.3	0.2	0.1	0	−0.1	−0.2	−0.3	−0.4	−0.5	−0.6	−0.7
Frequency	7	43	111	52	40	22	4	2	1	0	0	3	1	6	2	5	1

3.7 Goodness of fit and choice of q

If q is specified *a priori*, the goodness of fit of the factor model can be judged using the likelihood ratio statistic for testing the hypothesis $\boldsymbol{\Sigma} = \boldsymbol{\Lambda}\boldsymbol{\Lambda}' + \boldsymbol{\Psi}$ (H_0) against the alternative that $\boldsymbol{\Sigma}$ is unconstrained (H_1). The statistic is then

$$\begin{aligned} -2\{L(H_0) - L(H_1)\} &= n\{\log|\hat{\boldsymbol{\Sigma}}| + \text{trace}\,\hat{\boldsymbol{\Sigma}}^{-1}\mathbf{S} - \log|\mathbf{S}| - p\} \\ &= n\{\text{trace}\,\hat{\boldsymbol{\Sigma}}^{-1}\mathbf{S} - \log|\hat{\boldsymbol{\Sigma}}^{-1}\mathbf{S}| - p\} \end{aligned} \quad (3.35)$$

where $\hat{\boldsymbol{\Sigma}} = \hat{\boldsymbol{\Lambda}}\hat{\boldsymbol{\Lambda}}' + \hat{\boldsymbol{\Psi}}$ is the estimated covariance matrix. If $\boldsymbol{\Psi} > \mathbf{0}$ this statistic is asymptotically distributed as χ^2 with degrees of freedom

$$\nu = \frac{1}{2}p(p+1) - \left\{pq + p - \frac{1}{2}q(q-1)\right\} = \frac{1}{2}\{(p-q)^2 - (p+q)\}. \quad (3.36)$$

This is the difference between the number of parameters in $\boldsymbol{\Sigma}$ and the number of free parameters on the null hypothesis. Bartlett (1950) showed that the approximation can be improved by replacing n in (3.35) by $n - 1 - \frac{1}{6}(2p+5) - \frac{2}{3}q$. The behaviour of the test when n is small was investigated by Geweke and Singleton (1980), whose results suggest that it is adequate for n as low as 30 with one or two factors. However, Schönemann (1981) argues that this is too optimistic because they used untypical examples with high communalities.

Since q is not usually specified in advance the test is often made the basis of a procedure for choosing the best value. Starting with $q = 1$, we then take successive values in turn until the fit of the model is judged to be adequate. Viewed as a testing procedure this is not strictly valid because it does not adjust the significance levels to allow for the sequential character of the test. It rather depends on regarding the p-value of the test as a measure of the adequacy of the model.

3.7.1 Model selection criteria

The trouble with a procedure of this kind is that the larger we make q the better the fit, but it provides us with no criterion for judging when to stop. An alternative approach is provided by Akaike's *information criterion* for model selection. The situation is that we have a set of linear models indexed by q and a selection has to be made. Akaike (1983) proposed that method for use in factor analysis and showed that it required q to be chosen to make

$$\text{AIC} = -2L + 2\nu$$

a minimum, where ν is the number of free parameters in the model, and L is the maximised value of the log-likelihood function for the estimated model. Note that L and ν are both functions of q. The criterion can be justified in a variety of ways but, in essence, it effects a trade-off between the bias introduced by fitting the wrong number of factors and the precision with which the parameters are estimated—as q

is increased the bias decreases but the error increases. An alternative version, due to Schwarz (1978), also known as the *Bayesian information criterion* (BIC), replaces the term 2ν by $\nu \ln n$. A further criterion based on the residuals of the fitted correlation matrix is proposed by Bozdogan and Ramirez (1986), who also report a comparison of all three criteria using Monte Carlo methods applied to models used in the study of Francis (1974). All methods perform reasonably well, but for the examples considered no method is uniformly best. These model selection ideas show considerable promise and their performance over a wide range of models merits further investigation.

Other criteria for selecting q which do not involve distributional assumptions will be mentioned in Section 3.11.

3.8 Fitting without normality assumptions: least squares methods

If nothing is assumed about the distributions of $\mathbf{y}$ and $\mathbf{e}$, it remains true that the covariance matrix predicted by the model is given by $\boldsymbol{\Sigma} = \boldsymbol{\Lambda}\boldsymbol{\Lambda}' + \boldsymbol{\Psi}$. We could then aim to estimate $\boldsymbol{\Lambda}$ and $\boldsymbol{\Psi}$ in such a way that $\boldsymbol{\Sigma}$ was as close to $\mathbf{S}$, the sample covariance matrix, as is possible. For this we need some scalar measure of distance between $\boldsymbol{\Sigma}$ and $\mathbf{S}$ which must then be minimised with respect to the parameters. The distance function

$$\Delta(\boldsymbol{\Sigma}, \mathbf{S}) = -\text{trace}\, \boldsymbol{\Sigma}^{-1}\mathbf{S} + \log |\boldsymbol{\Sigma}^{-1}\mathbf{S}| \tag{3.37}$$

which arose in the course of maximum likelihood estimation is one possibility. It will only yield maximum likelihood estimators if the normal assumptions hold, but if (3.37) were regarded as a reasonable way of measuring distance the estimators would be justified on a much broader basis.

There are many other possible measures of distance which could be used in place of (3.37). It is natural to turn to least squares ideas. A simple unweighted least squares criterion would be

$$\Delta_1 = \sum_{i=1}^{p}\sum_{u=1}^{p}(s_{iu} - \sigma_{iu})^2 = \text{trace}(\mathbf{S} - \boldsymbol{\Sigma})^2. \tag{3.38}$$

Another, suggested by the role played by the matrix $\boldsymbol{\Psi}^{-1/2}\mathbf{S}\boldsymbol{\Psi}^{-1/2}$ in the Lawley and Maxwell approach to maximum likelihood estimation, is

$$\Delta_2 = \text{trace}\{[\boldsymbol{\Psi}^{-1/2}(\mathbf{S} - \boldsymbol{\Sigma})\boldsymbol{\Psi}^{-1/2}]^2\} = \text{trace}\{(\mathbf{S} - \boldsymbol{\Sigma})\boldsymbol{\Psi}^{-1}\}^2. \tag{3.39}$$

These are both special cases of a general class of measures stemming from fundamental work by Browne (1982, 1984) (see also Tanaka and Huba (1985)) which may be written

$$\Delta = \text{trace}\{(\mathbf{S} - \boldsymbol{\Sigma})\mathbf{V}\}^2. \tag{3.40}$$

These arise from a generalised least squares approach which allows the deviations $\{s_{ij} - \sigma_{ij}\}$ to be weighted in various ways. We shall not develop these ideas here, but reference to the authors mentioned above will show that these methods lead to robust methods of estimation under a range of distributional assumptions. The case $\mathbf{V} = \mathbf{S}^{-1}$ was investigated by Jöreskog and Goldberger (1972); see also Anderson (1984, Section 14.3.4). The attraction of Δ_1 and Δ_2 is that, like maximum likelihood, the optimisation requires the solution of an eigenvalue problem. In the case of Δ_1 the function to be minimised may be written

$$\Delta_1 = \sum_{i=1}^{p}\sum_{u=1}^{p}\left(s_{iu} - \delta_{iu}\psi_i - \sum_{j=1}^{q}\lambda_{ij}\lambda_{uj}\right)^2.$$

Differentiating with respect to λ_{rs},

$$\frac{\partial \Delta_1}{\partial \lambda_{rs}} = 4\left\{-\sum_{i=1}^{p}(s_{ri} - \delta_{ir}\psi_i)\lambda_{is} + \sum_{i=1}^{p}\lambda_{is}\sum_{j=1}^{q}\lambda_{ij}\lambda_{rj}\right\}$$

$$(r = 1, 2, \ldots, p;\ s = 1, 2, \ldots, q) \tag{3.41}$$

or

$$\frac{\partial \Delta_1}{\partial \mathbf{\Lambda}} = 4\{\mathbf{\Lambda}(\mathbf{\Lambda}'\mathbf{\Lambda}) - (\mathbf{S} - \mathbf{\Psi})\mathbf{\Lambda}\},$$

which gives the estimating equations

$$(\mathbf{S} - \mathbf{\Psi})\mathbf{\Lambda} = \mathbf{\Lambda}(\mathbf{\Lambda}'\mathbf{\Lambda}). \tag{3.42}$$

Differentiating with respect to ψ_r,

$$\frac{\partial \Delta_1}{\partial \psi_r} = -2\left(s_{rr} - \psi_r - \sum_{j=1}^{q}\lambda_{rj}^2\right)$$

or

$$\operatorname{diag}\frac{\partial \Delta_1}{\partial \mathbf{\Psi}} = -\operatorname{diag}\mathbf{S} + \mathbf{\Psi} + \operatorname{diag}\mathbf{\Lambda}\mathbf{\Lambda}',$$

leading to

$$\mathbf{\Psi} = \operatorname{diag}(\mathbf{S} - \mathbf{\Lambda}\mathbf{\Lambda}'). \tag{3.43}$$

These estimating equations are similar to those obtained for maximum likelihood estimation in (3.14) and (3.17) and are solved in a similar manner. Suppose first that $\mathbf{\Psi}$ is known and that $\mathbf{S} - \mathbf{\Psi}$ is positive definite. Then (3.42) will be satisfied if:

1. the columns of $\mathbf{\Lambda}$ consist of any q eigenvectors of $\mathbf{S} - \mathbf{\Psi}$;
2. $\mathbf{\Lambda}'\mathbf{\Lambda}$ is a diagonal matrix with elements equal to the eigenvalues of $\mathbf{S} - \mathbf{\Psi}$ associated with the vectors in $\mathbf{\Lambda}$.

Thus if we have a starting value for $\mathbf{\Psi}$, the solution of (3.42) will yield a first approximation to $\mathbf{\Lambda}$ which can then be inserted in (3.43) to give a second estimate of $\mathbf{\Psi}$, and then the cycle can be continued until convergence occurs. The question of which eigenvectors of $\mathbf{S} - \mathbf{\Psi}$ are to be included in $\mathbf{\Lambda}$ can be answered as follows:

$$\begin{aligned}\Delta_1 &= \text{trace}(\mathbf{S} - \mathbf{\Psi} - \mathbf{\Lambda}\mathbf{\Lambda}')^2 \\ &= \text{trace}(\mathbf{S} - \mathbf{\Psi})^2 - 2\text{trace}(\mathbf{S} - \mathbf{\Psi})\mathbf{\Lambda}\mathbf{\Lambda}' + \text{trace}(\mathbf{\Lambda}\mathbf{\Lambda}')^2 \\ &= \text{trace}(\mathbf{S} - \mathbf{\Psi})^2 - \text{trace}(\mathbf{\Lambda}\mathbf{\Lambda}')^2,\end{aligned}$$

using (3.42).

Now $\mathbf{\Lambda}\mathbf{\Lambda}'$ has $(p - q)$ zero eigenvalues because it is of rank q. The others are also eigenvalues of $\mathbf{S} - \mathbf{\Psi}$ since if we replace $\mathbf{S} - \mathbf{\Psi}$ in (3.42) by $\mathbf{\Lambda}\mathbf{\Lambda}'$ the equation is obviously satisfied. Let the eigenvalues which the two matrices have in common be $\theta_1, \theta_2, \ldots, \theta_q$ and the remaining eigenvalues of $\mathbf{S} - \mathbf{\Psi}$ be $\theta_{q+1}, \ldots, \theta_p$; then

$$\Delta_1 = \sum_{i=1}^{p} \theta_i^2 - \sum_{i=1}^{q} \theta_i^2 = \sum_{i=q+1}^{p} \theta_i^2. \tag{3.44}$$

For this to be a minimum, $\theta_{q+1}, \ldots, \theta_p$ must be the smallest eigenvalues and hence $\mathbf{\Lambda}$ must consist of the eigenvectors associated with the q largest eigenvalues.

In this method $\mathbf{\Psi}$ is chosen so that $\mathbf{S} - \mathbf{\Psi}$ is positive definite and so that the sum of squares of the $p - q$ smallest eigenvalues of $\mathbf{S} - \mathbf{\Psi}$ is as small as possible.

3.9 Other methods of fitting

The foregoing method is known as the *principal factor* (or *principal axis*) method because of its similarity to principal components analysis to which it is equivalent if $\mathbf{\Psi} = \mathbf{0}$. If we use the correlation matrix $\mathbf{R}$ instead of $\mathbf{S}$ the estimates obtained for the scale-invariant parameters $\mathbf{\Lambda}^*$ and $\mathbf{\Psi}^*$ defined in Section 3.12.2 will not be identical to those arrived at by first using $\mathbf{S}$ and then transforming as they were with the maximum likelihood method.

Estimation for Δ_2 using (3.39) proceeds in an exactly similar manner. In fact, since

$$\mathbf{\Psi}^{-1/2}\mathbf{\Sigma}\mathbf{\Psi}^{-1/2} = \mathbf{\Lambda}^*\mathbf{\Lambda}^{*\prime} + \mathbf{I} = (\mathbf{\Psi}^{-1/2}\mathbf{\Lambda})(\mathbf{\Psi}^{-1/2}\mathbf{\Lambda})' + \mathbf{I},$$

all we have to do is to replace $\mathbf{\Lambda}$ by $\mathbf{\Lambda}^* = \mathbf{\Psi}^{-1/2}\mathbf{\Lambda}$ in (3.42) and $\mathbf{S} - \mathbf{\Psi}$ by $\mathbf{S}^* - \mathbf{I}$. The estimating equation for $\mathbf{\Lambda}^*$ is then identical to that for maximum likelihood. Rather surprisingly, it turns out that the differing distance functions both lead to the eigenvalues and vectors of $\mathbf{S}^*$. However, this is true only for fixed $\mathbf{\Psi}$. If we bring $\mathbf{\Psi}$ into the picture its partial derivatives will be different from those in the maximum likelihood case and, in fact, a good deal more complicated. With this method the choice of $\mathbf{\Psi}$ is made so that $\mathbf{\Psi}^{-1/2}\mathbf{S}\mathbf{\Psi}^{-1/2}$ has eigenvalues all no less than 1 and so that the sum of the squared differences from 1 of the $p - q$ smallest eigenvalues is as small as possible.

For Δ given by (3.40) with $\mathbf{V} = \mathbf{S}^{-1}$ the same equation for $\mathbf{\Lambda}$ for fixed $\mathbf{\Psi}$ is obtained, but the (implicit) estimating equation for $\mathbf{\Psi}$ is

$$\operatorname{diag} \mathbf{S}^{-1}\{(\mathbf{\Lambda}\mathbf{\Lambda}' + \mathbf{\Psi}) - \mathbf{S}\}\mathbf{S}^{-1} = \operatorname{diag} \mathbf{0};$$

see Anderson (1984). Here the sum of the squared differences from 1 of the reciprocals of the $p - q$ smallest eigenvalues of $\mathbf{\Psi}^{-1/2}\mathbf{S}\mathbf{\Psi}^{-1/2}$ is minimised.

We have noted that if $\mathbf{\Psi}$ were known, these methods would be much simpler, being straightforward eigenproblems. Since $\mathbf{\Psi}$ only enters into the diagonal elements of $\mathbf{\Sigma}$, there is a prospect of avoiding the difficulties by eliminating these terms from Δ_1. We would then minimise

$$\Delta_1' = \sum_{i=1,\ i \neq u}^{p} \sum_{u=1}^{p} \left(s_{iu} - \sum_{j=1}^{q} \lambda_{ij}\lambda_{uj} \right)^2. \tag{3.45}$$

This approach is described in Harman and Jones (1966) and is usually known as the 'minres' method. Various methods of obtaining estimates have been given by Comrey (1962), Comrey and Ahumada (1964), Okamoto and Ihara (1983) and Zegers and ten Berge (1983). Unfortunately the omission of the diagonal terms destroys the structure which led to the easily solved eigenequations. Given that the minres method is unweighted and that the iterative methods for the Δ-family are well within the scope of modern computers, the method offers few advantages and is not included in modern computer packages.

There are yet other methods of estimating the normal linear factor model. Knott (2005) has a contribution to, and references for, the *minimum trace* method.

3.10 Approximate methods for estimating $\mathbf{\Psi}$

The iterative methods require a starting value for $\mathbf{\Psi}$, and although one can use an arbitrary value such as $\mathbf{\Psi} = \operatorname{diag} \mathbf{S}$ there are approximations which will reduce the number of iterations required. Apart from their practical value they give some insight into the interpretation of the analysis. We have a non-negative definite matrix

$$\mathbf{\Sigma} - \mathbf{\Psi} \tag{3.46}$$

and so, assuming now that $\mathbf{\Psi}$ is invertible,

$$\mathbf{\Psi}^{-1} - \mathbf{\Sigma}^{-1} \tag{3.47}$$

is non-negative definite, (see, for instance, Rao (1973, p. 70, Exercise 9)), from which it follows that

$$\sigma^{ii} \leq \psi_r^{-1}. \tag{3.48}$$

Since σ^{ii} can be estimated from the inverse of $\mathbf{S}$, we can estimate an upper bound for ψ_i and this may be used as a starting value for the iteration.

This result may be given another interpretation: s^{ii}, which we would use as an approximation, may be expressed using standard regression results as

$$s^{ii} = s_{ii}(1 - R_i^2), \tag{3.49}$$

where R_i^2 is the multiple correlation coefficient of x_i regressed on the remaining xs. Now

$$s_{ii}(1 - R_i^2) = \text{var}(x_i \mid x_1, x_2, \ldots, x_{i-1}, x_{i+1}, \ldots, x_p)$$

and

$$\psi_i = \text{var}(x_i \mid \mathbf{y}).$$

We would expect the latter to be smaller than the former because $\mathbf{y}$ is not precisely determined by any finite set of xs (see (3.6)). A simpler but less precise bound for ψ follows from the fact that $R_i^2 \geq \max_{j \neq i} r_{ij}^2$, where r_{ij} is the correlation between x_i and x_j. Hence an estimated upper bound for ψ_i is $s_{ii}(1 - \max r_{ij}^2)$.

The approximation $\hat{\mathbf{\Psi}} \approx \left(\text{diag}\,\mathbf{S}^{-1}\right)^{-1}$ derived from (3.48) is the basis of *image factor analysis*. This was proposed by Guttman (1953); see also Jöreskog (1969). Like other methods of fitting the NLFM which seem to have fallen into disuse, it offered a relatively simple method of fitting at a time when computing facilities were very modest. Treating $\mathbf{\Psi}$ as if it were known and proportional to $(\text{diag}\,\mathbf{S}^{-1})^{-1}$ means that the fitting can be carried out by the principal axis method without iteration. An iterative version can be used in which $\mathbf{\Psi}$ is updated using the latest values of $\hat{\mathbf{\Psi}}$, but Basilevsky (1994, Section 5.33) has shown that this leads to inconsistent estimators. With modern computing facilities there is no longer any need for such methods.

3.11 Goodness of fit and choice of q for least squares methods

Little appears to be known about the sampling behaviour of the methods of fitting discussed in Sections 3.8–3.10, but there is an important result due to Amemiya and Anderson (1985) which shows that the limiting χ^2 distribution of the goodness-of-fit statistic of (3.40) is valid under very general circumstances. They show that if the elements of $\mathbf{e}$ are independent and if $\mathbf{y}$ and $\mathbf{e}$ have finite second moments then (3.35) calculated using the maximum likelihood estimators has the same distribution as in the normal case. This result also holds for another goodness-of-fit statistic which is sometimes used, namely

$$\frac{1}{2}n \operatorname{trace}\{(\mathbf{S} - \hat{\boldsymbol{\Sigma}})\hat{\boldsymbol{\Sigma}}^{-1}\}^2$$

which, in the normal case, is asymptotically equivalent to (3.35). This is a further reason for using the maximum likelihood method of fitting, whether or not one makes the normality assumptions.

One consequence of these results is that we can use the same methods for choosing q as were proposed for the maximum likelihood method in Section 3.7. There are also two other methods which do not depend on distributional assumptions. Both are based on the eigenvalues of the sample correlation matrix and the role which they have in principal components analysis. The Kaiser–Guttman criterion chooses q equal to the number of eigenvalues greater than unity. The rationale is that the average contribution of a manifest variable to the total variation is 1, and that a principal component which did not contribute at least as much variation as a single variable represents no advantage. The carry-over of this argument from principal components to factor analysis rests on the similarity between the two techniques noted in Chapters 1 and 9. Simulation results obtained by Fachel (1986) suggest that if p is large this method is likely to overestimate q.

The second method, due to Cattell, is known as the 'scree test'. If the eigenvalues are plotted against their rank order they will lie on a decreasing curve. One then looks for an 'elbow' in the curve, as this would indicate the point at which the further addition of factors shows diminishing returns in terms of variation explained. A simulation study by Hakistian *et al.* (1982) comparing these two methods with the use of the likelihood ratio statistic does not lead to clear-cut conclusions.

3.12 Further estimation issues

3.12.1 Consistency

In the discussion of estimation methods we have tacitly assumed that the parameters could be consistently estimated. A necessary condition for this to be possible is that there shall be at least as many sample statistics as there are parameters to be estimated. The number of parameters in a q-factor model is $pq+p$, but in order to obtain a unique solution we have to impose $\frac{1}{2}q(q-1)$ constraints (usually by requiring the off-diagonal elements of $\boldsymbol{\Lambda}'\boldsymbol{\Psi}^{-1}\boldsymbol{\Lambda}$ to be zero). The number of free parameters is then

$$pq + p - \frac{1}{2}q(q-1).$$

The sample covariance matrix $\mathbf{S}$ has $\frac{1}{2}p(p+1)$ distinct elements, so for consistent estimation to be possible we must have

$$\frac{1}{2}p(p+1) - pq - p + \frac{1}{2}q(q-1) = \frac{1}{2}[(p-q)^2 - (p+q)] \geq 0. \tag{3.50}$$

Equation (3.50) implies that there is an upper bound to the number of factors which can be fitted which is given by

$$q \leq \frac{1}{2}\{2p + 1 - (8p+1)^{1/2}\}. \tag{3.51}$$

The condition in (3.50) is not sufficient because it does not guarantee that the estimates of the ψ_i will be non-negative. Building on earlier work by Anderson and Rubin (1956), Kano (1983, 1986a,b) has provided general conditions under which maximum likelihood and generalised least squares estimators are consistent. However, in Kano (1986a) he provides an example of a model which does not admit any consistent estimator. The question of consistency when p is not fixed is investigated in Kano (1986b).

3.12.2 Scale-invariant estimation

In practice the factor model is almost always fitted using the sample correlation matrix rather than the covariance matrix. The reason usually given for this is that the scaling adopted for the manifest variables is often arbitrary, especially in social science applications. Using correlations rather than covariances amounts to standardising the xs using their sample standard deviations, and this ensures that changing the scale of the xs has no effect on the analysis. However, the joint distribution of the correlation coefficients is not the same as that of the covariances and so it is not obvious that the estimators obtained in the way described above will be true maximum likelihood estimators or that the asymptotic goodness-of-fit test will be valid. A good deal of confusion has surrounded this topic so we shall approach the matter from first principles.

We start from the proposition that if the units of measurement of the xs are arbitrary then there can be no scientific interest in any aspect of the model which depends on the choice of units. The effect of a scale change on the model of (3.3) can be seen by transforming to $\mathbf{x}^* = \mathbf{C}\mathbf{x}$, where $\mathbf{C}$ is a diagonal matrix with positive elements. Expressed in terms of $\mathbf{x}^*$, the model becomes

$$\mathbf{x}^* = \mathbf{C}\boldsymbol{\mu} + \mathbf{C}\boldsymbol{\Lambda}\mathbf{y} + \mathbf{C}\mathbf{e}, \tag{3.52}$$

with

$$\text{var}(\mathbf{x}^*) = \mathbf{C}\boldsymbol{\Lambda}\boldsymbol{\Lambda}'\mathbf{C} + \mathbf{C}\boldsymbol{\Psi}\mathbf{C}.$$

In general, the parameters $\mathbf{\Lambda}$ and $\mathbf{\Psi}$ do not therefore meet our requirements since their estimated values and the interpretation to be put upon them will depend on the scales of the xs. Suppose, however, that we choose $\mathbf{C} = (\text{diag}\,\mathbf{\Sigma})^{-1/2}$ so that each x is divided by its theoretical standard deviation; then we may write (3.52) as

$$\mathbf{x}^* = \boldsymbol{\mu}^* + \mathbf{\Lambda}^*\mathbf{y} + \mathbf{e}^*, \tag{3.53}$$

with $\text{var}(\mathbf{x}^*) = \mathbf{\Lambda}^*\mathbf{\Lambda}^{*\prime} + \mathbf{\Psi}^*$, where $\mathbf{\Lambda}^* = \mathbf{C\Lambda}$ and $\mathbf{\Psi}^* = \mathbf{C\Psi C}$.

Clearly, any change in the scale of $\mathbf{x}$ now has no effect on the value of $\mathbf{x}^*$ and hence the parameter estimates obtained for $\mathbf{\Lambda}^*$ and $\mathbf{\Psi}^*$ will be independent of the scaling of $\mathbf{x}$. These parameters will be said to be *scale-invariant* and it is on them that our interest will be centred. We shall also refer to $\mathbf{\Lambda}^*$ and $\mathbf{\Psi}^*$ as the parameters of the model in *standard form*.

The important difference between this approach and that starting with the sample correlations is that in the latter case the xs would be standardised using the *sample* standard deviations. Their distribution would not then be normal, and that would be inconsistent with the assumptions we have made about the form and distribution of the right-hand side of (3.3).

To estimate $\mathbf{\Lambda}^*$ and $\mathbf{\Psi}^*$ one could first estimate $\mathbf{\Lambda}$ and $\mathbf{\Psi}$ for any arbitrary scaling of the xs using the sample covariance matrix and then transform them by

$$\hat{\mathbf{\Lambda}}^* = (\text{diag}\,\hat{\mathbf{\Sigma}})^{-1/2}\hat{\mathbf{\Lambda}} \quad \text{and} \quad \hat{\mathbf{\Psi}}^* = (\text{diag}\,\hat{\mathbf{\Sigma}})^{-1/2}\,\hat{\mathbf{\Psi}}(\text{diag}\,\hat{\mathbf{\Sigma}})^{-1/2}.$$

However, it turns out that the usual practice of treating the correlation matrix as if it were the covariance matrix does yield the maximum likelihood estimators of the scale-invariant parameters. This was shown by Krane and McDonald (1978) and an outline of the justification is as follows.

We reparameterise the likelihood by writing $\mathbf{T} = \mathbf{C}^{-1/2}\mathbf{\Sigma}\mathbf{C}^{-1/2}$, where $\mathbf{C} = \text{diag}\,\mathbf{\Sigma}$. From (3.11) we then have

$$\begin{aligned} -2L/n &= \text{constant} + \log|\mathbf{C}^{1/2}\mathbf{T}\mathbf{C}^{1/2}| + \text{trace}(\mathbf{C}^{1/2}\mathbf{T}\mathbf{C}^{1/2})^{-1}\mathbf{S} \\ &= \text{constant} + \log|\mathbf{C}| + \log|\mathbf{T}| + \text{trace}\,\mathbf{C}^{-1/2}\mathbf{T}^{-1}\mathbf{C}^{-1/2}\mathbf{S} \\ &= \text{constant} + \log|\mathbf{C}| + \log|\mathbf{T}| + \text{trace}\,\mathbf{T}^{-1}\mathbf{C}^{-1/2}\mathbf{S}\mathbf{C}^{-1/2}. \end{aligned} \tag{3.54}$$

The matrix $\mathbf{T}$ involves only the parameters $\mathbf{\Lambda}^*$, since its diagonal elements are now unity. The expression in (3.54) may be maximised in two stages, first with respect to $\mathbf{C}$ and then with respect to $\mathbf{T}$. Krane and McDonald showed, as one might have anticipated, that $\mathbf{C} = \text{diag}\,\mathbf{S}$, so at the second stage the quantity to be maximised is

$$\log|\mathbf{T}| + \text{trace}\,\mathbf{T}^{-1}\mathbf{R},$$

where $\mathbf{R} = (\text{diag}\,\mathbf{S})^{-1/2}\mathbf{S}(\text{diag}\,\mathbf{S})^{-1/2}$ which is exactly what we would have done if we had treated the sample correlation matrix as a covariance matrix. It is easily verified that the log-likelihood ratio statistic is the same whichever parameterisation is used, so the procedure for estimation and testing goodness of fit is fully justified.

3.12.3 Heywood cases

All the methods of estimating the parameters of the factor model involve minimising a measure of the distance between $\mathbf{S}$ and $\mathbf{\Sigma}$. However, the parameter space is restricted by the condition $\mathbf{\Psi} \geq \mathbf{0}$ so the usual procedure of setting all the partial derivatives equal to zero may yield a solution with a negative ψ. In such cases the minimum we seek will lie on a boundary of the admissible region at a point where one or more of the ψs is zero. When this happens we have what is called a Heywood case, after Heywood (1931). In practice this will be recognised either by the appearance of a negative estimate or by the convergence of an estimate to zero, depending on the algorithm used.

There is no inconsistency in the occurrence of a zero residual variance and, taken at its face value, it would simply mean that the variation of the manifest variable in question was wholly explained by the latent variables. In practice this rarely seems plausible and the rather frequent occurrence of Heywood cases has caused a good deal of unease among practitioners. Taken with the fact that it is known empirically that zero estimates can easily arise when the true parameter values are not zero, there is good reason for not taking such estimates at their face value.

There is a good deal of evidence, mainly of an empirical kind, about the circumstances under which Heywood cases are likely to occur. Much of this is based on simulation studies where there is no question of the model itself being invalid. The results are widely scattered in the literature but we have drawn heavily in what follows on van Driel (1978), Anderson and Gerbing (1984), Boomsma (1985) and Fachel (1986).

If a Heywood case arises when the data conform to the linear factor model it will probably be the result of sampling error. A key factor is therefore sample size. For a model with positive ψs the probability of a Heywood case tends to zero as n tends to infinity. We are therefore dealing with a small-sample phenomenon. The risk of a Heywood case depends on other factors mentioned below, but as a rough guide it appears that it is high with sample sizes of 100 or less and low with samples of 500 or more.

For a given sample size the risk decreases as p, the number of variables, increases. Obviously the risk will also be greater if one or more ψs is very small, but this is not something which would be known in advance. Nevertheless, one can obtain some clues from the data. For example, in Section 3.10 and the remarks that followed, we saw that an estimated upper bound for ψ could be obtained and, in particular, that ψ_i would be small if x_i was highly correlated with any other variable. The presence of one or more high correlations is therefore indicative of a potential Heywood case.

One of the commonest cause of Heywood cases is the attempt to extract more factors than are present. This is readily demonstrated by simulation but might have been anticipated on the grounds that artificially inflating the communality forces the residuals towards zero.

Once the causes of Heywood cases are understood, the ways of dealing with them become clearer. At the stage of designing an enquiry one should aim for a large sample with a good number of variables. But in selecting variables it is important to avoid introducing new variables which add little to those already there. This will merely create high correlations without contributing significantly to the information about the latent variables.

At the analysis stage the options are more limited. Over-factoring can be avoided by paying careful attention to the various criteria suggested in Section 3.7. If a highly correlated pair of variables appears to be the cause, one of them can be dropped with little loss. However, it does not follow that dropping a variable which is implicated in a Heywood case is always advisable. If it arises because of a single small residual variance we should then be omitting one of the more valuable variables which was a relatively pure indicator of the latent variables. In any case, experience shows that such a course often leads to a Heywood case involving another variable, which defeats the object.

If the foregoing precautions fail there are at least three courses still open:

1. We can use a Bayesian approach like that suggested by Martin and McDonald (1975). Here we maximise not the likelihood but the posterior density, using a prior distribution for the ψs which assigns zero probability to negative values. Martin and McDonald propose a form which is both tractable and plausible in that it implies a distribution which is almost uniform except that it decreases to zero at the point $\psi_i = 0$. Lee (1981) also investigated the form of the posterior density under different informative prior distributions, some of which have been designed to deal with Heywood cases. Furthermore, a simulation-based Bayesian estimation approach discussed in Sections 2.10 and 4.11 has been also used by Shi and Lee (1998) for the factor analysis model for continuous variables that can also handle categorical variables.

2. We may stop the iteration at some arbitrarily small value of ψ_i such as 0.05 or 0.01. In effect, this is a special case of course of action 1 with a uniform prior on the interval $(\epsilon, 1)$ where ϵ is the chosen cut-off point. It may be justified on the grounds that the likelihood (or other distance measure) will be very close to its optimum and all that matters for purposes of interpretation is to know that ψ_i is 'small'.

3. A third method which shows promise but needs further investigation rests on the following simple idea. If a variable x_i has a small ψ_i we could increase the latter by adding to x_i an independent random variable with known variance σ^2. Denoting the new variable by x_i', its variance would then be

$$\mathrm{var}(x_i') = \sum_{j=1}^{q} \lambda_{ij}^2 + \psi_i + \sigma^2,$$

but the covariances would be unchanged.

If x_i' were used instead of x_i in estimating the parameters, we would obtain an estimate of $\psi_i + \sigma^2$. Knowing σ^2, we could then obtain an estimate of ψ_i

by subtraction. If this still led to a negative ψ_i the procedure would need to be repeated with a larger value of σ^2. It is not, in fact, necessary to add the artificial variable to each value of x_i; we could simply add σ^2 to the appropriate diagonal element of **S**.

If the analysis is carried out on the correlation matrix the effect of replacing x_i by x_i' is to multiply the off-diagonal elements in the ith row of the sample correlation matrix by $(1+\sigma^2)^{-1/2}$. The relationship between the parameters of the original model and the modified one is then given by

$$\lambda_{ij}^* = \lambda_{ij}'(1+\sigma^2)^{1/2}, \quad \psi_i^* = 1 - (1+\sigma^2)\sum_{j=1}^{q} \lambda_{ij}'^2, \tag{3.55}$$

where the asterisk denotes the modified loading. The result holds for any value of σ^2 but it seems reasonable to choose a value just large enough to avoid the occurrence of a Heywood case. In practice $\sigma^2 = 1$ seems to be the right order of magnitude. It should be noted that the results on the standard errors of the estimates and on goodness of fit will be invalidated by this device.

3.13 Rotation and related matters

In Section 2.11 we introduced the idea of orthogonal and oblique rotations in the factor space. These transformations have the property that they leave the joint distribution of the manifest variables unchanged and so the fit of the model is unaffected. A choice of rotation thus has to be made on non-statistical grounds. We saw that patterns of loadings exhibiting simple structure were relatively easy to interpret. The rotations available in the major software packages for the NLFM all embody some algorithm for getting as close as possible to simple structure.

If only two factors have been fitted the position will usually be clear from a plot of the loadings. In the space of the factor loadings one is looking to position the axes so that the points lie close to one or other axis. This is often all that is necessary, and we shall illustrate the method on the examples in Section 3.17.

3.13.1 Orthogonal rotation

One characteristic of simple structure is that the loadings on any factor are either zero or large. Since the squares of standardised loadings lie between 0 and 1, an algorithm which seeks to maximise their variance will move them towards opposite ends of the range. This is the varimax approach which was proposed by Kaiser (1958). In an alternative version λ_{ij}^2 is replaced by $\lambda_{ij}^2/(1-\Psi_i)$ $(i = 1, 2, \ldots, p)$ in the criterion for maximisation. The maximisation is achieved iteratively, and details are given in Magnus and Neudecker (1988).

Another way of describing simple structure is in terms of the variance of the squares of the factor loadings across the rows of $\mathbf{\Lambda}$ rather than down the columns. This idea is implemented in the quartimax method. It would, of course, be possible to combine the characteristics of varimax and quartimax in a single criterion.

3.13.2 Oblique rotation

An oblique rotation offers a better chance of finding simple structure, but at the price of complicating the interpretation. In practical terms there is no obvious reason why factors of substantive interest should be uncorrelated as orthogonality implies. For example, if one is seeking to identify two factors called verbal ability and numerical ability in an educational test, there is every reason to expect them to be correlated and, if so, they will not be uncovered by a search restricted to orthogonal factors. Algorithms similar to those in the orthogonal case, but without the constraint of orthogonality, are available. Details of the options available may be found in the specialist literature, including Tabachnick and Fidell (1996) and McDonald (1985). We shall use the OBLIMIN routine provided by Mplus (Muthén and Muthén 2010).

When the factors are uncorrelated we saw that the standardised factor loadings could be interpreted as the correlation coefficients of the manifest variables and the factors. Under oblique rotation this is no longer true. The calculation in (a) of Section 3.2 now gives

$$E(\mathbf{x} - \boldsymbol{\mu})\mathbf{y}' = \boldsymbol{\Lambda\Phi}, \tag{3.56}$$

where $\boldsymbol{\Phi}$ is the correlation matrix of the ys. The matrix of correlations is sometimes called the *structure loading* matrix to distinguish it from the *pattern loading* matrix $\boldsymbol{\Lambda}$.

3.13.3 Related matters

There are other methods of fitting the factor model which are best regarded as ways of selecting a particular rotation. One of these is given by Rao (1955). This approach views the problem in the context of canonical correlation, where the object is to find linear combinations of one group of variables which are maximally correlated with linear combinations of a second group. In factor analysis the xs form one group and the ys the other. Using the methods of canonical correlation, we find that linear combination of the xs and of the ys which are most highly correlated. Then we find two further linear combinations, uncorrelated with the corresponding member of the first pair, and so on. This procedure leads to estimates of $\boldsymbol{\Lambda}$ and $\boldsymbol{\Psi}$ which satisfy the maximum likelihood equations and also the constraint that $\boldsymbol{\Lambda}'\boldsymbol{\Psi}^{-1}\boldsymbol{\Lambda}$ be diagonal. In effect, therefore, it is selecting the rotation with this property and, in the process, providing another reason for choosing that particular solution.

Alpha factor analysis can be regarded in the same light. This was proposed by Kaiser and Caffrey (1965) and based on the psychometric notion of generalisability which we considered in Section 2.17. It is based on the eigendecomposition of the correlation matrix of the common parts of the manifest variable. Thus for the model

$$\mathbf{x} = \boldsymbol{\mu} + \boldsymbol{\Lambda}\mathbf{y} + \mathbf{e}, \tag{3.57}$$

$\mathbf{c} = \boldsymbol{\mu} + \boldsymbol{\Lambda}\mathbf{y}$ is the vector of the common parts of $\mathbf{x}$. Its correlation matrix is

$$\mathbf{R}^* = (\mathbf{I} - \boldsymbol{\Psi})^{-1/2}(\mathbf{R} - \boldsymbol{\Psi})(\mathbf{I} - \boldsymbol{\Psi})^{-1/2}, \tag{3.58}$$

where $\mathbf{R}$ is the correlation matrix of the xs.

The standard least squares fit of principal axis factoring then yields

$$\hat{\mathbf{\Lambda}} = (\mathbf{I} - \mathbf{\Psi})^{1/2}\mathbf{M}\mathbf{\Theta}, \tag{3.59}$$

where $\mathbf{\Theta}^2$ is the diagonal matrix containing the q largest eigenvalues of $\hat{\mathbf{R}}^*$ and $\mathbf{M}$ the matrix of associated eigenvectors. The unknown $\mathbf{\Psi}$ can be estimated by an iterative procedure.

The solution is thus the rotation which maximises the generalisability as measured by the correlations of the common parts. Given the questionable nature of the psychometric measure of generalisability, other methods of selecting rotations are to be preferred. An alternative version of alpha factor analysis was proposed by Bentler (1968) which turned out to be equivalent to Rao's method.

The software packages CEFA (Browne *et al.* 1998) and Mplus (Muthén and Muthén 2010) allow a choice between very many different sorts of rotations and standardisations.

3.14 Posterior analysis: the normal case

Having determined the dimension of the factor space and established that the model is a reasonable fit, we may wish to assign scale values to individuals in the latent dimensions. We may need these for selection purposes as, for example, when we wish to choose those with the highest ability. Or we may wish to use scale values as an input to some further analysis to see how ability is related to some other variable. It is sometimes argued that this latter need can best be met within the framework of a structural equation model (see Chapter 8) where the role of the latent variables is implicit and does not need to be made explicit. In Chapter 8 we shall advance reasons for treating this recommendation with great caution. For the present we simply note that this does not eliminate the need for factor scores.

Within the framework we have adopted and set out in Chapter 2, the so-called *factor scores problem* is solved by the posterior distribution of $\mathbf{y}$ given $\mathbf{x}$. Scale values for the dimensions of $\mathbf{y}$ can be obtained by using some measure of location of the posterior distribution such as $E(\mathbf{y} \mid \mathbf{x})$. The precision of any predicted variable can be measured by the posterior variance. This procedure was set out in general terms in Chapter 2, where it was noted that the posterior expectation was close to a linear function of the sufficient statistic. We have already given the posterior distribution for the normal case in (3.6) (and also in (1.13)) which shows that the relationship for the normal linear latent variable model is precisely linear. Whether we use the expectation or the component is therefore of no significance since one is a linear function of the other.

The posterior covariance matrix of $\mathbf{y} \mid \mathbf{x}$ was given in (3.6) in the form $(\mathbf{I} + \mathbf{\Gamma})^{-1}$. It is much easier to interpret the posterior analysis when $\mathbf{\Gamma}$ is diagonal because this makes the ys independent.

The traditional way of approaching the factor scores problem has been to start with (3.3). This is an equation in *random* variables, but realised values of these random variables will satisfy the same equation. However, the only realised values we are able to observe are the xs. This leaves the ys and the es undetermined. The set of equations cannot, therefore, be solved for the ys alone. For this reason the factor scores are said to be indeterminate and various devices have been adopted to get best-fitting values, in some sense, for $\mathbf{y}$.

The case for adopting the posterior approach was argued in Bartholomew (1981) but had been anticipated by Dolby (1976). The 'indeterminacy' is reflected in the fact that $\mathbf{y}$ remains a random variable after $\mathbf{x}$ has been observed though its variance will have been reduced by the knowledge of $\mathbf{x}$. The choice between the posterior and the traditional approach remains a matter of lively debate, as reference to Maraun (1996) and the ensuing discussion shows. In purely practical terms there is little at stake and we shall confine our treatment to the posterior method. (It may be noted that it is only in relation to the linear factor model that this controversy arises. With other latent variable models, in particular the latent class model, the posterior approach has been used without question.)

The greatest apparent drawback of the posterior method is that the scale values depend on the choice of prior distribution. However, this is an inevitable consequence of the essential indeterminacy of the prior. Theorem 2.15.1 shows that, using the component scores, no higher level of scaling than ordinal is possible. Any assumption about the form of the prior distribution is therefore a matter of convention. By choosing a standard normal prior we are implicitly choosing to measure the latent variable on a scale which renders its distribution normal. The predicted value of y for a given individual is therefore with reference to that prior scaling.

3.15 Posterior analysis: least squares

If we are unwilling to make assumptions about the forms of the distributions in the linear model we can still aim to find functions of the xs which are, in some sense, as near as possible to the ys. In the case of the linear model it is sufficient to consider only linear functions, as we now show. Let us consider the q linear functions $\mathbf{X}$ obtained by pre-multiplying $\mathbf{x} - \boldsymbol{\mu}$ by a $q \times p$ matrix of the form $(\mathbf{B\Lambda})^{-1}\mathbf{B}$. Then from (3.3),

$$\mathbf{X} = (\mathbf{B\Lambda})^{-1}\mathbf{B}(\mathbf{x} - \boldsymbol{\mu}) = \mathbf{y} + (\mathbf{B\Lambda})^{-1}\mathbf{Be}, \tag{3.60}$$

and hence we may write

$$X_j = y_j + u_j \quad (j = 1, 2, \ldots, q). \tag{3.61}$$

We could then regard X_j as a factor score for y_j since u_j is a random error independent of y_j. Since the transformation in (3.60) involves the arbitrary matrix $\mathbf{B}$ we shall seek to choose the transformation so that $\text{var}(u_j)$ is as small as possible for each j. In this way X_j is made as close as possible to y_j in a mean square sense.

It can easily be shown that choosing $(\mathbf{B\Lambda})^{-1}\mathbf{B} = \mathbf{C} = \mathbf{\Gamma}^{-1}\mathbf{\Lambda}'\mathbf{\Psi}^{-1}$ leads to the smallest possible values for the variances of the u_j. These smallest values are the diagonal elements of $\mathbf{\Gamma}^{-1}$. One way to see this is to show that for fixed $\mathbf{a}$,

$$\text{var}\,\mathbf{a}'\mathbf{u} = \mathbf{a}'\mathbf{C\Psi C}'\mathbf{a} \geq \mathbf{a}'\mathbf{\Gamma}^{-1}\mathbf{a}.$$

Since $\mathbf{C\Lambda} = \mathbf{I}$,

$$\mathbf{a}'\mathbf{\Gamma}^{-1}\mathbf{a} = \mathbf{a}'\mathbf{C\Psi}^{1/2}\mathbf{\Psi}^{-1/2}\mathbf{\Lambda\Gamma}^{-1}\mathbf{a}.$$

The Cauchy inequality gives

$$\begin{aligned}(\mathbf{a}'\mathbf{\Gamma}^{-1}\mathbf{a})^2 &\leq (\mathbf{a}'\mathbf{C\Psi}^{1/2}\mathbf{\Psi}^{1/2}\mathbf{C}'\mathbf{a})(\mathbf{a}'\mathbf{\Gamma}^{-1}\mathbf{\Lambda}'\mathbf{\Psi}^{-1/2}\mathbf{\Psi}^{-1/2}\mathbf{\Lambda\Gamma}^{-1}\mathbf{a})\\ &= (\mathbf{a}'\mathbf{C\Psi C}'\mathbf{a})(\mathbf{a}'\mathbf{\Gamma}^{-1}\mathbf{a}).\end{aligned}$$

So,

$$\mathbf{a}'\mathbf{\Gamma}^{-1}\mathbf{a} \leq \mathbf{a}'\mathbf{C\Psi C}'\mathbf{a} = \text{var}(\mathbf{a}'\mathbf{u}).$$

We have

$$\mathbf{X} = \mathbf{\Gamma}^{-1}\mathbf{\Lambda}'\mathbf{\Psi}^{-1}(\mathbf{x} - \boldsymbol{\mu}) \quad \text{and} \quad \text{var}(\mathbf{u}) = \mathbf{\Gamma}^{-1}. \tag{3.62}$$

These are known as Bartlett's scores (Bartlett 1937), and the advantage often claimed for them is that $E(\mathbf{X} \mid \mathbf{y}) = \mathbf{y}$, as (3.61) shows. However, an expectation of $\mathbf{x}$ conditional on $\mathbf{y}$ is hardly relevant when it is $\mathbf{x}$ that is known and $\mathbf{y}$ that is to be predicted. Of more interest is the comparison with (3.9). The only difference is that $(\mathbf{I} + \mathbf{\Gamma})^{-1}$ in (3.9) is replaced by $\mathbf{\Gamma}^{-1}$ in (3.62). If $\mathbf{\Gamma}$ is diagonal the effect of this difference is merely to introduce different scaling of the Xs which is of little practical significance. The distribution-free argument can thus be viewed as establishing the result of (3.9) on a broader basis. However, a slightly different approach leads to scores which are identical to those of (3.9), as we now show.

In this method, due to Thomson (1951), we choose $\mathbf{X} = \mathbf{C}(\mathbf{x} - \boldsymbol{\mu})$, where $\mathbf{C}$ is a $q \times p$ matrix, and try to make X_j to as close to y_j as possible in the sense of minimising

$$\phi_j = E(X_j - y_j)^2 \tag{3.63}$$

where the expectation is with respect to both $\mathbf{y}$ and $\mathbf{x}$.

The choice $\mathbf{C} = \mathbf{\Lambda}'\mathbf{\Sigma}^{-1}$ leads to the smallest ϕ_j because for a fixed vector $\mathbf{a}$, $\mathbf{a}'(\mathbf{X} - \mathbf{y})$ has variance

$$\mathbf{a}'(\mathbf{C\Sigma C}' - \mathbf{C\Lambda} - \mathbf{\Lambda}'\mathbf{C}' + \mathbf{I})$$

which may be written

$$\mathbf{a}'[(\mathbf{C}-\mathbf{\Lambda}'\mathbf{\Sigma}^{-1})\mathbf{\Sigma}(\mathbf{C}-\mathbf{\Lambda}'\mathbf{\Sigma}^{-1})'+\mathbf{I}-\mathbf{\Lambda}'\mathbf{\Sigma}^{-1}\mathbf{\Lambda}]\mathbf{a}.$$

This is clearly no less than

$$\mathbf{a}'[\mathbf{I}-\mathbf{\Lambda}'\mathbf{\Sigma}^{-1}\mathbf{\Lambda}]\mathbf{a}$$

and that lower bound is achieved for

$$\mathbf{C}=\mathbf{\Lambda}'\mathbf{\Sigma}^{-1}=(\mathbf{I}+\mathbf{\Gamma})^{-1}\mathbf{\Lambda}'\mathbf{\Psi}^{-1}. \tag{3.64}$$

When defined in this way the scores are known as 'regression' scores. In this case it is *not* true that $E(\mathbf{X}\mid\mathbf{y})=\mathbf{y}$, but for the reason given above this is not a relevant objection.

3.16 Posterior analysis: a reliability approach

An entirely different approach to determining factor scores was proposed by Knott and Bartholomew (1993) (see also Bartholomew and Knott (1993)). This starts from the *reliability* of the score and aims to choose a score for which the reliability is a maximum. Let $\phi(\mathbf{x})$ denote any function of the xs which is proposed as a score for the latent variable y. Suppose we were able to make two independent determinations of $\phi(\mathbf{x})$, $\phi(\mathbf{x}_1)$ and $\phi(\mathbf{x}_2)$ say, for all members of a population. Then the better the function ϕ, the closer we would expect $\phi(\mathbf{x}_1)$ and $\phi(\mathbf{x}_2)$ to be. Closeness can be measured by the correlation of $\phi(\mathbf{x}_1)$ and $\phi(\mathbf{x}_2)$ – known as the *test–retest* correlation. Knott and Bartholomew showed that the function h which maximised this correlation satisfies

$$E\{\phi(\mathbf{x}_1)\mid\mathbf{x}_2\}=\lambda\phi(\mathbf{x}_2). \tag{3.65}$$

In the case of the NLFM it turns out that the function ϕ is identical to the components and hence proportional to the conditional expectation $E(y\mid\mathbf{x})$. For other models this is not necessarily so though it appears, on the limited evidence available, that the components are close to optimal. This approach therefore provides an interesting further justification for the existing methods rather than a different type of score. For a fuller account of the use of reliability in this context, see Bartholomew (1996, Section 1.2).

3.17 Examples

We illustrate the application of the foregoing theory on two examples. The first is small-scale and is designed to show some of the options available in a typical computer package. The second is on a larger scale and is more typical of examples encountered in the applied literature. Both examples exhibit features which are quite common but seldom remarked upon in textbook treatments.

Example 3.17.1 In the first case the data are taken from a study by Smith and Stanley (1983) on the relationship between reaction times and intelligence test scores. Here we consider only the factor analysis of the ability variables. Scores were available for 112 individuals on the following six variables:

1. a non-verbal measure of general intelligence (Spearman's g) using Cattell's culture-fair test;
2. picture completion test;
3. block design;
4. mazes
5. reading comprehension;
6. vocabulary.

Full details may be found in the original paper. The correlation coefficients, covariances and variances, supplied by the authors, are set out in Table 3.3. The data were first analysed using the maximum likelihood routine with the correlation matrix input.

Table 3.3 Correlation coefficients (right upper) and variances and covariances (left lower) for Smith and Stanley's (1983) data.

	1	2	3	4	5	6
1	26.641	0.466	0.552	0.340	0.576	0.510
2	5.991	6.700	0.572	0.193	0.263	0.239
3	33.520	18.137	149.831	0.445	0.354	0.356
4	6.023	1.782	19.424	12.711	0.184	0.219
5	20.755	4.936	31.430	4.757	52.604	0.794
6	29.701	7.204	50.753	9.075	66.762	135.292

Since the xs have no natural common scale it is sensible to base the interpretation on the standardised loadings and communalities, which are given in Table 3.4 for a two-factor fit. We discuss the reasons for choosing two factors later.

The interpretation of these results is not immediately clear. The communalities vary a good deal. In the case of reading comprehension, block design and vocabulary the high figure means that a large part of their variation is accounted for by the two factors. The other three variables, especially mazes, have smaller communalities and so are much poorer indicators of the factors.

The particularly high communality of 0.96 for reading comprehension indicates that we are close to a Heywood case. This output was obtained from Mplus (Muthén and Muthén 2010).

Table 3.4 Standardised loadings, with standard errors in brackets, and communalities of a two-factor model estimated by maximum likelihood for Smith and Stanley's (1983) data.

Variable (i)	$\hat{\lambda}_{i1}$	$\hat{\lambda}_{i2}$	$1 - \hat{\psi}_i$
1	0.53 (0.09)	0.52 (0.09)	0.54
2	0.19 (0.09)	0.61 (0.08)	0.41
3	0.25 (0.08)	0.85 (0.07)	0.78
4	0.13 (0.09)	0.46 (0.09)	0.23
5	0.97 (0.07)	0.13 (0.05)	0.96
6	0.79 (0.07)	0.18 (0.07)	0.66

One has to deduce from the factor loadings what these two factors might be. The loadings in the first column are all positive, indicating that the first factor contributes to all variables, and this might tentatively be identified with some general ability. In the second column the loadings indicate that the second factor is largely concerned with the first four items. These are all non-verbal items, and so this seems to point to a non-verbal dimension to ability.

It might be that the interpretation would become clearer if we look at rotations. Some idea of what to try can be gained from a plot of the loadings, and this is given in Figure 3.1.

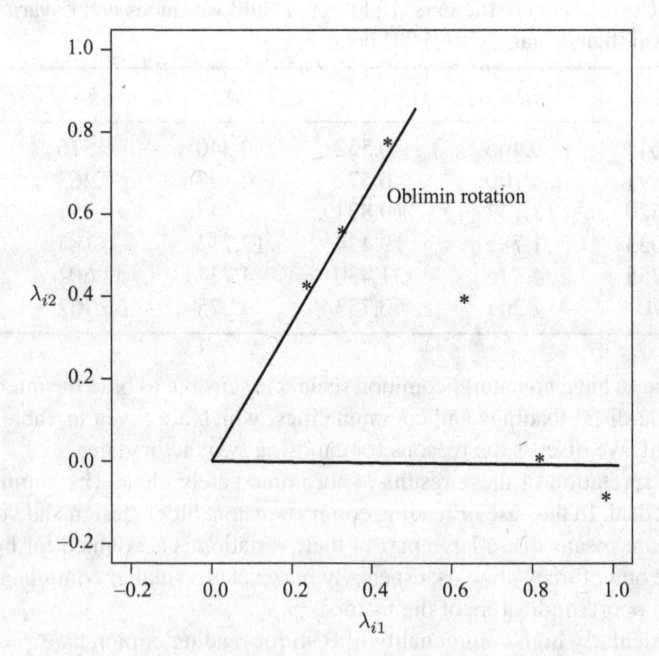

Figure 3.1 Factor loadings given in Table 3.4. The axes for OBLIMIN rotation are also shown.

The aim of rotation, we recall, is to achieve 'simple structure'. In geometrical terms this means seeking a rotation which makes the points lie as nearly as possible along the two axes. It is clear from the figure that this cannot be achieved with an orthogonal rotation. We could certainly rotate to make items 1, 2 and 3 lie very close to one axis, but then the remaining points would be nowhere near the other. Equally an axis passing close to 6 and 5 would leave the remaining points farther away from the other axis than before rotation.

The figure shows that something close to simple structure could be obtained by an oblique rotation, and one resulting from the OBLIMIN option of Mplus is illustrated. This still leaves item 1 with a substantial loading on both dimensions, but this is not surprising given that it is itself a composite measure. The rotated loadings are given in Table 3.5.

Table 3.5 Factor loadings for the OBLIMIN rotation, with standard errors in brackets, for Smith and Stanley's (1983) data (the communalities are unchanged by rotation and so are omitted).

Variable (i)	$\hat{\lambda}_{i1}$	$\hat{\lambda}_{i2}$
1	0.38 (0.10)	0.48 (0.10)
2	−0.01 (0.08)	0.65 (0.09)
3	−0.03 (0.03)	0.90 (0.07)
4	−0.02 (0.10)	0.49 (0.10)
5	1.00 (0.07)	−0.04 (0.02)
6	0.78 (0.09)	0.06 (0.08)

Leaving aside item 1 for the moment, we have factor 1 loading almost entirely on the verbal items, 5 and 6, and factor 2 loading on the remaining non-verbal items. This strongly suggests two dimensions of ability – one verbal and one non-verbal. We would expect the measure of general intelligence to contribute to both factors but more strongly, perhaps, to the non-verbal dimension because of the non-verbal character of that test.

We might thus conclude that the analysis identifies two dimensions of ability, one verbal and one non-verbal, but that these are not uncorrelated (because the rotation is oblique). The package gives an estimate of the correlation between the two factors of 0.462 with an estimated standard error of 0.086.

In view of this analysis we might expect the two-factor structure to be clearer if we omitted item 1 from the analysis, thus leaving items which fall clearly into verbal and non-verbal categories.

A repeat of maximum likelihood estimation for the last five items reveals a Heywood case. Mplus produces an unsatisfactory solution with a very large negative variance for item 2. In effect, we have a Heywood case with a communality of 1 for item 2. An R routine has been specifically written to cope with Heywood cases based on the maximum likelihood estimation presented in Section 3.4, the results of which are given in Table 3.6.

Table 3.6 Factor loadings and communalities of a two-factor model for Smith and Stanley's (1983) data (item 1 omitted).

Variable (i)	$\hat{\lambda}_{i1}$	$\hat{\lambda}_{i2}$	$1-\hat{\psi}_i$
2	−0.57	0.05	0.33
3	−1.00	0.00	1.00
4	−0.45	0.06	0.20
5	−0.35	0.78	0.73
6	−0.36	0.86	0.87

Heywood cases are not uncommon with small sample sizes of a hundred or so as we have already remarked. However, taking the loadings at their face values, the interpretation is essentially the same as before. The verbal ability factor emerges clearly in the second column whereas the first factor loads more heavily on the non-verbal items, with the dominating item being block design.

Given the small sample size, one should not push the interpretation beyond claiming some evidence for verbal and non-verbal dimensions of ability. If we wished to scale individuals on any of the dimensions we have uncovered, we would need the coefficients of the factor scores (or components), and these are provided by the standard packages. However, those coefficients can be very unstable in small samples because they involve the factor $\psi^{-1/2}$. If $\hat{\psi}_i$ is small, and subject to a large sampling error, the weight of x_i will be large and uncertain. If a Heywood case occurs, as in Table 3.6, the estimate becomes infinite. Posterior analysis for this example is thus not worthwhile.

All of the foregoing analysis presupposes that a model with two factors provides a satisfactory fit and that two is the best number of factors. In this case the default value provided by Mplus was two and we set this question on one side in order to proceed directly to questions of interpretation. We now return to the question of goodness of fit.

For the example with all six items the overall global test of goodness of fit yields: for the one-factor model, $\chi^2 = 78.95$ with 9 degrees of freedom ($P = 0.000$); and for the two-factor model, $\chi^2 = 6.36$ with 4 degrees of freedom ($P = 0.174$). There is little point in going beyond two factors because when $q = 3$ the model is only just identifiable with the number of parameters equal to the number of statistics. In any case the fit with two factors is good.

If the fit is not good, a comparison of the observed and predicted correlation (or covariance) matrix may help to identify the source of the deviation. Table 3.7 gives for the present example the differences between the observed and fitted correlation matrices. It appears that, with the possible exception of items 2 and 4, the 2-factor model predicts the observed correlation matrix very closely.

Table 3.7 Differences between the observed and fitted correlation matrices for Smith and Stanley's (1983) data.

Variable	1	2	3	4	5	6
1	0	0.05	−0.02	0.03	0.00	0.00
2	—	0	0.01	−0.11	0.00	−0.02
3	—	—	0	0.02	0.00	0.00
4	—	—	—	0	0.00	0.03
5	—	—	—	—	0	0.00
6	—	—	—	—	—	0

Example 3.17.2 The second example is taken from Harman (1976) and is based on eight physical variables measured on 305 individuals. The eight variables are: height, arm span, length of forearm, length of lower leg, weight, bitrochanteric diameter, chest girth and chest width. The first four variables were assumed to measure 'lankiness' and the last four 'stockiness'. Table 3.8 gives the sample correlation matrix of the eight variables.

Table 3.8 Sample correlation matrix for the physical data.

Variable	1	2	3	4	5	6	7	8
1	1.000							
2	0.846	1.000						
3	0.805	0.881	1.000					
4	0.859	0.826	0.801	1.000				
5	0.473	0.376	0.380	0.436	1.000			
6	0.398	0.326	0.319	0.329	0.762	1.000		
7	0.301	0.277	0.237	0.327	0.730	0.583	1.000	
8	0.382	0.415	0.345	0.365	0.629	0.577	0.539	1.000

Inspection of the correlation matrix shows a quite clear grouping of the first four variables and the last four. We would therefore expect to find that a two-factor solution adequately explains the correlation matrix. We first address the question of the choice of q, the number of factors. Inspection of the scree plot of the eigenvalues in Figure 3.2 shows that there are two eigenvalues greater than 1. If we look at the goodness-of-fit test and the AIC in Table 3.9 the position is less clear. The χ^2 test is highly significant for $q \leq 3$ and the AIC shows no sign of reaching a minimum value by $q = 4$. However, the normality assumption on which the χ^2 test is based is suspect, and we note that, whether the factors lower in the order are significant or not, they individually contribute very little to the overall variation in scores. When $q = 5$ no solution could be obtained due to convergence problems, indicating that the factoring has been taken too far. We shall, therefore, concentrate on the two-factor model for purposes of interpretation.

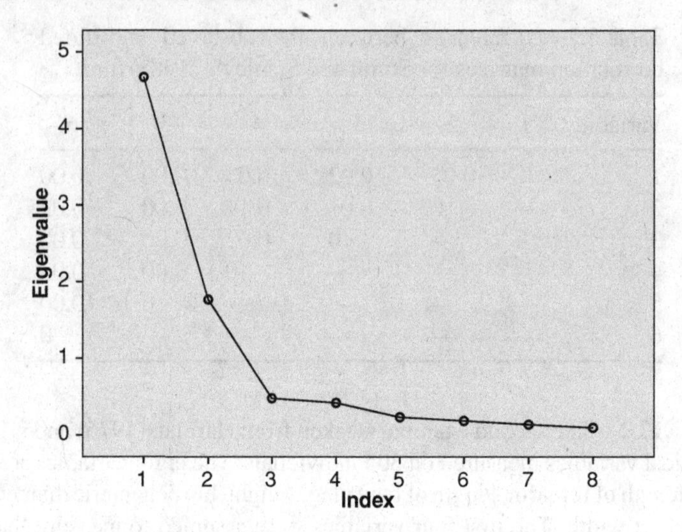

Figure 3.2 Scree plot for physical data.

Table 3.9 χ^2 and AIC for physical data.

Factors	χ^2	df	P	AIC
1	713.7	20	0.000	6254.5
2	88.6	13	0.000	5643.3
3	23.9	7	0.001	5590.6
4	0.84	2	0.658	5577.6

Table 3.10 Two-factor solution (unrotated and geomin rotated), with standard errors in brackets, for physical data.

	Unrotated				Rotated			
Variable	Factor 1		Factor 2		Factor 1		Factor 2	
1	0.885	(0.013)	0.219	(0.024)	**0.855**	(0.020)	0.115	(0.032)
2	0.939	(0.009)	0.108	(0.020)	**0.951**	(0.013)	−0.015	(0.022)
3	0.907	(0.011)	0.109	(0.022)	**0.917**	(0.012)	−0.010	(0.011)
4	0.876	(0.014)	0.184	(0.025)	**0.858**	(0.021)	0.078	(0.034)
5	0.304	(0.041)	0.905	(0.021)	0.003	(0.019)	**0.953**	(0.019)
6	0.254	(0.044)	0.756	(0.027)	0.002	(0.026)	**0.797**	(0.027)
7	0.192	(0.045)	0.739	(0.028)	−0.057	(0.043)	**0.787**	(0.033)
8	0.325	(0.046)	0.598	(0.037)	0.133	(0.047)	**0.612**	(0.040)

The parameter estimates obtained by maximum likelihood are given in Table 3.10, with standard errors in brackets, both for the unrotated and the rotated solution. The geomin oblique rotation, available in Mplus (Muthén and Muthén 2010), (Jennrich 2006) is used here. Both the unrotated and the oblique solution produced interpretable factors. However, the rotated solution produced a simple structure. In order to identify what the two factors are measuring, we have to ask what distinguishes the items with the higher loading from those with the lower. This, of course, requires some expertise in the field but the task might not be very difficult here because of the nature of the items. The loadings printed in bold type for factor 1 constitute a group which load heavily on that factor. On the second factor it is the complementary set of items which have the high loadings. The high loadings on the first four items are associated with length and height, and therefore factor 1 is related to aspects of physique that have to do with tallness and thinness. The high loadings on the last four items are related to weight, diameter and width, and so factor 2 measures the stockiness aspects of the physique. The two factors are significantly correlated ($r = 0.436$, (0.049)) where the standard error is given in brackets.

4

Binary data: latent trait models

4.1 Preliminaries

Binary responses are extremely common, especially in the social sciences. Individuals can be classified according to whether or not they belong to a trade union or take holidays abroad. They can be recorded as agreeing or disagreeing with some proposition or as getting some item in an educational test right or wrong. Such binary variables are often supposed to be indicators of more fundamental attitudes or abilities, and it is in these circumstances that latent variable modelling is relevant. Even when the observed responses fall into more than two categories, it is often useful to conflate them into two categories. Models for polytomous data are discussed in Chapter 5.

In this chapter we shall consider only the case where the latent variables are continuous. Because the greatest stimulus for the development of models in this area has come from the field of educational testing, where the latent variables are conceived of as 'traits', it is usual to speak of *latent trait models*. This contrasts with latent class models where the latent space is treated as categorical. Such models are the subject of Chapter 6.

If we code the possible responses as 0 or 1, the *response pattern* for any sample member consists of a string of zeros and ones. If there are p items the number of distinct response patterns is 2^p. The most economical way of setting out the data for a sample of size n will depend on the relative magnitudes of n and 2^p. If p is as large as 50, for example, the possible number of response patterns is approximately 1.1×10^{15}. In such a case only a tiny fraction of possible patterns will occur and the chance of any one being repeated is negligible. One would then simply list the response patterns in some convenient order. This has been done with the BSA sexual attitudes data used in Chapter 6, for example.

Latent Variable Models and Factor Analysis: A Unified Approach, Third Edition.
David Bartholomew, Martin Knott and Irini Moustaki.

If 2^p is not too large the data may be set out as a frequency distribution over response patterns listed in a systematic order. Examples are given in Tables 4.2 and 4.4 which appear in Section 4.13. The first, in Table 4.2, is taken from educational testing and relates to the Law School Admission Test (Section VI). It is included here because it has been analysed many times before and so allows us to compare the newer approaches with earlier work. The second, in Table 4.4, is taken from Schuessler's Social Life Feelings Scales (for some background see Schuessler (1982); Krebs and Schuessler (1987)). The scale was labelled 'economic self-determination' and was based on five items. It is immediately clear that, in both data sets, the different response patterns occur with very unequal frequencies. The main kind of question to be addressed in this chapter is how far these patterns can be explained by supposing that all items depend on one or more latent variables.

The relative frequencies from Tables 4.2 and 4.4 are the empirical counterparts of the joint probability distribution $f(\mathbf{x})$ which assigns probabilities to the 2^p response patterns $\mathbf{x}$. For some purposes, especially when testing goodness of fit, it is useful to specify the joint distribution in terms of a set of *marginal probabilities*. Thus the univariate distributions $f(x_i)$ $(i = 1, 2, \ldots, p)$ are the first-order marginal distributions; $f(x_i, x_j)$ $(i, j = 1, 2, \ldots, p)$ the second-order margins and so on. Suppose we consider the following set of marginal probabilities:

$$
\begin{aligned}
&P\{x_i = 1\} \quad (i = 1, 2, \ldots, p),\\
&P\{x_i = 1, x_j = 1\} \quad (i, j = 1, 2, \ldots, p \quad (i \neq j)),\\
&P\{x_i = 1, x_j = 1, x_k = 1\} \quad (i, j, k = 1, 2, \ldots, p \quad (i, j, k \text{ not equal})), \qquad (4.1)\\
&\ldots,\\
&P\{x_1 = 1, x_2 = 1, \ldots, x_p = 1\}.
\end{aligned}
$$

The total number of probabilities in this set is

$$\binom{p}{1} + \binom{p}{2} + \cdots + \binom{p}{p} = 2^p - 1.$$

Taking account of the fact that the 2^p response probabilities $f(\mathbf{x})$ sum to one, there are just sufficient marginal probabilities in the above set to allow the use of an inclusion–exclusion formula to uniquely determine the complete joint distribution. Either set thus provides a full specification of the joint distribution. The second-order margins, in particular, tell us about the pairwise associations between the xs, and the success with which these are predicted by a model will be an important part of assessing its goodness of fit.

4.2 The logit/normal model

If the dependence among the xs is wholly explained by a vector $\mathbf{y}$ of latent variables they may be regarded as mutually independent random variables with

$$P\{x_i = 1 \mid \mathbf{y}\} = \pi_i(\mathbf{y}) \quad (x_i = 0, 1; i = 1, 2, \ldots, p), \qquad (4.2)$$

where $\pi_i(\mathbf{y})$ is called the *response function*. Since it is a probability, $0 \leq \pi_i(\mathbf{y}) \leq 1$. In test theory, where y is usually a scalar, $\pi_i(y)$ is known as the *item response function* or *item characteristic curve*. In that context y represents an ability of some kind, in which case one would expect $\pi_i(y)$ to be a monotonic function.

The conditional distribution of x_i given $\mathbf{y}$ is thus

$$g_i(x_i \mid \mathbf{y}) = \{\pi_i(\mathbf{y})\}^{x_i}\{1 - \pi_i(\mathbf{y})\}^{1-x_i} \quad (x_i = 0, 1; i = 1, 2, \ldots, p). \tag{4.3}$$

We have met this distribution in Section 2.1 and, because it is a member of the exponential family, the general linear latent variable model takes the form

$$\text{logit}\,\pi_i(\mathbf{y}) = \alpha_{i0} + \sum_{j=1}^{q} \alpha_{ij} y_j \quad (i = 1, 2, \ldots, p). \tag{4.4}$$

If we complete the specification of the model by assuming that the prior distribution of $\mathbf{y}$ is standard normal with correlation matrix $\mathbf{I}$, then we have the *logit/normal model*. Figure 4.1 gives three examples showing the shape of the logit/normal response function when $q = 1$.

The parameter α_{i0} is sometimes called the 'intercept' because of its role in the linear plot of logit $\pi_i(y)$ against y. An alternative parameterisation uses

$$\pi_i(0) = \pi_i = \frac{1}{1 + e^{-\alpha_{i0}}}. \tag{4.5}$$

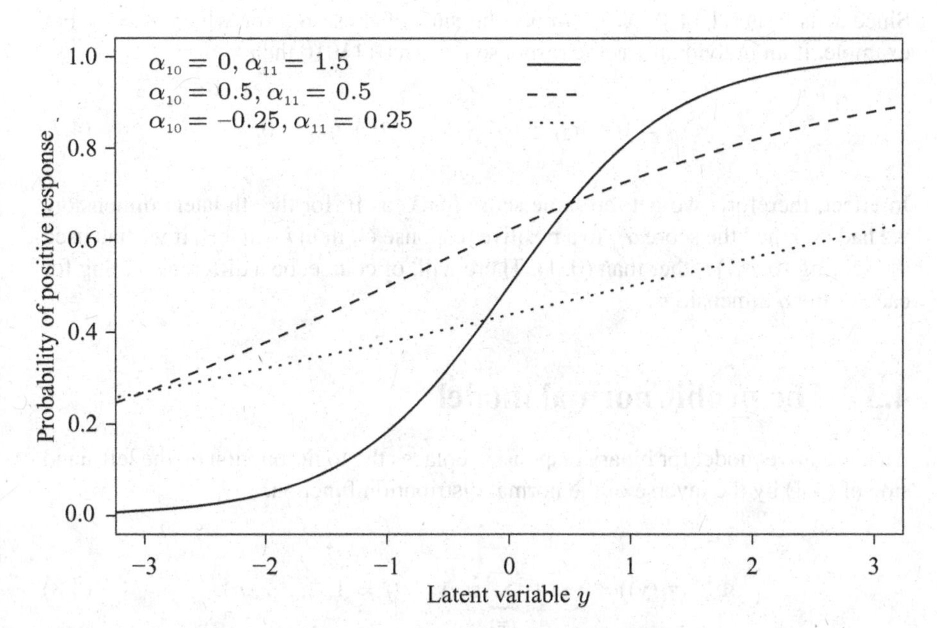

Figure 4.1 Several logistic response functions.

This is the probability of a positive response from an individual with $y = 0$ – that is, for someone at the median point of the latent scale.

The parameter α_{i1} governs the steepness of the curve; in educational testing it is known as the *discrimination* parameter because the bigger α_{i1} the easier it will be to discriminate between a pair of individuals a given distance apart on the latent scale. This is because the greater the difference between the value of $\pi_i(y)$ for the two individuals, the more likely it is that they will give a different response.

If the α_{i1} are equal the model reduces to the random effects Rasch model. The equality of the α_{i1}s implies equal discriminating power, which might be regarded as a desirable feature of a test on educational grounds. It certainly leads to greater simplification both conceptually and numerically. The fact that the minimal sufficient statistic becomes the total score supports the common practice of scaling individuals on that basis. However, we prefer to base our analysis on the more general case because, in our experience, it is unusual to find equal α_{i1} outside the educational field.

In the general case the α_{ij} $(j \geq 1)$ are effectively the same as factor loadings and may thus be interpreted according to the general principles laid out in Section 2.12. Later, in Section 7.8, we shall introduce a standardisation which makes the link with factor analysis even closer. For the moment we show how the special features of binary data allow us to approach the interpretation of the loadings from another angle.

From the general theory of the GLLVM we have that the components are

$$X_j = \sum_{i=1}^{p} \alpha_{ij} x_i \quad (j = 1, 2, \ldots, q). \tag{4.6}$$

Since x_i is either 1 or 0, X_j is simply the sum of those α_{ij} for which $x_i = 1$. For example, if an individual has the response pattern 0111010 then

$$X_j = 0 + \alpha_{2j} + \alpha_{3j} + \alpha_{4j} + 0 + \alpha_{6j} + 0. \tag{4.7}$$

In effect, therefore, we get the same score for X_j as if, for the jth latent dimension, we had assigned the score α_{ij} to a positive response on item i – that is, if we had used the scoring $(0, \alpha_{ij})$ rather than (0, 1). There will, of course, be a different scoring for each of the q dimensions.

4.3 The probit/normal model

An alternative model for binary responses replaces the logit function on the left-hand side of (4.4) by the inverse of the normal distribution function,

$$\Phi^{-1}(\pi_i(\mathbf{y})) = \alpha_{i0} + \sum_{j=1}^{q} \alpha_{ij} y_j \quad (i = 1, 2, \ldots, p). \tag{4.8}$$

Since

$$\text{logit}(u) \doteq (\pi/\sqrt{3})\Phi^{-1}(u), \tag{4.9}$$

it is clear that the two models are very close. It follows from (4.9) that the values of the αs for the probit/normal model will be less than those of the logit/normal model by a factor of approximately $\sqrt{3}/\pi$. In practice, therefore, the two models are virtually equivalent, though the probit/normal model lacks the sufficiency property of the linear component X.

The motivation for the introduction of the probit/normal model lies in its relationship to the standard normal linear model of factor analysis. This makes it possible to transfer much of the methodology and modes of interpretation which are already well established for that model.

The probit/normal model is an example of what we shall call the *underlying variable* approach to modelling categorical data. When applied to the case of binary manifest variables it supposes that, underlying the ith indicator, there is a continuous variable, say ξ_i. The observed binary variable, x_i, is then an indicator of whether or not ξ_i is above or below some threshold level τ_i. That is, we suppose (for $i = 1, 2, \ldots, p$) that

$$x_i = \begin{cases} 1 & \text{if } \xi_i \leq \tau_i, \\ 0 & \text{otherwise.} \end{cases} \tag{4.10}$$

The idea is then to set up a latent variable model for the continuous variable $\boldsymbol{\xi}$ and to find a way of fitting it using the partial information provided by the indicator variables, $\mathbf{x}$. In particular, if we assume $\boldsymbol{\xi}$ to be generated by a standard normal factor model, we shall have

$$\boldsymbol{\xi} = \boldsymbol{\mu} + \boldsymbol{\Lambda}\mathbf{y} + \mathbf{e}. \tag{4.11}$$

If we can find a way of estimating the parameters, the interpretation will then be exactly as in factor analysis. Before pursuing this matter, we need to consider whether postulating an underlying variable in this manner is substantively meaningful. In some cases it undoubtedly is. Indeed, the dichotomies may have been arrived at in just that way as, for example, when incomes are recorded as being below or above a tax threshold. Again, if a person is asked whether or not they are in favour of some proposition they may be imagined to have a quantifiable degree of enthusiasm which is imperfectly revealed by their response to a question of the agree/disagree variety. But there are other cases where the notion is less appealing. The question asked may be a simple matter of fact such as 'Were you born in this country?' In this case it is difficult to conceive of any relevant underlying variable. However, from a purely practical point of view, this is immaterial because there is an equivalence between underlying variable and response function models which we now demonstrate. As we shall see in Chapter 5, this breaks down if the indicators are polytomous nominal.

4.4 The equivalence of the response function and underlying variable approaches

To show that the two approaches are equivalent we must show that the joint probability distributions of $\mathbf{x}$ to which they lead are the same. For if this is so then no empirical information can distinguish between them. Instead of working with the probability function $f(\mathbf{x})$ we use the equivalent set of marginal probabilities $\{P\{x_i = 1, x_j = 1, \ldots\}\}$ defined by (4.1). Let S denote any non-empty subset of $\{1, 2, \ldots, p\}$; then

$$P\left\{\bigcap_{i \in S}(x_i = 1)\right\} = \int_{-\infty}^{\infty} \cdots \int P\left\{\bigcap_{i \in S}(x_i = 1 \mid \mathbf{y})\right\} h(\mathbf{y})d\mathbf{y}$$

$$= \int_{-\infty}^{\infty} \cdots \int \prod_{i \in S} P\{x_i = 1 \mid \mathbf{y}\} h(\mathbf{y})\, d\mathbf{y} \tag{4.12}$$

because of the conditional independence of the xs. In the response function approach $P\{x_i = 1 \mid \mathbf{y}\}$ is the response function and $h(\mathbf{y})$ is the prior distribution of $\mathbf{y}$. To show the equivalence with the underlying variable approach we have to show in this case that the left-hand probability of (4.12) admits the same representation. Thus

$$P\left\{\bigcap_{i \in S}(x_i = 1)\right\} = P\left\{\bigcap_{i \in S}(\xi_i \le \tau_i)\right\}$$

$$= \int_{-\infty}^{\infty} \cdots \int P\left\{\bigcap_{i \in S}(\xi_i \le \tau_i) \,\middle|\, \mathbf{y}\right\} h(\mathbf{y})d\mathbf{y}. \tag{4.13}$$

When $\mathbf{y}$ is fixed the ξs are independent (because the es are independent) and therefore

$$P\left\{\bigcap_{i \in S}(x_i = 1)\right\} = \int_{-\infty}^{\infty} \cdots \int \prod_{i \in S} P\{\xi_i \le \tau_i \mid \mathbf{y}\} h(\mathbf{y})d\mathbf{y}. \tag{4.14}$$

Now

$$P\{\xi_i \le \tau_i \mid \mathbf{y}\} = P\left\{\mu_i + e_i + \sum_{j=1}^{q} \lambda_{ij} y_j \le \tau_i \,\middle|\, \mathbf{y}\right\}$$

$$= P\left\{\frac{e_i}{\psi_i^{1/2}} \le \frac{\tau_i - \mu_i - \sum_{j=1}^{q} \lambda_{ij} y_j}{\psi_i^{1/2}} \,\middle|\, \mathbf{y}\right\}$$

$$= R\left(\frac{\tau_i - \mu_i - \sum_{j=1}^{q} \lambda_{ij} y_j}{\psi_i^{1/2}}\right), \tag{4.15}$$

where $R(\cdot)$ is the distribution function of $e_i/\psi_i^{1/2}$ (note that in the factor model this does not depend on i). For the response function model,

$$P\{x_i = 1 \mid \mathbf{y}\} = G\left(\alpha_{i0} + \sum_{j=1}^{q} \alpha_{ij} y_j\right), \tag{4.16}$$

where G is temporarily used to denote the inverse logit or probit function. Equivalence between the two models therefore exists if $G \equiv R$ and if $\alpha_{i0} = (\tau_i - \mu_i)/\psi_i^{1/2}$ and

$$\alpha_{ij} = -\lambda_{ij}/\psi_i^{1/2} \quad (i = 1, 2, \ldots, p). \tag{4.17}$$

Thus for any model of the one kind we can find a model of the other which is empirically indistinguishable from it. For example, if G^{-1} is the logit function,

$$G(u) = (1 + e^{-u})^{-1} = R(e_i/\psi_i^{1/2}).$$

The standard logistic random variable with distribution function $G(u)$ has mean zero and variance $\pi^2/3$; e_i therefore has a logistic distribution with mean zero and variance $\psi_i \pi^2/3$.

The equivalence of the two methods has also been noted by, for example, Takane and de Leeuw (1987) and by Moran in unpublished work.

An important corollary of the foregoing analysis is that the parameters $\{\tau_i\}$, $\{\mu_i\}$ and $\{\lambda_{ij}\}$ of the underlying variable model are not individually estimable. The likelihood is a function of the parameters $\{(\tau_i - \mu_i)/\psi_i^{1/2}\}$. This is a reflection of the fact that there can be no information about the standard deviation of the underlying ξs in the dichotomized data. In practice, when working with continuous variables, we usually throw this information away by first standardising the xs; that is, we assume that

$$\text{var}(\xi_i) = 1 = \sum_{j=1}^{q} \lambda_{ij}^2 + \psi_i.$$

The loadings that we estimate are then

$$\frac{\lambda_{ij}}{\left(1 - \sum_j \lambda_{ij}^2\right)^{1/2}} = \frac{\lambda_{ij}}{\psi_i^{1/2}} \quad (i, j = 1, 2, \ldots, p; \ i \neq j).$$

For later reference we emphasise that member of the underlying variable family which is equivalent to the logit model. In that case G^{-1} is the logit function, so

$$G(u) = \frac{1}{1 + e^{-u}}.$$

This is the distribution function of a symmetrically distributed random variable with zero mean and variance $\pi^2/3$. In the underlying variable model $e_i/\psi_i^{1/2}$ thus has to have a distribution of this form, which implies that e_i has a logistic distribution with zero mean and variance $\psi_i\pi^2/3$ for all i.

4.5 Fitting the logit/normal model: the E-M algorithm

The parameters of the logit/normal model can be estimated in various ways. Many of the earlier methods, some of which were given in the first edition (Bartholomew 1987), were devised more with an eye to ease of calculation than efficiency. Now one can use efficient methods like maximum likelihood or generalised least squares. In this section we give the E-M algorithm, and in Section 4.8 we outline a method based on generalised least squares. The E-M method was first used by Bock and Aitkin (1981).

We shall set out the method for the case of a single latent variable. This is notationally simpler and can be extended to the general case almost immediately. The approach also provides the basis for the general method applicable to a wider class of distributions to be given in Chapter 7.

Dempster *et al.* (1977) developed the iterative E-M method for maximum likelihood estimation when there are missing observations. In favourable applications of the E-M method, estimation is easy when there are no observations missing, but difficult when there is averaging over the distribution of the missing observations. The missing observations are the values of the latent variables y_h. If those were known, the logit/normal model would be p logistic response models for the $\mathbf{x}_{ih}$ for the explanatory variables y_h, and easy to estimate by maximum likelihood.

The E-M algorithm starts from the log-likelihood L_c for the complete data (x_{ih}, y_h) $(i = 1, \ldots, p, h = 1, \ldots, n)$, where y_h is the value of the latent variable for individual h:

$$L_c = \sum_{h=1}^{n} \ln f(\mathbf{x}_h, y_h) = \sum_{h=1}^{n} \left[\sum_{i=1}^{p} \ln g(x_{ih} \mid y_h) + \ln h(y_h) \right]. \tag{4.18}$$

The E-M algorithm alternates E-steps, which evaluate $E[L_c \mid \mathbf{x}_1, \ldots, \mathbf{x}_n]$, and M-steps, which maximise $E[L_c \mid \mathbf{x}_1, \ldots, \mathbf{x}_n]$ over the parameters. The expectation of the E-step uses current estimates for the parameters, and the M-step finds new estimates for the parameters that achieve the maximisation. Since $h(y)$ does not depend on the parameters, there is no need to use the complete data likelihood; we can replace L_c by

$$L = \sum_{h=1}^{n} \ln g(\mathbf{x}_h \mid y_h) = \sum_{h=1}^{n} \left[\sum_{i=1}^{p} \ln g(x_{ih} \mid y_h) \right], \tag{4.19}$$

and at each E-step find $E[L \mid \mathbf{x}]$.

From Section 2.4,

$$g(x_{ih} \mid y_h) = (1 - \pi_i(y_h)) \exp\{x_{ih}(\alpha_{i0} + \alpha_{i1} y_h)\},$$

and so

$$L = \sum_{h=1}^{n} \ln g(\mathbf{x}_h \mid y_h) = \sum_{h=1}^{n} \left[\sum_{i=1}^{p} (\ln(1 - \pi_i(y_h)) + x_{ih}(\alpha_{i0} + \alpha_{i1} y_h)) \right]$$
$$= \sum_{i=1}^{p} \left[\sum_{h=1}^{n} \ln(1 - \pi_i(y_h)) + \alpha_{i0} \sum_{h=1}^{n} x_{ih} + \alpha_{i1} \sum_{h=1}^{n} x_{ih} y_h \right].$$

The E-step is therefore

$$E[L \mid \mathbf{x}] = \sum_{i=1}^{p} \left[\sum_{h=1}^{n} E(\ln(1 - \pi_i(y_h)) \mid \mathbf{x}_1, \ldots, \mathbf{x}_n) \right.$$
$$\left. + \alpha_{i0} \sum_{h=1}^{n} x_{ih} + \alpha_{i1} \sum_{h=1}^{n} x_{ih} E(y_h \mid \mathbf{x}_1, \ldots, \mathbf{x}_n) \right]$$
$$= \sum_{i=1}^{p} \left[\sum_{h=1}^{n} E(\ln(1 - \pi_i(y_h)) \mid \mathbf{x}_h) + \alpha_{i0} \sum_{h=1}^{n} x_{ih} + \alpha_{i1} \sum_{h=1}^{n} x_{ih} E(y_h \mid \mathbf{x}_h) \right].$$

For the M-step one can maximise separately, for each item i,

$$\sum_{h=1}^{n} E(\ln(1 - \pi_i(y)) \mid \mathbf{x}_h) + \alpha_{i0} \sum_{h=1}^{n} x_{ih} + \alpha_{i1} \sum_{h=1}^{n} x_{ih} E(y \mid \mathbf{x}_h). \tag{4.20}$$

One can write (4.20) as

$$\int \left[\ln(1 - \pi_i(y)) \sum_{h=1}^{n} h(y \mid \mathbf{x}_h) + \alpha_{i0} \sum_{h=1}^{n} x_{ih} + \alpha_{i1} \sum_{h=1}^{n} x_{ih} h(y \mid \mathbf{x}_h) \right] dy,$$

or, simplifying the notation, as

$$\int \left[\ln(1 - \pi_i(y)) N(y) + \alpha_{i0} \sum_{h=1}^{n} x_{ih} + \alpha_{i1} R_i(y) \right] dy,$$

where

$$R_i(y) = \sum_{h=1}^{n} x_{ih} h(y \mid \mathbf{x}_h),$$

$$N(y) = \sum_{h=1}^{n} h(y \mid \mathbf{x}_h). \tag{4.21}$$

So the M-step is the same as maximum likelihood estimation in a continuous analogue of logistic regression with $R_i(y)$ positive responses out of $N(y)$ trials at explanatory variable value y.

Bock and Aitkin (1981) used Gauss–Hermite quadrature to approximate $h(y \mid x_h)$, which would work best if that distribution were standard normal. Suppose that the Gauss–Hermite quadrature uses the discrete distribution at values $y^1, \ldots, y^k$ with probabilities $h(y^1), \ldots, h(y^k)$. Then approximating the integrals in (4.21) gives

$$r_{it} = \sum_{h=1}^{n} x_{ih} h(y^t \mid \mathbf{x}_h), \tag{4.22}$$

$$n_t = \sum_{h=1}^{n} h(y^t \mid \mathbf{x}_h). \tag{4.23}$$

This leads to the M-step used by Bock and Aitkin, who used maximum likelihood estimation for a logistic regression with responses r_{it} from n_t trials at explanatory variable values y_t for $t = 1, \ldots, k$.

It may be helpful to give an interpretation of r_{it} and n_t. The quantity $h(y_t \mid \mathbf{x}_h)$ is the probability that an individual with response vector $\mathbf{x}_h$ is located at y_t; $\sum_{h=1}^{k} h(y_t \mid \mathbf{x}_h)$ is thus the expected number of individuals at y_t. By a similar argument r_{it} is the expected number of those predicted to be at y_t who will respond positively.

Note that $E[L \mid \mathbf{x}]$ is the log-likelihood of the observed data, and can be directly maximised, without fixing the parameters in the conditional expectation. That can allow an interplay between the E-M algorithm and a maximisation of the observed-data log-likelihood. Newton–Raphson algorithms may be used for direct maximisation. They are known for being fast but sensitive to initial values. On the contrary, the E-M is not sensitive to the initial values but convergence can be slow in high-dimensional problems. Combination of the two methods can give an improvement.

As noted above, the simple form of the equations in the estimation part of the algorithm depends on choosing the logit function on the left-hand side of (4.4). However, the same idea can be used for other members of the general family, and it was for the probit/normal version that Bock and Aitkin (1981) developed their method.

It should be noted that in both versions it is only necessary to evaluate the posterior probabilities and expectations for those response patterns that actually occur. Furthermore, these quantities will be the same for all sample members having the same response pattern. This is of particular importance when p is large.

4.5.1 Fitting the probit/normal model

The model can be fitted by the E-M algorithm using essentially the method described above.

A simple, though less efficient, method is to use the fact that the correlation matrix of the ξs can be estimated from the *tetrachoric correlations* obtained from the xs. The tetrachoric correlation is that value of the normal bivariate correlation ρ_{ij} which makes the volumes under the normal surface in the four quadrants formed by the thresholds equal to those in the fourfold association table between x_i and x_j. If the tetrachoric coefficients are treated as if they were product-moment correlations, then the parameters can be estimated using a standard method of fitting the factor model.

More efficient methods have been given based on generalised least squares and are discussed in Section 4.8.

4.5.2 Other methods for approximating the integral

Gauss–Hermite quadrature as described above is not always the most accurate of the many ways of approximating the integral required in the calculation of the posterior distribution. Other methods, which are particularly valuable for large q, are adaptive quadrature (Rabe-Hesketh *et al.* 2002; Schilling and Bock 2005), Monte Carlo methods (Sammel *et al.* 1997) and Laplace approximations (Huber *et al.* 2004).

Adaptive quadrature is an improved version of the Gauss–Hermite quadrature described above. The values $y^1, \ldots, y^k$ and probabilities $h(y^1), \ldots, h(y^k)$ are chosen so that the normal distributions used match the location and dispersion of the posterior distribution $h(y \mid \mathbf{x}_h)$ for each response pattern $\mathbf{x}_h$ during every E-step.

Schilling and Bock (2005) show that when there are several factors, this method of numerical integration can produce fast and accurate solutions using k as low as 2 for each factor. Assuming that $h(\mathbf{y} \mid \mathbf{x}_h)$ can be approximated by a multivariate normal distribution $N(\mathbf{u}, \boldsymbol{\Sigma})$, Schilling and Bock suggest that in the posterior distribution $\mathbf{y}$ are transformed to $\mathbf{y}^*$, which have approximately uncorrelated standard normal distributions. One way to do this is to use

$$\mathbf{y}^* = \mathbf{L}'(\mathbf{y} - \boldsymbol{\mu}),$$

where $\mathbf{L}$ is the lower triangle Cholesky factor of $\boldsymbol{\Sigma} = \mathbf{LL}'$. Then for each dimension of $\mathbf{y}^*$ the values and probabilities for the adaptive quadrature are those for Gauss–Hermite quadrature. Approximate averaging over the distribution $h(\mathbf{y} \mid \mathbf{x}_h)$ is done with a sum over the values for the $\mathbf{y}^*$ weighted by their probabilities, using $\mathbf{y} = \mathbf{L}\mathbf{y}^* + \boldsymbol{\mu}$, and multiplying by the determinant of $\mathbf{L}$. Schilling and Bock use the mode, and the inverse information matrix computed at the mode, of $h(\mathbf{y} \mid \mathbf{x}_h)$ for $\boldsymbol{\mu}$ and $\boldsymbol{\Sigma}$, respectively.

The Monte Carlo method known in the literature as the Monte Carlo E-M (MCEM) algorithm works by generating a large sample of $\mathbf{y}(1), \ldots, \mathbf{y}(R)$ values from $h(\mathbf{y} \mid \mathbf{x}_h)$, which distribution is not usually available in a closed form. Expectation over the latter is replaced by averaging over the large sample.

A large number of draws R are required to obtain precise estimates, which slows down convergence of the E-M algorithm. Meng and Schilling (1996) proposed drawing from the posterior distribution $h(\mathbf{y} \mid \mathbf{x}_h)$ by using a Gibbs sampler (see Section 4.11.1). They showed that, by retaining every fifth value generated, the Gibbs sampler can produce nearly independent samples from the posterior distribution. A Monte Carlo integration technique has also been successfully applied to structural equation models (see, for example, Lee and Song (2004a); Lee and Zhu (2002)).

4.6 Sampling properties of the maximum likelihood estimators

The determination of exact standard errors for the parameter estimates does not appear to be possible. The usual way to deal with this situation for maximum likelihood and generalised least squares estimation is to compute the asymptotic variance–covariance matrix using the information matrix. Even this proves to be impracticable for large p, and so we shall give a more tractable alternative.

Asymptotically the sampling variances and covariances of the maximum likelihood estimates of the $\boldsymbol{\alpha}$s are given by the elements of the inverse of the information matrix evaluated at the solution point. Thus if we have a set of parameters $\boldsymbol{\beta}$ then

$$\{\mathrm{var}(\hat{\boldsymbol{\beta}})\}^{-1} = E\left[-\frac{\partial^2 L}{\partial \beta_i \partial \beta_j}\right]_{\boldsymbol{\beta}=\hat{\boldsymbol{\beta}}},$$

in which

$$\frac{\partial^2 L}{\partial \beta_i \partial \beta_j} = \sum_{h=1}^{n} \left\{ \frac{1}{f} \frac{\partial^2 f}{\partial \beta_i \, \partial \beta_j} - \frac{1}{f^2} \frac{\partial f}{\partial \beta_i} \frac{\partial f}{\partial \beta_j} \right\} \tag{4.24}$$

where $f \equiv f(\mathbf{x}_h)$. On taking the expectation the first term vanishes leaving

$$\{\mathrm{var}(\hat{\boldsymbol{\beta}})\}^{-1} = n \left\{ E \frac{1}{f^2} \frac{\partial f}{\partial \beta_i} \frac{\partial f}{\partial \beta_j} \right\}_{\boldsymbol{\beta}=\hat{\boldsymbol{\beta}}}. \tag{4.25}$$

In our case $\mathbf{x}$ is a score pattern taking 2^p different values and the expectation in (4.25) is thus

$$\sum_{\text{all } \mathbf{x}} \frac{1}{f(\mathbf{x})} \frac{\partial f(\mathbf{x})}{\partial \beta_i} \frac{\partial f(\mathbf{x})}{\partial \beta_j}. \tag{4.26}$$

If p is small it is feasible to evaluate this sum for all i and j and then to invert the resulting matrix. However, if p is larger the computer resources required to compute (4.26) become prohibitive. Moreover, many of the probabilities $f(\mathbf{x}_h)$ will become so small that the computation of $1/f(\mathbf{x}_h)$ will cause overflow on most computing machines. In these circumstances an approximation can be obtained by replacing the expectation of the information matrix by its observed value. This requires the computation of (4.24) and the calculation of a generalised inverse of the resulting matrix. Since the first term has expectation zero, a further approximation may be obtained from the generalised inverse

$$\text{var}^*(\hat{\boldsymbol{\beta}}) = \left\{ \sum_{h=1}^{n} \frac{1}{f^2(\mathbf{x}_h)} \frac{\partial f(\mathbf{x}_h)}{\partial \beta_i} \frac{\partial f(\mathbf{x}_h)}{\partial \beta_j} \right\}^{-1}. \tag{4.27}$$

The dimension of this matrix is $p + qp$ (p α_{i0}s and p α_{ij}s for the q values of j). The number of distinct terms in the sum of (4.27) will usually be less than n since more than one individual may have the same score pattern.

The approximation given by (4.27) appears to be good for standard errors but less reliable for covariances. Some comparative figures for the case of a single latent variable are given in Table 4.1 for standard errors, and these are typical of the differences found in other cases. It therefore seems acceptable to use the simpler version from (4.27) which, in any case, tends to err on the safe side.

In the case of two or more latent variables the computation of standard errors requires more care. The position is essentially the same as for the normal linear model set out in Section 3.6 because of the rotational indeterminacy. If we impose sufficient constraints on the parameters to ensure that there is a unique maximum, then we can use the standard asymptotic approach as for the normal latent variable model. Without such constraints the maximum is not unique and the information matrix will be singular. This may not be true for the approximation of (4.27), but it is not clear that this then has any meaning.

We showed in Section 2.12 how to transform any solution for the maximum likelihood estimates into one satisfying some desired orthogonality condition. In principle, therefore, we could use the asymptotic theory to find standard errors for the rotated solution. Standard errors could be estimated by simulation, as in the one-factor case, but this is currently very time-consuming.

4.7 Approximate maximum likelihood estimators

In certain circumstances it is possible to approximate the likelihood function in a manner which allows us to obtain the maximum likelihood estimators explicitly. As the result for binary data is a special case of a more general result for polytomous data to be given in Section 5.6, we merely quote the result here. The method starts with the matrix of phi-coefficients between all pairs of the p variables. (The phi-coefficient is the product-moment correlation of the binary variables.) We then find the eigenvectors associated with all eigenvalues greater than one. Let $\mathbf{g}$ be the normalised vector associated with the largest eigenvalue; then there is a vector $\boldsymbol{\alpha}' = (\alpha_{11}, \alpha_{21}, \ldots, \alpha_{p1})$

given by

$$\boldsymbol{\alpha} \propto \hat{\mathbf{Q}}^{-1/2}\mathbf{g}, \tag{4.28}$$

where $\hat{\mathbf{Q}}$ is a diagonal matrix with ith diagonal element equal to $\hat{\pi}_i(1-\hat{\pi}_i)$, which maximises the approximate likelihood. The constant of proportionality is determined in a manner to be explained later. The vector associated with the second largest eigenvalue provides a set of estimators for a second factor, and so on for all eigenvalues exceeding one. There is an obvious similarity between this method and a principal components analysis of the phi-coefficient matrix. However, the methods give different loadings because of the matrix $\hat{\mathbf{Q}}$ and the constant of proportionality in (4.28). Like all of the approximate methods, this is concerned only with the α_{ij} for $j \geq 1$; the α_{i0} or π_i can easily be estimated approximately using a method given in the first edition (Bartholomew 1987, pp. 118–119) as

$$\hat{\alpha}_{i0} = \left(1 + \frac{3}{\pi}\Sigma_{j\geq 1}\alpha_{ij}^2\right)^{1/2} \Phi^{-1}(P_i), \tag{4.29}$$

where P_i is the proportion with the score 1 on item i.

4.8 Generalised least squares methods

Before the introduction of the E-M algorithm for obtaining maximum likelihood estimators, two other methods were devised for circumventing the heavy calculations required for full maximum likelihood. These were based on weighted least squares, and, as they have the advantage of being applicable when there are several latent variables, we now give a brief outline of them. They were proposed for the probit/normal model, but the idea could readily be extended to the other members of the family.

Both methods start with the presumption that most of the relevant information in the sample data is contained in the first- and second-order margins. A heuristic justification for this is obtained by considering the underlying variable model. If the ξs are known, then we know that the sample covariance matrix is sufficient for the model parameters, and these require only a knowledge of the bivariate distributions. One would expect the bivariate distributions of the indicator variables, $\mathbf{x}$, to contain almost all of the information about the underlying bivariate distributions. This is supported by an approximation to the likelihood function, given below, which depends only on the first- and second-order margins. Retrospectively, of course, one can see that the sampling errors of the least squares estimators are virtually the same as those obtained by maximum likelihood.

The Christoffersson (1975) method was to choose parameter estimates which minimise the distance, in a least squares sense, between the observed and expected first- and second-order marginal proportions. Let $\hat{P}_i$ be the observed proportion who respond positively on variable i, and $\hat{P}_{ij}$ the proportion for variables i and j, with P_i and P_{ij} being the corresponding population values. The parameters are then estimated by minimising

$$SS = (\hat{\mathbf{P}} - \mathbf{P})'\boldsymbol{\Sigma}_p^{-1}(\hat{\mathbf{P}} - \mathbf{P}), \tag{4.30}$$

where $\mathbf{P}' = (P_1, P_2, \ldots, P_p, P_{12}, P_{13}, \ldots, P_{p,p-1})$ (of dimension $p + \frac{1}{2}p(p-1)$) and $\boldsymbol{\Sigma}_p$ is the covariance matrix of $\mathbf{P}$. Christoffersson (1975) showed that a consistent estimator of $\boldsymbol{\Sigma}_p$ could be obtained and hence that efficient estimators could be found. ($\boldsymbol{\Sigma}_p$ involves third- and fourth-order marginal proportions.)

Although this method is much faster than full maximum likelihood, Muthén (1978) showed that a further improvement was possible. At each stage of the iterative minimisation of (4.30) one has to evaluate the integrals

$$P(\tau_i) = \int_{-\infty}^{\tau_i} \phi(u)\,du, \quad P(\tau_i, \tau_j) = \int_{-\infty}^{\tau_i}\int_{-\infty}^{\tau_j} \phi(u_1, u_2, : l_{ij})\,du_1\,du_2, \tag{4.31}$$

where ϕ is the standard univariate (or bivariate) normal density function. Muthén's method was to invert equations (4.31) to give

$$\tau_i = \Phi^{-1}(P(\tau_i)), \quad l_{ij} = f(P(\tau_i, \tau_j), P(\tau_i), P(\tau_j)) \quad (i, j = 1, 2, \ldots, p; i \neq j), \tag{4.32}$$

and then to fit the model by minimising the weighted squared distance between the parameters of (4.32) and their sample estimates. To obtain the weights, Muthén used a Taylor expansion expressing the estimator $\hat{\mathbf{t}}$ of $\mathbf{t} = (\tau_i, \ldots, \tau_p, l_{12}, \ldots, l_{p-1,p})$ in the form

$$\hat{\mathbf{t}} = \mathbf{t} + \boldsymbol{\delta}$$

and showing that the covariance matrix of $\boldsymbol{\delta}$ could be consistently estimated. This method avoids the repeated integration involved in Christoffersson's approach while retaining the same asymptotic efficiency. The sample estimators of $\{\hat{l}_{ij}\}$ are, of course, the tetrachoric correlations, so the method is analogous to the weighted least squares method for normal models. The difference lies in the use of tetrachoric correlations, with consequent differences in the weighting.

4.9 Goodness of fit

The theoretical hypothesis is $H_0 : \boldsymbol{\pi} = \boldsymbol{\pi}(\boldsymbol{\alpha})$, where $\boldsymbol{\pi}$ is the true probability and $\boldsymbol{\pi}(\boldsymbol{\alpha})$ is the probability computed from the model. In practice, the sample proportions $\mathbf{P}$ are compared with the estimated ones which are $\boldsymbol{\pi}(\hat{\boldsymbol{\alpha}})$, where $\hat{\boldsymbol{\alpha}}$ can be either the maximum likelihood estimator or any other. If n is large compared with 2^p, the expected frequency for each response pattern is likely to be large enough to carry out a valid chi-squared (χ^2) or log-likelihood ratio (G^2) test to compare the observed and expected frequencies. Up to a point, a few small expected frequencies can be dealt with by pooling response patterns so that all are greater than 5, say. In the unpooled case the number of degrees of freedom will be $2^p - p(q+1) - 1$. As 2^p becomes larger the amount of pooling needed may be such as to reduce the degrees of freedom to zero, at which point no test can be done. It is common to have values of $p \geq 10$, large enough to make this a common problem. We can think of the data as set out in a 2^p contingency table and so our problem is an example of how to test goodness of fit in sparse contingency tables. We shall meet the problem again in relation to the latent class model in Chapter 6. The problem has been investigated in that case by Collins *et al.* (1993) and, in a more general context, by Langeheine *et al.* (1996). The specific case of the fit of the latent trait model is treated in Bartholomew and Tzamourani (1999). The main conclusion of this work is that the best way to do a global test of fit is to generate the empirical sampling distribution of the statistic (χ^2 or G^2) using the parametric bootstrap method. This proceeds as follows:

1. Fit the desired model in the usual way.
2. Generate a random sample (of the same size) from the population in which the parameter values are equal to those estimated for the actual sample.
3. Fit the model in each case and compute the chosen test of fit.
4. Compare the actual value of the statistic with the bootstrap sampling distribution.

The number of bootstrap samples needs to be large enough to give a reasonable estimate of the sampling distribution, but the time needed to do the calculation will increase proportionally.

A global test is a useful first step but, on its own, it may fail to reveal much about the nature of any departure from the model. In particular, it may happen that just two or three items are responsible for a poor fit because, perhaps, they also depend on some factor peculiar only to them. The effect of this may be diffused across all response patterns and so be difficult to detect by looking at the individual contributions to χ^2. Therefore, we recommend carrying out supplementary tests designed to detect the effects of such inappropriate items.

These tests are based on residuals calculated from the marginal frequencies of various orders. In Section 4.1 we showed that the set of response patterns was equivalent to a set of marginal probabilities. The first-order margins, $P\{x_i = 1\}$ $(i = 1, 2, \ldots, p)$ contain no information about the dependencies among the xs. The higher-order margins do contain that information, beginning with the simple pairwise associations derivable from probabilities of the form $P\{x_i = 1,\ x_j = 1\}$. A model which successfully predicts these associations may be useful even if it fails to capture the higher-order dependencies. We note that we can obtain all other pairwise probabilities using results such as $P\{x_i = 0,\ x_j = 1\} = P\{x_j = 1\} - P\{x_i = 1,\ x_j = 1\}$. Similarly, all second-order marginal probabilities for $x_i,\ x_j,\ x_k$ can be derived from $P\{x_i = 1,\ x_j = 1,\ x_k = 1\}$ and the lower-order margins. In practice the second- and third-order margins usually suffice.

If O denotes the observed frequency for any marginal probability and E the corresponding expected frequency, we define the residual as $R = (O - E)^2/E$. (Note that such a residual is not one of the terms making up the global χ^2 test.) Large values of R for the second-order margins will identify pairs of xs which are more (or less) strongly associated than the model predicts. Such an excess of association might point to some extraneous factor affecting only those two variables. In the binary case, Reiser and VandenBerg (1994) and Bartholomew *et al.* (2008) used residuals extensively for identifying pairs and triplets of items where the fit was poor. However, only recently have proper goodness-of-fit test statistics based on the lower-order observed and expected margins probabilities been developed. These are also known as limited-information goodness-of-fit tests (Reiser 1996; Bartholomew and Leung 2002; Cai *et al.* 2006; Maydeu-Olivares and Joe 2005).

Let us denote by $\mathbf{P}' = (P_1, \ldots, P_{2^p})$ the vector of sample (observed) proportions for all possible distinct response patterns with a corresponding vector of estimated probabilities under the model by $\boldsymbol{\pi}(\hat{\boldsymbol{\alpha}})' = (\pi_1(\hat{\boldsymbol{\alpha}}), \ldots, \pi_{2^p}(\hat{\boldsymbol{\alpha}}))$. In general, the fit of a model is judged by how close the estimated proportions $\boldsymbol{\pi}(\hat{\boldsymbol{\alpha}})$ are to the sample proportions $\mathbf{P}$. One can obtain the lower-order probabilities by pre-multiplying the overall probabilities by an indicator matrix.

A test statistic based on the lower-order residuals is

$$n(\mathbf{P}^l - \boldsymbol{\pi}^l(\hat{\boldsymbol{\alpha}}))'\boldsymbol{\Sigma}_l^{-1}(\hat{\boldsymbol{\alpha}})(\mathbf{P}^l - \boldsymbol{\pi}^l(\hat{\boldsymbol{\alpha}})), \tag{4.33}$$

where l indicates residuals obtained either from lower-order margins such as univariate ($l = 1$) or bivariate ($l = 2$) or up to the lth margins, and $\mathbf{P}^l$ and $\boldsymbol{\pi}^l(\hat{\boldsymbol{\alpha}})$ are the sample and estimated proportions under the model respectively for the lth-order margins. The covariance matrix $\boldsymbol{\Sigma}_l(\hat{\boldsymbol{\alpha}})$ can be obtained from

$$\boldsymbol{\Sigma}_l = \Xi_l - \Delta_l \mathcal{I}^{-1} \Delta_l',$$

evaluated at $\hat{\boldsymbol{\alpha}}$, where

$$\Delta_l = \frac{\partial \boldsymbol{\pi}^l(\boldsymbol{\alpha})}{\partial \boldsymbol{\alpha}'},$$

$\mathcal{I}$ is the Fisher information matrix and Ξ_l is the covariance matrix of $\sqrt{n}(P_l - \boldsymbol{\pi}_l)$. Using known results from multivariate theory and under the null hypothesis (H_0 : $\boldsymbol{\pi} = \boldsymbol{\pi}(\boldsymbol{\alpha})$), (4.33) follows a χ^2 distribution with degrees of freedom equal to the rank of the asymptotic variance–covariance matrix given by $\boldsymbol{\Sigma}_l$. Maydeu-Olivares and Joe (2005) suggested using the Moore–Penrose inverse of the covariance matrix; however, they noted that its calculation might not be always feasible due to the presence of eigenvalues close to zero. To overcome this problem an alternative test statistic has been proposed by Maydeu-Olivares and Joe (2005) given by

$$n(\mathbf{P}^l - \boldsymbol{\pi}^l(\hat{\boldsymbol{\alpha}}))'\hat{\mathbf{M}}_l(\mathbf{P}^l - \boldsymbol{\pi}^l(\hat{\boldsymbol{\alpha}})), \tag{4.34}$$

where

$$\hat{\mathbf{M}}_l(\hat{\boldsymbol{\alpha}}) = \hat{\Delta}_l^c([\hat{\Delta}_l^c]'\hat{\Xi}_l\hat{\Delta}_l^c)^{-1}[\hat{\Delta}_l^c]'$$

and $\hat{\Delta}_l^c$ is a full-rank orthogonal complement to $\hat{\Delta}_l$. Under the null hypothesis, (4.34) follows a χ^2 distribution with degrees of freedom equal to the number of sample moments up to order l minus the total number of free parameters.

A closely related problem to that discussed here has been the subject of a good deal of research in item response theory. This is concerned with testing for unidimensionality. If one wishes to construct a unidimensional scale using a latent variable model, one needs to know whether a single latent variable is sufficient to explain the distribution of response patterns. Here we are looking for a particular kind of departure from the logit/normal model. Therefore it is relevant to know the best way of detecting the effect of additional latent variables. The approach initiated by Holland (1981) and continuing with Rosenbaum (1984) and Holland and Rosenbaum (1985) is concerned with the restraints which unidimensionality imposes on the joint distribution $f(\mathbf{x})$. In particular, these authors arrive at a requirement of conditional association. Roughly speaking, this says that the pairwise associations, conditional on all other variables, must be positive. Looking at these conditional associations is somewhat similar to inspection of the residuals. A different approach to testing unidimensionality is found in Stout (1987, 1990). He bases a test on the proportion of positive responses and then extends the idea to what he terms essential unidimensionality. This condition allows minor variations from strict unidimensionality. Junker (1993) reviews these approaches and establishes the links between them.

An important feature of these methods is that they are non-parametric in the sense that they make no assumption about the form of the item response functions or of the prior distribution. They are not, therefore, tests of the logit/normal model but the information they give is relevant to that question.

4.10 Posterior analysis

We have given some general results applicable to all members of the GLLVM family of models in Section 2.15. Here we illustrate them for the case of binary data and add one or two further results. In principle, all the information about $\mathbf{y}$ is contained in the posterior distribution $h(\mathbf{y} \mid \mathbf{x})$. In the case of the logit/normal model we have seen that this distribution depends on $\mathbf{x}$ only through the q-variate sufficient statistic $\mathbf{X} = \mathbf{A}\mathbf{x}$, where $\mathbf{A} = \{\alpha_{ij}\}$. Much of the interest in posterior analysis lies in its use for scaling, and this aspect has been especially prominent in the measurement of abilities. In that context one is usually dealing with a single latent dimension, and for the remainder of this section we shall concentrate on that case. Some of the results can easily be extended to the case of q latent variables. We quoted the result of Knott and Albanese (1993) that if the posterior distribution was normal for one response pattern, say $(0, 0, \ldots, 0)$, then it would be normal for all response patterns. In that event, they showed that the posterior mean was a linear function of $\mathbf{X}$. Chang and Stout (1993) subsequently showed that the posterior distribution tends to normality as

p, the number of items, increases without limit. In practice the normal approximation and the consequent linearity of the posterior mean in $\mathbf{X}$ appear to be good if p is quite small. This can be checked in the case of Example 4.13.2 below.

The posterior variance, or standard deviation, of y is a useful measure of the precision of the predicted scale point. In the context of test theory this is usually known as *reliability* since it has to do with the reproducibility of a measurement. The posterior variance serves this purpose. By convention we assume the prior variance to be unity. After we have observed $\mathbf{x}$ we shall be less uncertain about the location of the subject on the scale of y, and this will be reflected in the smaller value obtained for $\text{var}(y \mid X)$. If this turns out to be around 0.35, as in Example 4.13.2, then we can say that 65% of the original variation in y has been removed by the information contained in X.

The traditional measure of reliability has been based on the variance of X given y, and not the other way round as proposed here. Since y is never known, the average has been taken with respect to y. Thus the coefficient is

$$r = 1 - \frac{E\,\text{var}(X \mid y)}{\text{var}(X)}. \tag{4.35}$$

This tells us by how much we would expect the variance of X to be reduced by knowledge of y. This coefficient is also equivalent to the test–retest correlation (see, for example, Bartholomew and Schuessler (1991)). If we were to take a random sample of individuals and obtain two independent measurements of X on each, then r is the correlation between the first and second determinations. A reliable method of scaling should yield a close correspondence between the two measures. The use of (4.35) for determining the reliability of measures derived from the logit/normal model was illustrated by Bartholomew and Schuessler (1991). A disadvantage of r is that it is based on an average variance, and this variance might depend strongly on y. More importantly, it inverts the natural way of regarding X and y. When X has been observed and y is unknown, it is more natural to condition on X rather than to consider the variation of something which is known conditional on something which is not.

One can thus define an average posterior variance in an analogous way, giving

$$r^* = 1 - \frac{E\,\text{var}(y \mid X)}{\text{var}(y)}, \tag{4.36}$$

as was done in Bartholomew (1995a) (see also Bartholomew (1996, section 2.2)). In that paper a common approximation to both r and r^* was obtained for the case where x_i has a conditional exponential family distribution. This implies that r and r^* are approximately equal, and numerical estimates of both show that this is usually the case.

However, once the relevance of $\text{var}(y \mid X)$ is accepted, averaging over X seems unnecessary. It is more useful to attach the appropriate conditional variance (or standard deviation) to each prediction. In actual fact the dependence of $\text{var}(y \mid X)$ on X is usually quite slight. Such departure as there is from constancy occurs for the extreme response patterns where most responses are the same.

4.11 Fitting the logit/normal and probit/normal models: Markov chain Monte Carlo

The main features of Bayesian estimation methods and, in particular, the MCMC methodology are outlined in Section 2.9. As already mentioned in Chapter 2, Bayesian inference is based on the posterior distribution of the vector of all random parameters conditionally on the data, and MCMC is primarily used for helping to make draws from that posterior distribution. MCMC methods, such as the Metropolis–Hastings algorithm and its special case, Gibbs sampling (Gelfand and Smith 1990), can be used to estimate the parameters of latent variable models. In particular, Metropolis–Hastings within Gibbs sampling is used for estimating the logit/normal model in (4.4), whereas Gibbs sampling can be used for the probit/normal model in (4.8).

4.11.1 Gibbs sampling

Gibbs sampling (Geman and Geman 1984; Gelfand and Smith 1990), is a way of sampling from complex joint posterior distributions. Consider the posterior distribution $h(\mathbf{v} \mid \mathbf{x})$, where it will be convenient here to consider partitioning the vector of unknown parameters $\mathbf{v}$ into two components, $\mathbf{v}_1 = \mathbf{y}$ and $\mathbf{v}_2 = \boldsymbol{\alpha}$. Other partitions might be more convenient for other problems. When it is computationally difficult to take draws from the joint posterior of all parameters, it might be easier to simulate from the full conditional distributions, here $h(\mathbf{y}^1 \mid \boldsymbol{\alpha}^0, \mathbf{x})$ and $h(\boldsymbol{\alpha}^1 \mid \mathbf{y}^1, \mathbf{x})$. Gibbs sampling produces a sequence of iterations $\mathbf{v}^0, \mathbf{v}^1, \ldots, \mathbf{v}^k$ that form a Markov chain that eventually converges to its stationary distribution, which is the posterior distribution. It can be summarised in the following two steps:

1. Start with initial guesses of $\mathbf{y}^0, \boldsymbol{\alpha}^0$.
2. Then simulate in the following order:

   (a) draw $\mathbf{y}^1$ from $h(\mathbf{y} \mid \boldsymbol{\alpha}^0, \mathbf{x})$,

   (b) draw $\boldsymbol{\alpha}^1$ from $h(\boldsymbol{\alpha} \mid \mathbf{y}^1, \mathbf{x})$,

   where the conditional distributions are given by

$$h(\mathbf{y} \mid \boldsymbol{\alpha}, \mathbf{x}) = \frac{g(\mathbf{x} \mid \mathbf{y}, \boldsymbol{\alpha})h(\mathbf{y}, \boldsymbol{\alpha})}{\int g(\mathbf{x} \mid \mathbf{y}, \boldsymbol{\alpha})h(\mathbf{y}, \boldsymbol{\alpha})d\mathbf{y}}, \tag{4.37}$$

$$h(\boldsymbol{\alpha} \mid \mathbf{y}, \mathbf{x}) = \frac{g(\mathbf{x} \mid \mathbf{y}, \boldsymbol{\alpha})h(\mathbf{y}, \boldsymbol{\alpha})}{\int g(\mathbf{x} \mid \mathbf{y}, \boldsymbol{\alpha})h(\mathbf{y}, \boldsymbol{\alpha})d\boldsymbol{\alpha}}. \tag{4.38}$$

Step 2 is iterated t times generating the sample $(\mathbf{v}^t = \mathbf{v}_1^t, \mathbf{v}_2^t)$. For suitably large t, the joint distribution of $\mathbf{v}^t$ will be close to the joint distribution of $\mathbf{v}$. So that will be considered as one simulated value from the posterior distribution of $\mathbf{v}$. Note that all the draws up to t are discarded (burn-in period) and t should be large enough to guarantee convergence. Repeating the above whole process s times gives a sample of size s from the posterior distribution. That sample can be used to approximate the mean and variance of the posterior distribution of each model parameter.

Note that each distribution used at step 2 is called a full conditional distribution; it is multivariate and, depending on the number of items and number of sample members, can be heavily multidimensional. That problem can be easily solved by arranging the blocks of parameters to have smaller dimensions. In latent trait models, one might block the parameters $(\boldsymbol{\alpha}_i, i = 1, \ldots, p)$ separately for each item (p blocks), and further form n blocks, one for each subject's latent vector $(\mathbf{y}_h, h = 1, \ldots, n)$. In that case, the vector of parameters $\mathbf{v}$ consists of $n + 2p$ components and the number of transitions required under step 2 will be equal to n+p. The elements of each of those components will not then usually be bigger than three or four dimensions. Gibbs sampling is particularly useful when the full conditional distributions given in (4.37) and (4.38) exist in closed form. However, the normalising constants that appear in the denominator of the posterior distributions are not often easy to compute. Methods based on data augmentation, rejection sampling and Metropolis–Hastings have been used with Gibbs sampling to simplify or avoid the computation of the normalising constants. Albert (1992) proposed a data augmentation scheme within Gibbs for estimating the parameters of the probit/normal of (4.8). The data augmentation scheme borrows the idea of the underlying variable approach discussed in Section 4.3. The observed binary variable, x_i, is assumed to be a manifestation of an underlying continuous variable ξ_i. Suppose that when the underlying variable ξ_i is positive the observed binary response is 1, and 0 otherwise. Furthermore, suppose that the underlying variable ξ conditional on the latent variable is normally distributed with mean $\alpha_{i0} + \alpha_{i1} y_h$ and variance 1. That leads to the known linear factor model,

$$\xi_{ih} = \alpha_{i0} + \alpha_{i1} y_h + \epsilon_{ih} \quad (i = 1, \ldots, p; h = 1, \ldots, n), \tag{4.39}$$

where the error term $\epsilon_{ih} \sim N(0, 1)$. We would like to draw samples from the joint posterior

$$h(\boldsymbol{\xi}, \mathbf{y}, \boldsymbol{\alpha} \mid \mathbf{x}) \propto g(\mathbf{x} \mid \boldsymbol{\xi}) h(\boldsymbol{\xi} \mid \mathbf{y}, \boldsymbol{\alpha}) h(\mathbf{y})(\boldsymbol{\alpha}), \tag{4.40}$$

but instead we draw samples from the full conditional posteriors, $h(\boldsymbol{\xi} \mid \mathbf{y}, \boldsymbol{\alpha}, \mathbf{x})$, $h(\mathbf{y} \mid \boldsymbol{\xi}, \boldsymbol{\alpha}, \mathbf{x})$ and $h(\boldsymbol{\alpha} \mid \boldsymbol{\xi}, \mathbf{y}, \mathbf{x})$, where for the one-factor model the vector $\boldsymbol{\xi}$ is of dimension $n \times p$, the vector $\mathbf{y}$ is of dimension $n \times 1$ and the vector of model parameters $\boldsymbol{\alpha}$ is of dimension $p \times 2$. Prior information needs to be provided for all model parameters. Usually, $y_h \sim N(0, 1)$, $\alpha_{i1} \sim N_{(0,\infty)}(\mu_{a_{i1}}, S^2_{\alpha_{i1}})$ and $\alpha_{i0} \sim N(\mu_{a_{i0}}, S^2_{\alpha_{i0}})$, where $(\mu_{a_{i1}}, \mu_{a_{i0}})$ and $(S^2_{\alpha_{i1}}, S^2_{\alpha_{i0}})$ are means and variances respectively chosen by the user. The steps of the Gibbs sampling at a given step $t - 1$ are as follows:

1. Suppose that at iteration $t - 1$ the values of all model parameters, including latent and underlying variables, are $\boldsymbol{\xi}^{(t-1)}, \mathbf{y}^{(t-1)}, \boldsymbol{\alpha}^{(t-1)}$.

2. Then simulate in the following order:

   (a) Draw $\xi_{ih}^{(t)}$ from $h(\boldsymbol{\xi} \mid \mathbf{y}^{(t-1)}, \boldsymbol{\alpha}^{(t-1)}, \mathbf{x})$. The random variables ξ_{ih} are assumed to be independent conditional on the latent variable and model parameters, and to have a truncated normal distribution with mean $\alpha_{i0} + \alpha_{i0} y_h$ and variance 1. The distribution is truncated to the right at 0 when the observed binary response is 0 and to the left at 0 when the observed response is 1.

   (b) Conditional on current values of the underlying variables and item parameters, the model in (4.39) becomes

$$\xi_{ih}^{(t)} - \alpha_{i0}^{(t-1)} = \alpha_{i1}^{(t-1)} y_h + \epsilon_{ih} \quad (i = 1, \ldots, p; h = 1, \ldots, n),$$

   with unknown parameters the latent variables y_h. We continue by drawing $y_h^{(t)}$ from $h(y_h \mid \boldsymbol{\xi}^{(t)}, \boldsymbol{\alpha}^{(t-1)}, \mathbf{x})$, which turns out to be normally distributed with mean

$$\frac{\sum_{i=1}^{p} (\xi_{ih}^{(t)} - \alpha_{i0}^{(t-1)}) \alpha_{i1}^{(t-1)}}{\sum_{i=1}^{p} \alpha_{i1}^{2(t-1)} + 1}$$

   and variance

$$\frac{1}{\sum_{i=1}^{p} \alpha_{i1}^{2(t-1)} + 1}.$$

   (c) Finally, conditional on current values of the underlying variables and latent variables, (4.39) becomes

$$\xi_{ih}^{(t)} = \alpha_{i0}^{(t-1)} + \alpha_{i1}^{(t-1)} y_h^{(t)} + \epsilon_{ih} \quad (i = 1, \ldots, p; h = 1, \ldots, n),$$

   with unknown item parameters. That model resembles the linear regression model. The draws $(\alpha_{i0}^{(t)}, \alpha_{i1}^{(t)})$ come from the posterior distribution of the vector of item parameters $(\alpha_{i0}, \alpha_{i1})$, conditional on the latent and underlying variables that can be shown to be normal with mean

$$[\mathbf{X}'\mathbf{X} + \boldsymbol{\Sigma}^{-1}]^{-1} [\mathbf{X}'\boldsymbol{\xi} + \boldsymbol{\Sigma}^{-1}[\mu_{a_{i1}}, \mu_{a_{i0}}]']$$

   and covariance matrix

$$[\mathbf{X}'\mathbf{X} + \boldsymbol{\Sigma}^{-1}]^{-1},$$

   where

$$\boldsymbol{\Sigma}^{-1} = \begin{bmatrix} S_{\alpha_{i1}}^2 & 0 \\ 0 & S_{\alpha_{i0}}^2 \end{bmatrix}$$

   and $\mathbf{X}$ is the covariate vector with elements $(y_h, 1)$.

4.11.2 Metropolis–Hastings

A method that altogether avoids the computation of the normalising constant is the Metropolis–Hastings (M-H) algorithm (Metropolis *et al.* 1953; Hastings 1970). For the M-H algorithm at each step, the next state is chosen by sampling a candidate from a proposal distribution $q(\cdot)$ that can be of any form. Independently of the form of the proposal distribution, the chain will converge to the desired distribution $\pi(\mathbf{y}, \boldsymbol{\alpha})$.

The transition kernel for the M-H algorithm is given by

$$
\begin{aligned}
t[(\mathbf{y}^0, \boldsymbol{\alpha}^0), (\mathbf{y}^1, \boldsymbol{\alpha}^1)] &= q[(\mathbf{y}^0, \boldsymbol{\alpha}^0), (\mathbf{y}^1, \boldsymbol{\alpha}^1)]a[(\mathbf{y}^0, \boldsymbol{\alpha}^0), (\mathbf{y}^1, \boldsymbol{\alpha}^1)] + I[(\mathbf{y}^1, \boldsymbol{\alpha}^1) \\
&= (\mathbf{y}^0, \boldsymbol{\alpha}^0)]\left[1 - \int q[(\mathbf{y}^*, \boldsymbol{\alpha}^*), (\mathbf{y}^0, \boldsymbol{\alpha}^0)]a[(\mathbf{y}^0, \boldsymbol{\alpha}^0), (\mathbf{y}^*, \boldsymbol{\alpha}^*)]d(\mathbf{y}^*, \boldsymbol{\alpha}^*)\right], \quad (4.41)
\end{aligned}
$$

where $I(\cdot)$ is an indicator function taking the value 1 when the statement in brackets is true and 0 otherwise. The first component of (4.41) occurs when the candidate state $(\mathbf{y}^1, \boldsymbol{\alpha}^1)$ has been accepted and the second term occurs from a rejection probability for all possible candidates $(\mathbf{y}^*, \boldsymbol{\alpha}^*)$. The proposal distribution q is a function of the current state of the Markov chain $(\mathbf{y}^0, \boldsymbol{\alpha}^0)$ and the candidate state $(\mathbf{y}^*, \boldsymbol{\alpha}^*)$, and a is the acceptance probability that determines whether the candidate state will be accepted as the next state of the chain, defined as

$$
a[(\mathbf{y}^0, \boldsymbol{\alpha}^0), (\mathbf{y}^*, \boldsymbol{\alpha}^*)] = \min\left(1, \frac{\pi(\mathbf{y}^*, \boldsymbol{\alpha}^*)q[(\mathbf{y}^*, \boldsymbol{\alpha}^*), (\mathbf{y}^0, \boldsymbol{\alpha}^0)]}{\pi(\mathbf{y}^0, \boldsymbol{\alpha}^0)q[(\mathbf{y}^0, \boldsymbol{\alpha}^0), (\mathbf{y}^*, \boldsymbol{\alpha}^*)]}\right). \quad (4.42)
$$

The normalising constant cancels out in the computation of the ratio of the $\pi(\cdot)$ densities of (4.42).

For a given value $(\mathbf{y}^0, \boldsymbol{\alpha}^0)$, the proposal distribution $q[(\mathbf{y}^0, \boldsymbol{\alpha}^0), (\mathbf{y}^*, \boldsymbol{\alpha}^*)]$ is a density function of $(\mathbf{y}^1, \boldsymbol{\alpha}^1)$. If the candidate is accepted, then $(\mathbf{y}^1, \boldsymbol{\alpha}^1) = (\mathbf{y}^*, \boldsymbol{\alpha}^*)$, otherwise we generate another candidate. One can show that under general conditions the sequence $(\mathbf{y}^0, \boldsymbol{\alpha}^0), (\mathbf{y}^1, \boldsymbol{\alpha}^1), (\mathbf{y}^2, \boldsymbol{\alpha}^2), \ldots$ is a Markov chain with stationary distribution $\pi(\mathbf{y}, \boldsymbol{\alpha})$.

In theory one can choose any proposal distribution. Depending on the choice of the proposal distribution q, the algorithm might converge to its stationary distribution after very few or very many steps. Usually, known distributions such as the uniform, beta, gamma, normal or t distribution are chosen. Common forms of the transition kernel are the *independence* chain and the *random walk* chain. For the independence chain, the proposal distribution $q[(\mathbf{y}^0, \boldsymbol{\alpha}^0), (\mathbf{y}^1, \boldsymbol{\alpha}^1)]$ does not depend on the current state $(\mathbf{y}^0, \boldsymbol{\alpha}^0)$; it will work effectively when the proposal is close to the target stationary distribution π since the acceptance probability will be close to 1 and accept nearly all candidates. In contrast, in the random walk chain the proposal distribution depends on the current state of the chain. The candidate value is considered to be the current state plus some noise that might be coming from a known distribution such as the normal with some variance σ^2.

The number of iterations until convergence or, in other words, the number of candidate states being rejected will increase with the number of parameters.

To avoid the slow convergence of the M-H algorithm and the problem of computing the normalising constant required in the Gibbs sampling, one can use both algorithms together. Patz and Junker (1999b) discuss rejection sampling techniques, such as the M-H, within Gibbs sampling. In essence, an M-H sampling scheme is used within the Gibbs for drawing samples from the complete conditionals, which need only be defined up to a normalising constant. M-H avoids the calculation of the normalising constants by performing draws using rejection sampling.

We now provide the steps of an M-H within Gibbs sampling algorithm for estimating the parameters of the logit/normal model of (4.4). For binary response the complete conditionals required for Gibbs sampling are not in closed form, and so M-H will be used to avoid computations of the normalising constants. To proceed we make the following assumptions:

1. All model parameters, including the latent variable, are assumed to be *a priori* independent,

$$h(y, \boldsymbol{\alpha}) = \prod_{h=1}^{n} h(y_h) \prod_{i=1}^{p} h(\alpha_{i0}) h(\alpha_{i1}).$$

2. Prior distributions are defined for all the random components of the model as follows: $y_h \sim N(0, \sigma_y^2)$, $h = 1, \ldots, n$; $\alpha_{0i} \sim N(0, \sigma_{\alpha_{i0}}^2)$ and $\alpha_{i1} \sim N(0, \sigma_{\alpha_{i1}}^2)$, $i = 1, \ldots, p$. Often a log-normal is assumed for the α_{i1} parameters to guarantee all positive loadings. Alternatively, $\boldsymbol{\alpha}_i \sim N(0, \boldsymbol{\Sigma}_{\alpha_i})$, where $\boldsymbol{\Sigma}_{\alpha_i}$ is the variance–covariance matrix of the parameters related to item i. To simplify things, assume that all variances are the same across items, giving $\sigma_{\alpha_{i0}}^2 = \sigma_{\alpha_0}^2$ and $\sigma_{\alpha_{i1}}^2 = \sigma_{\alpha_1}^2$.

3. The draws will be done separately for each y_h, $h = 1, \ldots, n$, and for each vector of parameters $\boldsymbol{\alpha}_i = (\alpha_{i0}, \alpha_{i1})$, $i = 1, \ldots, p$. We need two proposal distributions, namely, $q_y = (y_h^*, y_h^{k-1})$ and $q_\alpha = (\boldsymbol{\alpha}_i^*, \boldsymbol{\alpha}_i^{k-1})$. Furthermore, as stated earlier, proposal distributions are taken to have known forms. Here, the normal is assumed: $q_y = (y_h^*, y_h^{k-1}) \sim N(y_h^{k-1}, \sigma_y^2)$, $h = 1, \ldots, n$, and $q_\alpha = (\boldsymbol{\alpha}_i^*, \boldsymbol{\alpha}_i^{k-1}) \sim N(\boldsymbol{\alpha}_i^{k-1}, \sigma_\alpha^2)$, $i = 1, \ldots, p$, with means equal to the value at the current state $k - 1$.

4. We assume symmetrical proposal distributions and therefore the proposal distribution need not be included in the calculation of the acceptance probabilities.

The details of an M-H within Gibbs sampling algorithm for estimating the parameters of the two-parameter logistic model are given below. For iteration k, do the following:

1. Draw y^k from the complete conditional, $h(y \mid \boldsymbol{\alpha}^{k-1}, \mathbf{x})$. Since the complete conditional is not given in a closed form we use the M-H algorithm:

   (a) Draw y_h^* from $N(y_h^{k-1}, \sigma_y^2)$ independently for each $h = 1, \ldots, n$.

(b) Compute the acceptance probability a,

$$a[y_h^{k-1}, y_h^*] = \min\left(1, \frac{g(\mathbf{x}_h \mid y_h^*, \boldsymbol{\alpha}^{k-1})}{g(\mathbf{x}_h \mid y_h^{k-1}, \boldsymbol{\alpha}^{k-1})} \frac{h(y_h^*, \boldsymbol{\alpha}^{k-1})}{h(y_h^{k-1}, \boldsymbol{\alpha}^{k-1})}\right)$$
$$= \min\left(1, \frac{\prod_{i=1}^{p} g(x_{ih} \mid y_h^*, \boldsymbol{\alpha}_i^{k-1})}{\prod_{i=1}^{p} g(x_{ih} \mid y_h^{k-1} \boldsymbol{\alpha}_i^{k-1})} \frac{h(y_h^*)}{h(y_h^{k-1})}\right), \tag{4.43}$$

where

$$g(x_{ih} \mid y_h^*, \boldsymbol{\alpha}_i^{k-1}) = [\pi_i(y_h^*, \boldsymbol{\alpha}_i^{k-1})]^{x_{ih}}[1 - \pi_i(y_h^*, \boldsymbol{\alpha}_i^{k-1})]^{(1-x_{ih})}$$

and

$$\frac{h(y_h^*)}{h(y_h^{k-1})} = \frac{\exp(-\frac{1}{2\sigma_y^2}(y_h^*)^2)}{\exp(-\frac{1}{2\sigma_y^2}(y_h^{k-1})^2)}.$$

2. Draw $\boldsymbol{\alpha}_i^k$ from the complete conditional, $h(\boldsymbol{\alpha}_i \mid y^k, \mathbf{x})$. Again here, the complete conditionals are not given in a closed form and therefore the M-H algorithm is used:

(a) Draw α_{0i}^* from $N(\alpha_{0i}^{k-1}, \sigma_{\alpha_0}^2)$, $i = 1, \ldots, p$.

(b) Draw α_{1i}^* from $N(\alpha_{1i}^{k-1}, \sigma_{\alpha_1}^2)$, $i = 1, \ldots, p$.

(c) Compute the acceptance probability a for each item i,

$$a_i[\boldsymbol{\alpha}_i^{k-1}, \boldsymbol{\alpha}_i^*] = \min\left(1, \frac{g(\mathbf{x}_h \mid y^k, \boldsymbol{\alpha}_i^*)}{g(\mathbf{x}_h \mid y^k, \boldsymbol{\alpha}_i^{k-1})} \frac{h(y^k, \boldsymbol{\alpha}_i^*)}{h(y^k, \boldsymbol{\alpha}_i^{k-1})}\right)$$
$$= \min\left(1, \frac{\prod_{h=1}^{n} g(x_{ih} \mid y_h^k, \boldsymbol{\alpha}_i^*)}{\prod_{h=1}^{n} g(x_{ih} \mid y_h^k, \boldsymbol{\alpha}_i^{k-1})} \frac{h(\boldsymbol{\alpha}_i^*)}{h(\boldsymbol{\alpha}_i^{k-1})}\right), \tag{4.44}$$

where

$$g(x_{ih} \mid y_h^k, \boldsymbol{\alpha}^{k-1}) = [\pi_i(y_h^k, \boldsymbol{\alpha}^{k-1})]^{x_{ih}}[1 - \pi_i(y_h^k, \boldsymbol{\alpha}^{k-1})]^{(1-x_{ih})}$$

and

$$\frac{h(\boldsymbol{\alpha}_i^*)}{h(\boldsymbol{\alpha}_i^{k-1})} = \frac{\exp(-\frac{1}{2\sigma_\alpha^2}(\alpha_{0i}^*)^2)}{\exp(-\frac{1}{2\sigma_\alpha^2}(\alpha_{0i}^{k-1})^2)} \frac{\exp(-\frac{1}{2\sigma_\alpha^2}(\alpha_{1i}^*)^2)}{\exp(-\frac{1}{2\sigma_\alpha^2}(\alpha_{1i}^{k-1})^2)}.$$

More details on the M-H within Gibbs for latent variable models can be found in Patz (1996) and Patz and Junker (1999b).

4.11.3 Choosing prior distributions

For all the parameters defined in $\mathbf{v}$ we usually assume independent prior distributions. Diffuse or vague priors are needed in most applications, so that the likelihood is emphasised rather than the prior, unless one would like to impose some restrictions on the parameters such as positive loadings or values within a specific range. The latent variables $\mathbf{y}$ are a random sample from a known population, usually independent standard normal distributions.

4.11.4 Convergence diagnostics in MCMC

A main concern in MCMC estimation methods is to assess whether and at what stage the distribution of the parameter values produced by the Markov chain may be considered to be from the stationary distribution of the chain, which is the posterior distribution of the parameters given the data. Most of the work developed in this direction is based on diagnostic tools that rely for those conclusions on the MCMC output. The most popular diagnostic tools are that of Heidelberger and Welch (1983), Geweke (1992), Raftery and Lewis (1992b) and Gelman and Rubin (1992). Their popularity greatly depends on their availability in freely distributed software such as BUGS (Spiegelhalter *et al.* 1996). For a review and comparison of all the currently available diagnostic tools one should read Cowles and Carlin (1996) and Brooks and Roberts (1998). We briefly discuss a few of those diagnostics.

Geweke's (1992) criterion compares the mean of the parameter values from two sections of the Markov chain using a z test. The criterion uses two sections of the Markov chain. The first section contains, for example, the first 10% of the iterations and the second section contains the last 50% of the iterations. If the whole chain is stationary then the mean of the parameter values in the early stage and in the late stage should be similar. The test is based on a z-score that compares the difference in the two means divided by the asymptotic standard error of the difference. As the chain length tends to infinity the distribution of z tends towards standard normal. Absolute values of z greater than 2 indicate that the chain has not converged. In the case of non-convergence, the first 10% is discarded and Geweke's diagnostic score is recomputed to compare the first 20% with the last 50%.

The Raftery and Lewis (1992b) diagnostic tool aims to check the accuracy of the estimation of quantiles and computes the number of iterations required in order to achieve a certain level of accuracy in estimating quantiles. A statistic computed within that criterion is the dependence factor, computed as $I = N/N_{\text{min}}$, where N is the maximum number of iterations needed to achieve convergence and N_{min} is the minimum number. The dependence factor measures the increase in the number of iterations needed to reach convergence due to dependence between the samples in the chain. Values of I greater than 1.0 indicate high within-chain correlations and probable convergence failure. Raftery and Lewis (1992a) suggest that $I \geq 5.0$ indicates that a reparameterisation of the model might be needed.

Heidelberger and Welch (1983) developed a convergence test that uses the Cramér–von Mises statistic to test the null hypothesis that the sampled values (either the whole chain, or after discarding some percentage of the values of the chain) come from a stationary distribution. Gelman and Rubin (1992) monitor many parallel chains that are initialised with different starting values. Convergence is diagnosed when the chains have 'forgotten' their initial values, and the output from all chains is the same.

4.12 Divergence of the estimation algorithm

As with the normal model (see Section 3.12.3), it sometimes happens that the iterative estimation routine fails to give a sensible result. In the normal case this happened when one of the residual variances converged to zero – known as a Heywood case. The analogous situation with the present model occurs when the iterative estimation procedure fails to converge. This can happen when one of the αs appears to be increasing without limit. Experience shows that this situation is particularly prone to occur:

(a) when the sample size is small – a few hundred or less;

(b) when the number of variables is small;

(c) when the α_{ij} $(i \geq 1)$ are very unequal.

In practice this behaviour is less serious than might appear at first sight since, though the estimate may become very large, the likelihood changes very little and the fit of the model is hardly affected. The situation can be dealt with most simply by adopting a cut-off value of 10, say, which terminates the algorithm if any estimate reaches this value. Knott and Albanese (1992) investigated the situation in greater detail and pointed out that large estimated values may arise because the true α is, in fact, infinite. This means that the response function degenerates into a step function with all those above some critical value of y responding with probability 1 and those below with probability 0. This is not an unreasonable supposition, and we shall see in Chapter 6 that latent class models involve an implicit assumption of precisely this kind.

In addition to the above, in a Bayesian estimation framework, rotational indeterminacy often leads to computational problems such as non-convergence or label switching among the factors in the multidimensional case. Lopez and West (2004) suggested constraints for the classical factor analysis model, but no systematic work has been done for the factor analysis model for categorical manifest variables. Reparameterisations of the model are in some cases necessary to avoid lack of convergence when large discrimination parameters occur.

4.13 Examples

We give three examples chosen to illustrate many of the uses of the logit/normal model. The first, based on data from the Law School Admission Test, is a classical example of the investigation of latent traits in educational testing. The second is taken from a study of attitude scaling where the items are the responses to questions in a survey. The third example, taken from a fertility survey in Bangladesh, is more typical of the use of the model in the general exploration of relationships between binary variables in multivariate analysis. Some of the results are set out in the following tables, but much more detail will be found in the output files produced by the IRTPRO software (Cai *et al.* 2011). In our examples, we have not discussed questions of missing values in the data. All of the data sets are presented in complete form. In some cases this has been achieved by deleting cases or imputing values for missing observations. Missing values could be included in the analysis by adding a response propensity latent dimension that is correlated with the attitudinal latent variables (O'Muircheartaigh and Moustaki 1999; Moustaki and Knott 2000b; Glas and Pimental 2008). Such matters are very important in practice but are not relevant to our more limited objective here, which is to illustrate the theory of this chapter. There are many examples in the literature of fitting latent trait models to data on tests of ability. Attempts to use the models in the manner of exploratory factor analysis are more recent and less common. Those given here provide an introduction to what is possible.

Example 4.13.1 The data in Table 4.2 relate to Section VI of the Law School Admission Test and have appeared in the latent trait literature on many occasions (for example, in Muthén (1978)). The maximum likelihood estimates of the parameter are given in the second column of Table 4.1 and the fitted frequencies in Table 4.2. The latter are close to the observed frequencies and the chi-squared goodness-of-fit test gives $\chi^2 = 18.14$ on 21 degrees of freedom ($P = 0.6412$). We may thus conclude that the associations among the five items are adequately explained by their common dependence on a single latent variable which, in this case, may be identified with the underlying ability which the test is designed to measure.

In Table 4.1 we have also given three estimates of the standard errors. The results appear to be fairly typical for data sets of this size. The asymptotic estimates derived

Table 4.1 Parameter estimates and standard errors for the logit/normal model fitted to the Law School Admission (VI) data.

Parameter	Maximum likelihood estimator	Asymptotic by (4.25)	standard error by (4.27)	Standard error estimated by 1000 bootstrap samples
$\hat{\alpha}_{11}$	0.83	0.25	0.26	0.29
$\hat{\alpha}_{21}$	0.72	0.17	0.19	0.22
$\hat{\alpha}_{31}$	0.89	0.20	0.23	0.31
$\hat{\alpha}_{41}$	0.69	0.17	0.19	0.20
$\hat{\alpha}_{51}$	0.66	0.19	0.21	0.22
$\hat{\pi}_1$	0.94	0.01	0.01	0.01
$\hat{\pi}_2$	0.73	0.02	0.02	0.02
$\hat{\pi}_3$	0.56	0.02	0.02	0.02
$\hat{\pi}_4$	0.78	0.01	0.02	0.02
$\hat{\pi}_5$	0.89	0.01	0.01	0.01

Table 4.2 Frequencies and scores for the logit/normal model fitted to the Law School Admission Test (VI) data.

Response pattern	Frequency	Expected frequency	Total score	Component score
00000	3	2.3	0	0
00001	6	5.9	1	0.66
00010	2	2.6	1	0.69
01000	1	1.8	1	0.72
10000	10	9.5	1	0.83
00100	1	.7	1	0.89
00011	11	8.9	2	1.35
01001	8	6.4	2	1.38
01010	0	2.9	2	1.41
10001	29	34.6	2	1.48
10010	14	15.6	2	1.51
00101	1	2.6	2	1.55
11000	16	11.3	2	1.55
00110	3	1.2	2	1.58
01100	0	.9	2	1.61
10100	3	4.7	2	1.71
01011	16	13.6	3	2.07
10011	81	76.6	3	2.17
11001	56	56.1	3	2.21
00111	4	6.0	3	2.24
11010	21	25.7	3	2.24
01101	3	4.4	3	2.27
01110	2	2.0	3	2.30
10101	28	25.0	3	2.37
10110	15	11.5	3	2.40
11100	11	8.4	3	2.44
11011	173	173.3	4	2.89
01111	15	13.9	4	2.96
10111	80	83.5	4	3.06
11101	61	62.5	4	3.10
11110	28	29.1	4	3.13
11111	298	296.7	5	3.78
Total	1000	1000.0	—	—

from the inverse of the information matrix of (4.25) are close to but smaller than the approximation proposed in (4.27). This suggests that the approximation is adequate and will, if anything, err on the safe side. The bootstrap estimates suggest that the asymptotic theory tends to underestimate the true standard error though not, in this case, by very much. Our experience shows that this is often the case and it arises from the fact that the actual sampling distribution tends to be positively skewed with a few unusually large estimates.

The estimates of the πs show that the items are of varying difficulty, and are relatively easy in the sense that the median individual stands a better than even chance of getting any item correct. We note from the standard errors that the πs are estimated with little error. For scaling purposes it is the relative values of the αs which matter and, having regard to the large standard errors, they are obviously close. Little would be lost by weighting them equally and simply using the total score. This is shown in Table 4.2 alongside the component score given by $\sum \hat{\alpha}_{i1} x_i$. It is clear that the two scalings are consistent. The component score allows some distinctions to be made among those with the same total score, but it is doubtful whether much significance should be attributed to them.

Example 4.13.2 This example is taken from a major study of 'social life feelings' reported in Schuessler (1982) and Krebs and Schuessler (1987). The aim was to establish scales of measurement comparable to those used in ability testing. Further analyses of these data will be found in Bartholomew and Schuessler (1991), Bartholomew *et al.* (1993, 1997) and de Menezes and Bartholomew (1996). Here we take scale 10, called 'economic self-determination'. We use the 'intercultural version' derived from a sample of 1490 Germans. Yes or no responses were obtained to the following five questions.

1. Anyone can raise his standard of living if he is willing to work at it.
2. Our country has too many poor people who can do little to raise their standard of living.
3. Individuals are poor because of the lack of effort on their part.
4. Poor people could improve their lot if they tried.
5. Most people have a good deal of freedom in deciding how to live.

The frequency distribution of response patterns is given in Table 4.4. Table 4.3 gives the maximum likelihood with the E-M parameter estimates and their asymptotic standard errors as well as the MCMC estimates and standard deviations obtained from WinBUGS (Spiegelhalter *et al.* 1996) after 10 000 iterations. WinBUGS uses Gibbs sampling within M-H. There is little to choose between the two sorts of estimates. For the E-M solution, we have given both the $\hat{\alpha}_{i0}$ and the $\hat{\pi}_{i0}$, but with standard errors for the $\hat{\alpha}_{i0}$ only. The $\hat{\pi}$ vary much more in this example, as one might expect. The typical respondent is unlikely to agree that anyone can raise their standard of living (question 1) but is much more likely to believe that individuals are

Table 4.3 Parameter estimates and standard errors for the logit/normal model fitted to Social Life Feelings scale 10 with the E-M and MCMC algorithms.

	E-M			MCMC	
Item	$\hat{\alpha}_{i0}$	$\hat{\pi}_{i0}$	$\hat{\alpha}_{i1}$	$\hat{\alpha}_{i0}$	$\hat{\alpha}_{i1}$
1	−2.35 (0.13)	0.09	1.20 (0.15)	−2.37 (0.14)	1.21 (0.15)
2	0.80 (0.06)	0.69	0.71 (0.09)	0.79 (0.06)	0.70 (0.09)
3	0.99 (0.09)	0.73	1.53 (0.17)	0.99 (0.09)	1.52 (0.17)
4	−0.67 (0.12)	0.34	2.55 (0.39)	−0.74 (0.20)	2.86 (0.81)
5	−1.10 (0.07)	0.25	0.92 (0.10)	−1.10 (0.07)	0.92 (0.11)

poor because of their own lack of effort (question 3). The $\hat{\alpha}$s also show considerable variation, with questions 3 and 4 having the greatest discriminating power.

Goodness of fit as measured by χ^2 (= 38.92) or G^2 (= 39.09), both on 21 degrees of freedom, indicates a rather poor fit ($P \doteq 0.01$). This should be interpreted with some caution for two reasons. First, Bartholomew and Tzamourani (1999) have shown that both χ^2 and G^2 tend to overestimate significance in this context. Secondly, none of the marginal residuals is large. Almost all of the pairwise residuals are less than 1, and the largest is only 2.7. The three-way margins likewise yield no residual for the response (1, 1, 1) greater than 2.20, so the two- and three-way associations are well predicted by the model. This seems sufficient grounds for believing that a single latent dimension provides an adequate explanation of the dependencies among the item responses. However, the limited-information goodness-of-fit test based on the one- and two-way margins as measured by expression (4.34) is 15.71 on 5 degrees of freedom, indicating a poor fit ($P = 0.0077$).

The remaining columns of Table 4.4 are concerned with posterior analysis, which is relevant if we wish to construct a scale of economic self-determination. In this example the rankings given by the total score and the component score are not so close. For example, the response pattern 00010 has a component score which is substantially higher than the other cases where there was only one positive response. This is because item 4 has a much higher weight than the others, and so a single positive response on that item carries more weight than, say, two positive responses on items 1 and 2. Other inversions in the rankings of the two measures can be explained similarly.

The values in the column headed 'posterior mean' ($E(y \mid X)$) give the predicted positions on a standard normal scale for each of the response patterns. In Section 4.10 we quoted results about the approximate linear relationship between X and $E(y \mid X)$, which can be confirmed for this example by plotting one against the other. Apart from the extremes, the linearity is almost exact. The posterior standard deviation, given in the final column, is remarkably constant over all response patterns. The average variance with respect to the distribution of X will be close to the test–retest correlation of (4.35), but this adds nothing to what we can learn from inspection of the individual values. With a typical value of $\sigma(y \mid X)$ around 0.60 there would be considerable

Table 4.4 Frequencies and scores for the logit/normal model fitted to Social Life Feelings scale 10.

Response pattern	Observed	Expected	Total score	Component score	Posterior mean	Posterior standard deviation
00000	156	162.0	0	0.00	−1.19	0.72
01000	174	174.0	1	0.71	−0.85	0.67
00001	26	22.2	1	0.92	−0.76	0.66
10000	8	5.2	1	1.20	−0.64	0.64
00100	127	122.3	1	1.53	−0.50	0.63
01001	35	31.9	2	1.64	−0.46	0.62
11000	8	8.1	2	1.91	−0.36	0.61
10001	2	1.1	2	2.12	−0.28	0.61
01100	208	208.2	2	2.24	−0.23	0.60
00101	26	30.2	2	2.45	−0.16	0.60
00010	14	16.9	1	2.55	−0.12	0.59
10100	4	8.3	2	2.73	−0.06	0.59
11001	2	2.3	3	2.83	−0.02	0.59
01101	65	65.3	3	3.17	0.09	0.59
01010	36	37.6	2	3.26	0.12	0.58
11100	19	19.3	3	3.44	0.18	0.58
00011	9	5.8	2	3.47	0.19	0.58
10101	4	3.0	3	3.65	0.25	0.58
10010	1	1.8	2	3.74	0.29	0.58
00110	66	56.5	2	4.08	0.40	0.59
01011	13	16.2	3	4.19	0.44	0.59
11101	10	8.8	4	4.36	0.50	0.59
11010	5	5.3	3	4.46	0.53	0.59
10011	3	0.9	3	4.67	0.61	0.60
01110	195	182.7	3	4.79	0.65	0.60
00111	16	31.8	3	5.00	0.73	0.61
10110	18	11.2	3	5.27	0.83	0.62
11011	3	3.4	4	5.38	0.88	0.62
01111	129	130.9	4	5.72	1.01	0.64
11110	31	49.9	4	5.99	1.12	0.65
10111	9	9.6	4	6.20	1.21	0.66
11111	68	56.7	5	6.91	1.54	0.70

uncertainty about the location of any individual on the latent scale. Put another way, the observation of the score pattern **x** reduces the variance in our knowledge of the location of the individual from 1 to $0.6^2 = 0.36$.

Example 4.13.3 The third example analyses the UK data on Eysenck's Personality Questionnaire–Revised (EPQ-R); see Eysenck *et al.* (1985). The sample consists of

610 males. The EPQ-R is a widely used personality inventory that consists of three scales, each one measuring a broad personality trait: Extraversion (E, 23 items), Neuroticism or Emotionality (N, 24 items), and Psychoticism or Tough-Mindness (P, 32 items). We analyse here the revised P-scale on Psychoticism. The 32 items can be found in Eysenck *et al.* (1985). The revised P-scale was a result of criticisms of the psychometric properties of the original scale related to low reliability, low range of scoring and grossly skewed score distributions. Furthermore, the Psychoticism scale had been constructed to be unidimensional. The AIC and BIC favour the two-factor model (see Table 4.5). We also looked at the standardised χ^2 local dependence (LD) index (Chen and Thissen 1997) both for the one- and two-factor solutions.

Table 4.5 Standardised χ^2 local dependence index for item pairs greater than 3, and AIC and BIC for the one- and two-factor logit/normal model fitted to the Psychoticism scale.

Item pairs	Index one-factor	Index two-factor
(10, 4)	16.4	
(12, 9)	8.5	4.9
(13, 4)	16.1	
(13, 10)	15.3	
(13, 12)	3.0	
(15, 2)	4.0	3.8
(18, 15)	3.7	4.2
(21, 19)	6.7	
(21, 20)	7.4	
(22, 19)	4.8	
(22, 21)	73.0	
(23, 15)	3.8	4.2
(26, 7)	7.4	7.3
(26, 9)	14.7	11.5
(28, 14)	30.3	29.3
(29, 4)	6.2	
(29, 10)	6.9	
(29, 13)	32.1	6.1
(29, 17)	16.3	10.7
(30, 2)	10.3	7.5
(30, 20)	15.2	6.9
(30, 21)	5.2	
(32, 12)	3.0	
(32, 18)	21.9	21.5
AIC	18 243.77	17 739.10
BIC	18 526.23	18 354.79

Table 4.6 Unstandardised and standardised parameter estimates with standard errors in brackets for the one-factor logit/normal model, Psychoticism scale.

Item	$\hat{\alpha}_{i0}$ (s.e.)	$\hat{\pi}_{i0}$	$\hat{\alpha}_{i1}$ (s.e.)	st$\hat{\alpha}_{i1}$
1	−1.69 (0.14)	0.16	1.13 (0.16)	0.75
2	1.12 (0.10)	0.75	0.84 (0.13)	0.64
3	−3.93 (0.36)	0.02	1.24 (0.30)	0.78
4	−1.65 (0.13)	0.16	0.80 (0.15)	0.62
5	−2.36 (0.19)	0.09	1.36 (0.21)	0.81
6	0.53 (0.09)	0.63	0.93 (0.13)	0.68
7	−1.70 (0.13)	0.15	0.76 (0.14)	0.61
8	−0.05 (0.09)	0.49	0.63 (0.11)	0.53
9	−2.47 (0.16)	0.08	0.53 (0.18)	0.47
10	−2.13 (0.15)	0.11	0.87 (0.17)	0.66
11	−0.57 (0.09)	0.36	0.87 (0.12)	0.66
12	−2.36 (0.20)	0.09	1.54 (0.22)	0.84
13	−1.46 (0.12)	0.19	0.74 (0.14)	0.59
14	−1.84 (0.16)	0.14	1.25 (0.18)	0.78
15	−0.69 (0.09)	0.33	0.53 (0.11)	0.47
16	−1.19 (0.11)	0.23	0.78 (0.13)	0.62
17	−0.90 (0.09)	0.29	0.50 (0.11)	0.45
18	−3.67 (0.34)	0.02	1.45 (0.29)	0.82
19	−1.09 (0.10)	0.25	0.58 (0.12)	0.50
20	−0.21 (0.09)	0.45	0.99 (0.13)	0.70
21	−1.41 (0.12)	0.20	0.85 (0.14)	0.65
22	−1.93 (0.15)	0.13	1.11 (0.17)	0.74
23	−2.11 (0.14)	0.11	0.67 (0.16)	0.56
24	−1.63 (0.13)	0.16	1.02 (0.16)	0.71
25	−1.48 (0.11)	0.19	0.52 (0.13)	0.46
26	−2.65 (0.18)	0.07	0.65 (0.19)	0.54
27	−2.23 (0.18)	0.10	1.32 (0.20)	0.80
28	−1.40 (0.13)	0.20	1.19 (0.16)	0.77
29	−0.72 (0.09)	0.33	0.65 (0.12)	0.54
30	0.02 (0.09)	0.50	0.92 (0.13)	0.68
31	−1.95 (0.15)	0.12	0.94 (0.16)	0.68
32	−2.76 (0.21)	0.06	0.99 (0.21)	0.70

The standardised χ^2 statistic compares the observed and expected frequencies in the two-way marginal tables for pairs of items. The standardised version of LD produced by the IRTPRO program (Cai *et al.* 2011) is $(\chi^2 - df)/\sqrt{2df}$, where $df = 1$ in the binary case. Table 4.5 gives the standardised χ^2 LD index for those item pairs that exhibit positive dependencies that have not been well explained by the one- and two-factor models. Positive values indicate that the observed covariation between pairs of items exceeds that predicted by the fitted model. Therefore an additional factor might be needed to accommodate those dependencies. Values greater than 3 are reported here. There are few bad fits in both the one- and two-factor models. Certainly, the two-factor model is shown to be an improvement. However, there only a few large values and no obvious clustering of offensive items can be identified. Although the goodness of fit measures suggest that a two-factor solution is a better fit, it has been difficult to produce an interpretable two-factor solution even after an orthogonal or oblique rotation.

We give in Table 4.6 the unstandardised and standardised loadings for the one-factor logit/normal model. The standardised loadings are obtained from equation (7.52). All factor loadings are statistically significant and show moderate to strong positive correlations between the 32 items and the underlying construct. More analysis might be necessary to improve the 32-item scale. However, Eysenck *et al.* (1985) point out in their discussion that the Psychoticism scale accounts for several different aspects such as hostility, cruelty, lack of empathy, and non-conformism, whereas the rest of the scales constructed for Extraversion, Neuroticism, and Social Desirability are much more homogeneous in what they try to measure.

5

Polytomous data: latent trait models

5.1 Introduction

In conceptual terms it seems a small step from problems with binary manifest data to polytomous data, but this is deceptive. In the binary case we identified two approaches to the problem. The first, via the sufficiency principle, led to the logit model in which the response function turned out to be a logistic curve. The second started by supposing that the binary observations were formed by dichotomising an underlying continuous variable and then assuming that these underlying variables were generated by a normal linear factor model. The two models arrived at from these very different starting points proved to be equivalent. Which one we chose to adopt was therefore largely a non-empirical matter. In the polytomous case this equivalence breaks down. The natural extension of the binary underlying variable model is not equivalent to the response function model derived from the sufficiency argument. There is, however, another model based on the idea of underlying variables to which it is equivalent. This alerts us to the need to consider the choice of model more carefully and to determine the circumstances under which different choices might be appropriate.

There are also new practical questions which arise with polytomous data. In most models the number of parameters increases in line with the number of categories, and this can lead to problems of estimation and inference. Even with binary data we found that standard errors could be quite large and that there might be problems with the convergence of the E-M algorithm. These features are magnified when we move to polytomous data, as is the problem of sparseness when testing goodness of fit. More research is necessary on some of these matters and, for that reason, there will be a somewhat greater emphasis in this chapter on the theoretical aspects of the models.

Latent Variable Models and Factor Analysis: A Unified Approach, Third Edition.
David Bartholomew, Martin Knott and Irini Moustaki.

We shall give a number of examples to illustrate the different model strategies for polytomous variables.

5.2 A response function model based on the sufficiency principle

The general approach to choosing factor models set out in Chapter 2 yields a family of models for polytomous data which generalise those for binary data in Chapter 4. We first need to extend the notation used there. Let c_i denote the number of categories of variable i which are labelled $0, 1, \ldots, c_i - 1 \quad (i = 1, 2, \ldots, p)$ and indexed by s. The indicator variable x_i is now replaced by a vector-valued indicator function with its sth element defined by

$$x_i(s) = \begin{cases} 1 & \text{if the response falls in category } s, \\ 0 & \text{otherwise.} \end{cases} \tag{5.1}$$

The c_i-vector with these elements is denoted by $\mathbf{x}_i$ and, obviously, $\sum_s x_i(s) = 1$. The full response pattern for an individual is denoted by $\mathbf{x}' = (\mathbf{x}'_1, \mathbf{x}'_2, \ldots, \mathbf{x}'_p)$, of dimension $\sum_i c_i$. Note that if we specialise to the binary case by putting $c_i = 2$ for all i, this notation does not coincide exactly with that in Chapter 4. The two-element vector $\mathbf{x}_i$ was there represented by the single element which here is denoted by $x_i(1)$. Since $x_i(0) + x_i(1) = 1$, it is sufficient in the binary case to record only $x_i(1)$ and to delete alternate elements in $\mathbf{x}$; in the general case we could, in a similar way, delete $x_i(0)$ for each i. However, the symmetry which results from retaining the redundancy outweighs any advantage of the more parsimonious description.

The single response function $\pi_i(\mathbf{y})$ is now replaced by a set of functions defined by

$$P\{x_i(s) = 1 \mid \mathbf{y}\} = \pi_{is}(\mathbf{y}) \quad (s = 0, 1, \ldots, c_i - 1;\ i = 1, 2, \ldots, p). \tag{5.2}$$

The requirement that $\sum_s \pi_{is}(\mathbf{y}) = 1$ again means that there is some redundancy in the notation. We may now suppose the conditional probability function of $\mathbf{x}_i$ given $\mathbf{y}$ to be multinomial so that

$$g_i(\mathbf{x}_i \mid \mathbf{y}) = \prod_{s=0}^{c_i-1} \{\pi_{is}(\mathbf{y})\}^{x_i(s)} \quad (i = 1, 2, \ldots, p). \tag{5.3}$$

The posterior density function of $\mathbf{y}$ is then given by

$$\begin{aligned} h(\mathbf{y} \mid \mathbf{x}) &= h(\mathbf{y}) \left[\prod_{i=1}^{p} \prod_{s=0}^{c_i-1} \{\pi_{is}(\mathbf{y})\}^{x_i(s)} \right] / f(\mathbf{x}) \\ &= h(\mathbf{y}) \exp \left[\sum_{i=1}^{p} \sum_{s=0}^{c_i-1} x_i(s) \ln \pi_{is}(\mathbf{y}) \right] / f(\mathbf{x}). \end{aligned} \tag{5.4}$$

In Chapter 2 we introduced the GLLVM for the model in which the conditional distribution of a manifest variable, x_i, had a distribution in the exponential family with canonical parameter θ_i. Both x_i and θ_i were there supposed to be scalars. In the present case, both are vectors but the arguments presented in Sections 2.2 and 2.3 generalise almost immediately to provide a vector-valued GLLVM. In fact, if x_i and θ_i in (2.1) and the αs in (2.2) are vectors, the sufficiency property and the linearity of the components still hold. In applying this result to polytomous responses we must allow for the fact that $\sum_s \pi_{is}(\mathbf{y}) = 1$. We did this in the binary case by implicitly reparameterising the problem, replacing $\pi_i(\mathbf{y})$ and $1 - \pi_i(\mathbf{y})$ (here denoted by $\pi_{i0}(\mathbf{y})$ and $\pi_{i1}(\mathbf{y})$) by $\operatorname{logit} \pi_i(\mathbf{y})$ and $\{1 - \pi_i(\mathbf{y})\}$ respectively. The new parameter $\theta_i = \operatorname{logit} \pi_i(\mathbf{y})$ was then equated to a linear combination of the latent variables. In the general case the equivalent operation involves writing (1.9) in the form

$$
\begin{aligned}
g_i(\mathbf{x}_i \mid \mathbf{y}) &= \prod_{s=0}^{c_i-1} \{\pi_{i0}(\mathbf{y})\}^{x_i(s)} \{\pi_{is}(\mathbf{y})/\pi_{i0}(\mathbf{y})\}^{x_i(s)} \\
&= \pi_{i0}(\mathbf{y}) \exp \sum_{s=0}^{c_i-1} x_i(s) \ln \{\pi_{is}(\mathbf{y})/\pi_{i0}(\mathbf{y})\} \\
&= \pi_{i0}(\mathbf{y}) \exp \boldsymbol{\theta}_i' \mathbf{x}_i, \qquad (5.5)
\end{aligned}
$$

where

$$
\boldsymbol{\theta}_i' = \left\{0, \ln \frac{\pi_{i1}(\mathbf{y})}{\pi_{i0}(\mathbf{y})}, \ \ln \frac{\pi_{i2}(\mathbf{y})}{\pi_{i0}(\mathbf{y})}, \ldots, \ln \frac{\pi_{i(c_i-1)}(\mathbf{y})}{\pi_{i0}(\mathbf{y})}\right\}. \qquad (5.6)
$$

Which category is labelled 0 is, of course, quite arbitrary and any one may be chosen. The appropriate model may now be written

$$
\boldsymbol{\theta}_i = \boldsymbol{\alpha}_{i0} + \boldsymbol{\alpha}_{i1} y_1 + \boldsymbol{\alpha}_{i2} y_2 + \cdots + \boldsymbol{\alpha}_{iq} y_q, \qquad (5.7)
$$

where $\boldsymbol{\alpha}_{ij} = (\alpha_{ij}(0) = 0, \alpha_{ij}(1), \ldots, \alpha_{ij}(c_i - 1))$ for $j = 0, 1, \ldots, q$. It is easily shown that the components are given by

$$
X_j = \sum_{i=1}^{p} \sum_{s=0}^{c_i-1} \alpha_{ij}(s) x_i(s) \qquad (5.8)
$$

or

$$
\mathbf{X} = \mathbf{A}\mathbf{x},
$$

say. For complete consistency with the binary case we should have chosen the last probability, $\pi_{c_i-1}(\mathbf{y})$, as the divisor in (5.6) but it is notationally simpler to take the first. We shall refer to category 0 as the *reference* category and continue to refer to the elements of θ_i as logits and the model as the *logit model*.

The requirement that $\sum_s \pi_{is}(\mathbf{y}) = 1$, together with (5.5)and (5.6), implies that

$$\pi_{is}(\mathbf{y}) = \frac{\exp\left\{\alpha_{i0}(s) + \sum_{j=1}^{q} \alpha_{ij}(s) y_j\right\}}{\sum_{r=0}^{c_i-1} \exp\left\{\alpha_{i0}(r) + \sum_{j=1}^{q} \alpha_{ij}(r) y_j\right\}} \tag{5.9}$$

for $i = 1, 2, \ldots, p$ and $s = 0, 1, \ldots, c_i - 1$.

Figure 5.1 illustrates the forms of the response functions for the set of categories on a given manifest variable for a single latent variable measured on the y scale. As we move from left to right along the latent dimension the probability of a positive response is decreasing in some categories and increasing in others. The model thus determines an ordering of the categories such that the higher a subject is on the latent scale the greater is the tendency for the response to fall in the higher categories. This behaviour is plausible in many contexts. It would be appropriate, for example, if the categorisation were arrived at by grouping individuals according to an imperfect perception of their place on the scale. In educational testing it would be plausible if the possible answers were of varying degrees of 'rightness'. Thissen and Steinberg (1984, 1986) argue that the model would not apply if one response is correct and the others equally wrong. However, this can be accommodated as a limiting case. Nevertheless, the point made in connection with binary models bears repetition: one should *not* assume that the possession of attractive mathematical properties will

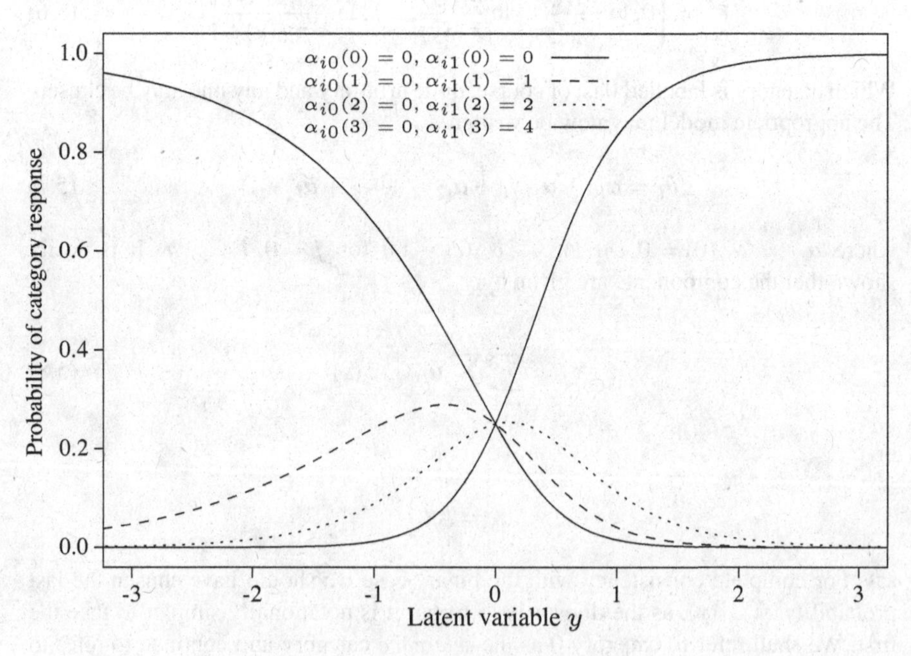

Figure 5.1 Probabilities of response categories.

guarantee the applicability of a model in any particular circumstances. Thissen and Steinberg (1984) discuss alternative models.

In (5.9), we have, as in the binary case of Chapter 4, $\alpha_{ij}(0) = 0$. The logit model for polytomous data is thus

$$\ln\left\{\frac{\pi_{is}(\mathbf{y})}{\pi_{i0}(\mathbf{y})}\right\} = \alpha_{i0}(s) + \sum_{j=1}^{q} \alpha_{ij}(s)y_j \quad (s = 0, 1, 2, \ldots, c_i - 1;\ i = 1, 2, \ldots, p). \tag{5.10}$$

We shall assume that the ys are mutually independent with standard normal distributions. Under this assumption we have a logit/normal model. It no longer makes sense to talk of a probit/normal model since there is no simple comparable way of introducing a probit function on the left-hand side. The logit model was first proposed by Bock (1972), though with a rather different rationale.

An interesting property of the polytomous logit model arises as follows. Suppose we consider the conditional probability that a positive response is in category s given that it is in r or s. This is clearly $\pi_{is}(\mathbf{y})/\{\pi_{is}(\mathbf{y}) + \pi_{ir}(\mathbf{y})\} = \pi_{i,s\,|\,s,r}(\mathbf{y})$, say. A simple manipulation then reveals that

$$\operatorname{logit} \pi_{i,s\,|\,s,r}(\mathbf{y}) = \{\alpha_{i0}(s) - \alpha_{i0}(r)\} + \sum_{j=1}^{q} \{\alpha_{ij}(s) - \alpha_{ij}(r)\}y_j. \tag{5.11}$$

In other words, if we reduce the data to binary form by extracting only those responses where the positive response is in r or s, then we have the standard binary logit model. This property provides an alternative route to the polytomous logit model. If we wish (5.11) to hold for any pair of categories, then the model (5.10) must hold for them all. This was the approach adopted by Thissen and Steinberg (1984, 1986), who used the term *divide-by-total* to refer to this model.

An alternative parameterisation which is useful later and which facilitates the interpretation of the $\alpha_{i0}(s)$ is obtained as follows. If we put $y_j = 0$ for all j, we obtain the response probability for the 'median' individual. Let this be denoted by

$$\pi_{is} = \frac{\exp \alpha_{i0}(s)}{\sum_{r=0}^{c_i-1} \exp \alpha_{i0}(r)}. \tag{5.12}$$

Remembering that $\alpha_{i0}(0) = 0$,

$$\pi_{is} = \frac{\exp \alpha_{i0}(s)}{1 + \sum_{r=1}^{c_i-1} \exp \alpha_{i0}(r)} \quad (s = 1, 2, \ldots, c_i - 1). \tag{5.13}$$

Substituting into (5.9) then gives

$$\pi_{is}(\mathbf{y}) = \frac{\pi_{is} \exp \sum_{j=1}^{q} \alpha_{ij}(s)y_j}{\sum_{r=0}^{c_i-1} \pi_{ir} \exp \sum_{j=1}^{q} \alpha_{ij}(r)y_j} \tag{5.14}$$

or, from (5.10) ,

$$\ln\{\pi_{is}(\mathbf{y})/\pi_{i0}(\mathbf{y})\} = \ln\{\pi_{is}/\pi_{i0}\} + \sum_{j=1}^{q} \alpha_{ij}(s) y_j \quad (s = 1, 2, \ldots, c_i - 1). \tag{5.15}$$

This is an alternative way of defining the polytomous logit model. When, later, we refer loosely to the 'αs' we shall mean the αs of (5.14) or (5.15) which are coefficients of ys. The other parameters will be referred to, collectively, as the πs.

5.3 Parameter interpretation

The interpretation of the πs is clear from their definition as 'median' response probabilities. The αs may be interpreted in ways similar to those of the binary case, but now we need to consider them, separately, as functions of s and i.

First, the discriminating power of the latent variable y_j is indicated by the *spread* of the $\alpha_{ij}(s)$ considered as a function of s. A large spread produces larger differences between the corresponding response probabilities and so a better chance of discriminating between individuals a given distance apart on the y scale on the evidence of $\mathbf{x}_i$.

Secondly, the αs are also weights in the components of (5.8). Here we are looking at the relative influence which each manifest variable has in determining the value of the component. A variable will thus be an important determinant if all of the $\alpha_{ij}(s)$ for given i are large. That is, it is the *average* level of the αs rather than their dispersion which counts.

The distinction between these two aspects disappears when we specialise to the binary case. The fact that $\alpha_i(0) = 0$ means that the dispersion and the mean of $\alpha_i(0)$ and $\alpha_i(1)$ are essentially the same thing.

The third way of looking at the αs is as category scores. If we think of $\alpha_{ij}(s)$ as a score associated with category s of variable i then X_j is the total score for an individual. This interpretation, already introduced in Section 4.2, will be particularly relevant when we come to consider the relationship with correspondence analysis.

All of these interpretations can be given a new aspect in relation to an equivalent model based on underlying variables described in Section 5.9.

5.4 Rotation

In Chapter 4 we favoured the choice of the standard normal distribution for y on the grounds that it left the likelihood unchanged under orthogonal transformation of the αs. The same advantage is gained in the polytomous case. To show this we redefine $\mathbf{A}$ to be the matrix

$$\mathbf{A} = (\boldsymbol{\alpha}_1, \boldsymbol{\alpha}_2, \ldots, \boldsymbol{\alpha}_q),$$

where $\boldsymbol{\alpha}'_j = (\alpha_{1j}(0), \alpha_{1j}(1), \ldots, \alpha_{1j}(c_1 - 1), \alpha_{2j}(0), \ldots, \alpha_{2j}(c_2 - 1), \ldots, \alpha_{pj}(c_p - 1))$. If we now make an orthogonal transformation to

$$\mathbf{A}^* = \mathbf{AM}, \tag{5.16}$$

where $\mathbf{M}$ is $q \times q$ with $\mathbf{MM}' = \mathbf{I}$, then $\sum_j \alpha_{ij}(s)y_j = \mathbf{A}_i(s)\mathbf{y}$, where $\mathbf{A}_i(s)$ is the row of $\mathbf{A}$ beginning with the element $\alpha_{i1}(s)$. As before,

$$\mathbf{A}_i(s)\mathbf{y} = \mathbf{A}_i^*(s)\mathbf{M}^{-1}\mathbf{y}$$

and the previous argument can be repeated making the transformation $\mathbf{y}^* = \mathbf{M}^{-1}\mathbf{y}$ to show that the likelihood is unchanged.

5.5 Maximum likelihood estimation of the polytomous logit model

The E-M algorithm given in Section 4.5 for binary data is easily generalised to the polytomous case as follows. As before, we work with only one factor to simplify the notation. Then (4.19) becomes

$$L = \sum_{h=1}^{n} \ln g(\mathbf{x}_h \mid y_h) = \sum_{h=1}^{n} \left[\sum_{i=1}^{p} \ln g(\mathbf{x}_{ih} \mid y_h) \right], \tag{5.17}$$

where at each E-step we have to find $E[L \mid \mathbf{x}]$. Substituting (5.5) in (5.17), we have

$$\begin{aligned} L &= \sum_{h=1}^{n} \ln g(\mathbf{x}_h \mid y_h) = \sum_{h=1}^{n} \left[\sum_{i=1}^{p} \left(\ln \pi_{i0}(y_h) + \sum_{s=0}^{c_i-1} x_{ih}(s)(\alpha_{i0}(s) + \alpha_{i1}(s)y_h) \right) \right] \\ &= \sum_{i=1}^{p} \left[\sum_{h=1}^{n} \ln \pi_{i0}(y_h) + \sum_{s=0}^{c_i-1} \left(\alpha_{i0}(s) \sum_{h=1}^{n} x_{ih}(s) + \alpha_{i1}(s) \sum_{h=1}^{n} x_{ih}(s) y_h \right) \right]. \end{aligned}$$

The E-step is therefore

$$\begin{aligned} E[L \mid \mathbf{x}] = &\sum_{i=1}^{p} \left[\sum_{h=1}^{n} E(\ln \pi_{i0}(y_h) \mid \mathbf{x}_1, \ldots, \mathbf{x}_n) \right. \\ &\left. + \sum_{s=0}^{c_i-1} \left(\alpha_{i0}(s) \sum_{h=1}^{n} x_{ih}(s) + \alpha_{i1}(s) \sum_{h=1}^{n} x_{ih}(s) E(y_h \mid \mathbf{x}_1, \ldots, \mathbf{x}_n) \right) \right] \\ = &\sum_{i=1}^{p} \left[\sum_{h=1}^{n} E(\ln \pi_{i0}(y_h) \mid \mathbf{x}_h) \right. \\ &\left. + \sum_{s=0}^{c_i-1} \left(\alpha_{i0}(s) \sum_{h=1}^{n} x_{ih}(s) + \alpha_{i1}(s) \sum_{h=1}^{n} x_{ih}(s) E(y_h \mid \mathbf{x}_h) \right) \right]. \end{aligned}$$

For the M-step one can maximise separately for each item i and each category s,

$$\sum_{h=1}^{n} E(\ln \pi_{i0}(y) \mid \mathbf{x}_h) + \alpha_{i0}(s) \sum_{h=1}^{n} x_{ih}(s) + \alpha_{i1}(s) \sum_{h=1}^{n} x_{ih}(s) E(y \mid \mathbf{x}_h). \qquad (5.18)$$

So, referring to Section 4.5, the M-step is the same as maximum likelihood estimation in a continuous analogue of logistic regression.

As in Section 4.5, we approximate the averages in (5.18) with Gauss–Hermite quadrature, giving

$$\begin{aligned} &\sum_{h=1}^{n} \sum_{t=1}^{k} x_{ih}(s) \ln \pi_{i0}(y^t) h(y^t \mid \mathbf{x}_h) + \alpha_{i0}(s) \sum_{h=1}^{n} x_{ih}(s) \\ &\quad + \alpha_{i1}(s) \sum_{t=1}^{k} y^t \sum_{h=1}^{n} x_{ih}(s) h(y^t \mid \mathbf{x}_h) \\ &= \sum_{t=1}^{k} n_t \ln \pi_{i0}(y^t) + \alpha_{i0}(s) \sum_{h=1}^{n} x_{ih}(s) + \alpha_{i1}(s) \sum_{t=1}^{k} y^t r_{it}(s), \end{aligned} \qquad (5.19)$$

where

$$r_{it}(s) = \sum_{h=1}^{n} x_{ih}(s) h(y^t \mid \mathbf{x}_h) \quad (s = 1, 2, \ldots, c_i - 1; \; i = 1, 2, \ldots, p) \qquad (5.20)$$

and

$$n_t = \sum_{h=1}^{n} h(y^t \mid \mathbf{x}_h). \qquad (5.21)$$

This leads to the M-step. For a given value of s, these equations are essentially the same as those of (4.22) and (4.23) where, again, $r_{it}(s)$ and n_t can be interpreted as pseudo-observed and expected frequencies. The simple form of the equations in the estimation process is because the logit link has been used on the left-hand side of (5.10).

At this juncture in the binary case we noted the similarity with dosage response analysis and hence that the conditional maximum likelihood estimators could be obtained by an iterative weighted least squares analysis. The same approach can be used here.

5.6 An approximation to the likelihood

We shall now derive an approximation to the likelihood function for the logit model which can easily be maximised with respect to the parameters, thus yielding approximate estimators. The reason for doing this is not to provide an easier way of finding maximum likelihood estimators but to bring out a link with correspondence analysis. It turns out that the approximations are rather poor if judged by the absolute differences. However, it is the profile of the set of weights which counts (i.e. their relative values) and in this respect the approximations appear to be good. This fact is illustrated by the example to be given in Table 5.2. The exact and approximate estimates are not precisely proportional but are near enough to lead to essentially the same interpretation. The result is particularly interesting because of the equivalence of the approximate estimates and the category scores of correspondence analysis to which we shall draw attention at the end of Section 5.7.

5.6.1 One factor

We treat this case first in order to establish the method and then, more briefly, extend the results to cover any q. The joint probability function of $\mathbf{x}$ for the logit model is

$$f(\mathbf{x}) = \frac{1}{\sqrt{2\pi}} \int_{-\infty}^{+\infty} \exp\left\{-\frac{1}{2}y^2 + \sum_{i=1}^{p}\sum_{r=0}^{c_i-1} x_i(r) \ln \pi_{ir}(y)\right\} dy. \tag{5.22}$$

Using the version of $\pi_{ir}(\mathbf{y})$ given in (5.14), (5.22) becomes

$$\begin{aligned} f(\mathbf{x}) = & \left[\exp \sum_{i=1}^{p}\sum_{r=0}^{c_i-1} x_i(r) \ln \pi_{is}\right] \\ & \times \left[\frac{1}{\sqrt{2\pi}} \int_{-\infty}^{+\infty} \exp\left\{-\frac{1}{2}y^2 + y\sum_{i=1}^{p}\sum_{r=0}^{c_i-1} x_i(r)\alpha_i(r) \right.\right. \\ & \left.\left. - \sum_{i=1}^{p} \ln\left(\sum_{r=0}^{c_i-1} \pi_{ir} \exp \alpha_i(r)y\right)\right\} dy\right]. \end{aligned} \tag{5.23}$$

We begin with the $\alpha_i(r)$ which appear only in the second factor of (5.23). The method is first to approximate this factor as described below and then to obtain estimators by maximising the relevant part of the likelihood. We replace the last term in the exponent by its Taylor expansion taken as far as terms of the second degree in α. Thus

$$\begin{aligned} \sum_{i=1}^{p} \ln\left(\sum_{r=0}^{c_i-1} \pi_{ir} \exp \alpha_i(r)y\right) \doteq & \; y\sum_{i=1}^{p}\sum_{r=0}^{c_i-1} \pi_{ir}\alpha_i(r) \\ & + \frac{1}{2}y^2 \sum_{i=1}^{p}\left\{\sum_{r=0}^{c_i-1} \pi_{ir}\alpha_i^2(r) - \left(\sum_{r=0}^{c_i-1} \pi_{ir}\alpha_i(r)\right)^2\right\}. \end{aligned} \tag{5.24}$$

Recalling that the probability is unaffected by translations of the $\alpha_i(r)$, there is no loss of generality if we impose the constraints

$$\sum_{r=0}^{c_i-1} \pi_{ir}\alpha_i(r) = 0 \quad (i = 1, 2, \ldots, p). \tag{5.25}$$

The right-hand side of (5.24) then simplifies to

$$\frac{1}{2}y^2 \sum_{i=1}^{p} \sum_{r=0}^{c_i-1} \pi_{ir}\alpha_i^2(r) = \frac{1}{2}y^2 \boldsymbol{\alpha}'\mathbf{P}\boldsymbol{\alpha}, \tag{5.26}$$

where $\boldsymbol{\alpha}' = (\alpha_1(0), \alpha_1(1), \ldots, \alpha_1(c_1 - 1), \alpha_2(0), \ldots, \alpha_p(c_p - 1))$ and $\mathbf{P}$ is the diagonal matrix with elements π_{ir} listed along the diagonal in dictionary order. Denoting the second factor of (5.23) by I, we now have

$$\begin{aligned} I &= \frac{1}{\sqrt{(2\pi)}} \int_{-\infty}^{+\infty} \exp\left\{-\frac{1}{2}y^2 + Yy - ay^2\right\} dy \\ &= (2a + 1)^{-1/2} \exp\left\{\frac{1}{2}\frac{Y^2}{2a + 1}\right\}, \end{aligned} \tag{5.27}$$

where $a = \frac{1}{2}\boldsymbol{\alpha}'\mathbf{P}\boldsymbol{\alpha}$ and $Y = \sum_{i=1}^{p} \sum_{r=0}^{c_i-1} x_i(r)\alpha_i(r) = \mathbf{x}'\boldsymbol{\alpha}$.

That part of the likelihood in which the αs occur may now be written

$$l = (2a + 1)^{-1/2} \exp\left\{\frac{1}{2}\sum_{h=1}^{n} \frac{Y_h^2}{2a + 1}\right\}, \tag{5.28}$$

where Y_h is the value of Y for the hth sample member. Now

$$\begin{aligned} \sum_{h=1}^{n} Y_h^2 &= \sum_{h=1}^{n} \sum_{i=1}^{p} \sum_{j=1}^{p} \sum_{r=0}^{c_i-1} \sum_{s=0}^{c_j-1} x_{ih}(r)x_{jh}(s)\alpha_i(r)\alpha_j(s) \\ &= \alpha'\mathbf{X}'\mathbf{X}\alpha. \end{aligned} \tag{5.29}$$

The hth row of the matrix $\mathbf{X}$ consists of the set of indicator variables for the hth sample member as defined in Section 5.2. Since each member falls into exactly one category on each variable, each row of $\mathbf{X}$ sums to p. The column sums of $\mathbf{X}$ give the numbers of sample members falling into each category on each variable. The diagonal elements of $\mathbf{X}'\mathbf{X}$ are $\{\sum_{h=1}^{n} x_{ih}^2(s)\} = \{\sum_{h=1}^{n} x_{ih}(s)\}$ and are thus the same as the column sums. The off-diagonal elements give the two-way marginal frequencies; thus

$$\sum_{h=1}^{n} x_{ih}(r)x_{jh}(s)$$

is the number of individuals in category r of variable i and category s of variable j. $\mathbf{X}'\mathbf{X}$ thus contains the first- and second-order marginal frequencies, and since our approximate likelihood depends on the data only through this matrix we may deduce that the estimates to be derived will also be functions of first- and second-order marginal frequencies only. Furthermore, this would also be true if the exponent in the probability function were approximated by any other quadratic function of $\boldsymbol{\alpha}$ – for example, a Taylor expansion about some point other than zero.

The relevant part of the log-likelihood may now be written

$$L = -\frac{n}{2}\ln(\boldsymbol{\alpha}'\mathbf{P}\boldsymbol{\alpha} + 1) + \frac{1}{2}\boldsymbol{\alpha}'\mathbf{X}'\mathbf{X}\boldsymbol{\alpha}/(\boldsymbol{\alpha}'\mathbf{P}\boldsymbol{\alpha} + 1) \tag{5.30}$$

which has to be maximised subject to the constraints of (5.25). These may be written $\boldsymbol{\alpha}'\mathbf{P}^* = \mathbf{0}$, where the ith column of $\mathbf{P}^*$ has elements $\pi_{i0}, \pi_{i1}, \ldots, \pi_{ic_i-1}$ in rows $\sum_{i_0<i} c_{i_0} + 1$ to $\sum_{i_0\leq i} c_{i_0}$ and zeros elsewhere ($i = 1, 2, \ldots, p$).

We maximise (5.30) in three stages. First choose the πs, then make $\boldsymbol{\alpha}'\mathbf{X}'\mathbf{X}\boldsymbol{\alpha}$ as large as possible for given πs and fixed $\boldsymbol{\alpha}'\mathbf{P}\boldsymbol{\alpha}$, and then choose the best value for $\boldsymbol{\alpha}'\mathbf{P}\boldsymbol{\alpha}$.

To choose the best πs we use only the first factor from (5.23), ignoring terms of second order in $\boldsymbol{\alpha}$. That factor is a multinomial probability, so the maximising πs are

$$\hat{\pi}_{is} = \sum_{h=1}^{n} x_{ih}(s)/n \quad (i = 1, 2, \ldots, p;\ s = 0, 1, \ldots, c_i - 1)$$

or

$$\hat{\boldsymbol{\pi}} = \mathbf{X}'\mathbf{1}/n, \tag{5.31}$$

where $\mathbf{1}$ is an n-vector of 1s and $\boldsymbol{\pi}$ is the vector of the π_{is} in dictionary order. The πs are thus estimated by corresponding marginal proportions. We assume for the rest of the maximisation that $\boldsymbol{\pi}$ is always replaced by $\hat{\boldsymbol{\pi}}$, but do not use the 'hat' notation.

The second stage of the maximisation is to find the best value for $\boldsymbol{\alpha}'\mathbf{X}'\mathbf{X}\boldsymbol{\alpha}$ for fixed $\boldsymbol{\alpha}'\mathbf{P}\boldsymbol{\alpha}$. It is easier to use $\boldsymbol{\beta}$, defined by

$$\boldsymbol{\beta} = \mathbf{P}^{1/2}\boldsymbol{\alpha}. \tag{5.32}$$

Then

$$\boldsymbol{\alpha}'\mathbf{X}'\mathbf{X}\boldsymbol{\alpha} = \boldsymbol{\beta}'\mathbf{P}^{-1/2}\mathbf{X}'\mathbf{X}\mathbf{P}^{-1/2}\boldsymbol{\beta}. \tag{5.33}$$

We maximise the right-hand side of (5.33) over all choices of $\boldsymbol{\beta}$ such that $(\mathbf{P}^*)'\boldsymbol{\alpha} = (\mathbf{P}^*)'\mathbf{P}^{-1/2}\boldsymbol{\beta} = \mathbf{0}$. This is a simple eigenvalue problem. It is happily the case that the constraints span a space that is spanned also by p eigenvectors of $\mathbf{P}^{-1/2}\mathbf{X}'\mathbf{X}\mathbf{P}^{-1/2}$. To maximise the quadratic form it is enough to find the eigenvector not in this spanning set that has the largest eigenvector. The constraints themselves are not eigenvectors, but one can rotate them to obtain eigenvectors. For $\mathbf{H}$ a $p \times p$ orthogonal matrix with all elements in its first column equal, the columns of $\mathbf{P}^{-1/2}\mathbf{P}^*\mathbf{H}$ are eigenvectors of $\mathbf{P}^{-1/2}\mathbf{X}'\mathbf{X}\mathbf{P}^{-1/2}$. This is because, for $\mathbf{1}_n$ an n-vectors of 1s,

$$\begin{aligned}\mathbf{P}^{-1/2}\mathbf{X}'\mathbf{X}\mathbf{P}^{-1/2}\mathbf{P}^{-1/2}\mathbf{P}^*\mathbf{H} &= \mathbf{P}^{-1/2}\mathbf{X}'\mathbf{X}\mathbf{P}^{-1}\mathbf{P}^*\mathbf{H}\\ &= \mathbf{P}^{-1/2}\mathbf{X}'\mathbf{X}(\mathbf{1}_{\sum c_i} \otimes \mathbf{I}_p)\mathbf{H}\\ &= \mathbf{P}^{-1/2}\mathbf{X}'\mathbf{1}_n\mathbf{1}_p'\mathbf{H}\\ &= n\mathbf{P}^{-1/2}\boldsymbol{\pi}\mathbf{1}_p'\mathbf{H},\end{aligned} \tag{5.34}$$

where (5.34) follows from (5.31). Since $\mathbf{1}_p$ is orthogonal to all but the first column of $\mathbf{H}$, (5.34)can be written

$$\begin{aligned}\mathbf{P}^{-1/2}\mathbf{X}'\mathbf{X}\mathbf{P}^{-1/2}\mathbf{P}^{-1/2}\mathbf{P}^*\mathbf{H} &= n\mathbf{P}^{-1/2}\boldsymbol{\pi}(\sqrt{p}, 0, \dots, 0)\\ &= \mathbf{P}^{-1/2}(n\sqrt{p}\boldsymbol{\pi}, \mathbf{0}, \dots, \mathbf{0})\\ &= \mathbf{P}^{-1/2}\mathbf{P}^*\mathbf{H}\begin{bmatrix} np & 0 & \dots & 0\\ 0 & 0 & \dots & 0\\ & \dots & & \\ 0 & 0 & \dots & 0\end{bmatrix}.\end{aligned} \tag{5.35}$$

So we have identified p eigenvectors of $\mathbf{P}^{-1/2}\mathbf{X}'\mathbf{X}\mathbf{P}^{-1/2}$, the first being $\boldsymbol{\pi}$, which by the Perron–Frobenius theorem has the maximum eigenvalue of size np. The other $p-1$ eigenvectors have eigenvalues 0. To maximise $\boldsymbol{\beta}'\mathbf{P}^{-1/2}\mathbf{X}'\mathbf{X}\mathbf{P}^{-1/2}\boldsymbol{\beta}$ subject to $(\mathbf{P}^*)'\mathbf{P}^{-1/2}\boldsymbol{\beta} = \mathbf{0}$ is the same as maximising subject to $\mathbf{H}'(\mathbf{P}^*)'\mathbf{P}^{-1/2}\boldsymbol{\beta} = \mathbf{0}$, and this is achieved by selecting $\boldsymbol{\beta}$ as an eigenvector $\hat{\boldsymbol{\beta}}$ of $\mathbf{P}^{-1/2}\mathbf{X}'\mathbf{X}\mathbf{P}^{-1/2}$ attached to its second largest eigenvalue, say λ_2. Then the maximum value of $\boldsymbol{\beta}'\mathbf{P}^{-1/2}\mathbf{X}'\mathbf{X}\mathbf{P}^{-1/2}\boldsymbol{\beta}$ is $\lambda_2\hat{\boldsymbol{\beta}}'\hat{\boldsymbol{\beta}}$ which, using (5.32), can be written as, say, $\lambda_2\hat{\boldsymbol{\alpha}}'\mathbf{P}\hat{\boldsymbol{\alpha}}$.

Substituting this maximum value back in the likelihood, it remains to carry out the third stage of the maximisation over the choice of size for $\hat{\boldsymbol{\alpha}}'\mathbf{P}\hat{\boldsymbol{\alpha}}$ (we can choose an eigenvector $\hat{\boldsymbol{\alpha}}$ of any length). The relevant part of the log-likelihood may now be written, putting $\hat{\boldsymbol{\alpha}}'\mathbf{P}\hat{\boldsymbol{\alpha}} = \delta$,

$$L = -\frac{n}{2}\ln(\delta + 1) + \frac{1}{2}\lambda_2\delta/(\delta + 1). \tag{5.36}$$

Assuming that $\lambda_2 > 0$, a weak assumption, the likelihood in (5.36) is easily shown to be maximised for $\delta = \max(\lambda_2/n - 1, 0)$.

5.6.2 More than one factor

The treatment of the general case follows the same lines, so we shall treat it more briefly by concentrating on the points at which it differs from the case $q = 1$. The second factor in (5.23) becomes

$$(2\pi)^{-q/2}\int_{-\infty}^{+\infty}\dots\int_{-\infty}^{\infty}\exp\left[-\frac{1}{2}\sum_{j=1}^{q}y_j^2-\sum_{j=1}^{q}y_jY_j-\sum_{i=1}^{p}\ln\sum_{r=0}^{c_i-1}\pi_{ir}\right.$$
$$\left.\times\exp\left\{\sum_{j=1}^{q}\alpha_{ij}(r)y_j\right\}\right]d\mathbf{y}. \tag{5.37}$$

The approximation now consists of replacing the last term in the exponent by a quadratic form derived from the Taylor expansion, giving

$$\sum_{i=1}^{p}\ln\sum_{r=0}^{c_i-1}\pi_{ir}\exp\left\{\sum_{j=1}^{q}\alpha_{ij}(r)y_j\right\}\doteq\mathbf{y}'\mathbf{B}\mathbf{y}+\mathbf{b}'\mathbf{y}, \tag{5.38}$$

where $\mathbf{b}$ is a q-vector with

$$b_j=-\sum_{i=1}^{p}\sum_{r=0}^{c_i-1}\pi_{ir}\alpha_{ij}(r)\quad(j=1,2,\dots,q) \tag{5.39}$$

and $\mathbf{B}$ is a $q\times q$ matrix with elements b_{jk} given by

$$2b_{jk}=\sum_{i=1}^{p}\left\{\sum_{r=0}^{c_i-1}\pi_{ir}\alpha_{ij}(r)\alpha_{ik}(r)-\sum_{r=0}^{c_i-1}\pi_{ir}\alpha_{ij}(r)\sum_{r=0}^{c_i-1}\pi_{ir}\alpha_{ik}(r)\right\}. \tag{5.40}$$

If, as before, we centre the αs for each variable at zero, then

$$\sum_{r=0}^{c_i-1}\pi_{ir}\alpha_{ij}(r)=0\quad(i=1,2,\dots,p;\ j=1,2,\dots,q)$$

and

$$\mathbf{b}=\mathbf{0}\quad\text{and}\quad 2\mathbf{B}=\mathbf{A}'\mathbf{P}\mathbf{A},$$

where $\mathbf{A}=(\boldsymbol{\alpha}_1,\boldsymbol{\alpha}_2,\dots,\boldsymbol{\alpha}_q)$, $\boldsymbol{\alpha}_q$ being the vector of αs for the jth latent variable.

The exponent in (5.37) can thus be written as

$$\frac{1}{2}\{\mathbf{y}'\mathbf{y}-2\mathbf{Y}'\mathbf{y}+\mathbf{y}'\mathbf{A}'\mathbf{P}\mathbf{A}\mathbf{y}\},$$

where $\mathbf{Y}' = (Y_1, Y_2, \ldots, Y_q)$. Completing the square, the integral reduces to

$$|\mathbf{I} + \mathbf{A}'\mathbf{P}\mathbf{A}|^{-1/2} \exp\left\{\frac{1}{2}\mathbf{Y}'(\mathbf{A}'\mathbf{P}\mathbf{A} + \mathbf{I})^{-1}\mathbf{Y}\right\}. \tag{5.41}$$

Noting that $\mathbf{Y} = \mathbf{A}'\mathbf{x}$ and that

$$\mathbf{Y}'(\mathbf{A}'\mathbf{P}\mathbf{A} + \mathbf{I})\mathbf{Y} = \text{trace}\{(\mathbf{A}'\mathbf{P}\mathbf{A} + \mathbf{I})^{-1}\mathbf{Y}\mathbf{Y}'\},$$

the relevant part of the log-likelihood becomes

$$L = -\frac{n}{2} \ln |\mathbf{A}'\mathbf{P}\mathbf{A} + \mathbf{I}| + \frac{1}{2}\text{trace}\{(\mathbf{A}'\mathbf{P}\mathbf{A} + \mathbf{I})^{-1}\mathbf{A}'\mathbf{X}'\mathbf{X}\mathbf{A}\}. \tag{5.42}$$

Our approximate estimators are obtained by maximising this expression with respect to $\mathbf{A}$. However, a new feature is that there is no unique maximum, for if we make an orthogonal transformation to $\mathbf{A}^* = \mathbf{A}\mathbf{N}$ ($\mathbf{N}'\mathbf{N} = \mathbf{I}$), the value of L is unchanged. The proof of this result depends on the two following well-known theorems:

(a) The determinant of a product of two square matrices equals the product of the determinants.

(b) trace $\mathbf{N}'\mathbf{Z}\mathbf{N}$ = trace $\mathbf{Z}$, where $\mathbf{N}$ is orthogonal.

Taking the two parts of (5.42) in turn,

$$\begin{aligned} \ln |(\mathbf{A}^*)'\mathbf{P}\mathbf{A}^* + \mathbf{I}| &= \ln |\mathbf{N}'\mathbf{A}'\mathbf{P}\mathbf{A}\mathbf{N} + \mathbf{I}| = \ln |\mathbf{N}'(\mathbf{A}'\mathbf{P}\mathbf{A} + \mathbf{I})\mathbf{N}| \\ &= \ln |\mathbf{N}'\mathbf{N}| + \ln |\mathbf{A}'\mathbf{P}\mathbf{A} + \mathbf{I}| = \ln |\mathbf{A}'\mathbf{P}\mathbf{A} + \mathbf{I}| \end{aligned} \tag{5.43}$$

and

$$\begin{aligned} \text{trace}[\{(\mathbf{A}^*)'\mathbf{P}\mathbf{A}^* + \mathbf{I}\}^{-1}(\mathbf{A}^*)'\mathbf{X}'\mathbf{X}\mathbf{A}^*] &= \text{trace}\{(\mathbf{N}'\mathbf{A}'\mathbf{P}\mathbf{A}\mathbf{N} + \mathbf{I})^{-1}\mathbf{N}'\mathbf{A}'\mathbf{X}'\mathbf{X}\mathbf{A}\mathbf{N}\} \\ &= \text{trace}[\{\mathbf{N}'(\mathbf{A}'\mathbf{P}\mathbf{A} + \mathbf{I})^{-1}\mathbf{N}\}^{-1}\mathbf{N}'\mathbf{A}'\mathbf{X}'\mathbf{X}\mathbf{A}\mathbf{N}] \\ &= \text{trace}[\mathbf{N}'(\mathbf{A}'\mathbf{P}\mathbf{A} + \mathbf{I})^{-1}\mathbf{N}\mathbf{N}'\mathbf{A}'\mathbf{X}'\mathbf{X}\mathbf{A}\mathbf{N}] \\ &= \text{trace}\{(\mathbf{A}'\mathbf{P}\mathbf{A} + \mathbf{I})^{-1}\mathbf{A}'\mathbf{X}'\mathbf{X}\mathbf{A}\}. \end{aligned} \tag{5.44}$$

To obtain a unique solution we must therefore impose sufficient constraints; other solutions can then be obtained by orthogonal rotation.

Since $\mathbf{A}'\mathbf{P}\mathbf{A}$ is a symmetric positive definite matrix, it can be written in the form $\mathbf{M}\boldsymbol{\Delta}\mathbf{M}'$, where $\boldsymbol{\Delta}$ is a diagonal matrix of its eigenvalues and $\mathbf{M}$ is an orthogonal matrix composed of the eigenvectors. It follows that

$$\boldsymbol{\Delta} = (\mathbf{A}\mathbf{M})'\mathbf{P}(\mathbf{A}\mathbf{M})$$

and therefore that the matrix $\mathbf{A}'\mathbf{P}\mathbf{A}$ can be diagonalised by the transformation $\mathbf{A}^* = \mathbf{A}\mathbf{M}$. We may thus obtain the unique maximiser of L, except for sign changes over whole columns of $\mathbf{A}$ and reorderings of those columns, when $\mathbf{A}'\mathbf{P}\mathbf{A}$ is diagonal, and then derive all others by orthogonal transformation. Since the determinant of a diagonal matrix is the product of its diagonal elements, (5.42) may now be written

$$L = \sum_{j=1}^{q} \left[-\frac{n}{2} \ln\{\boldsymbol{\alpha}'_j \mathbf{P} \boldsymbol{\alpha}_j + 1\} + \frac{\boldsymbol{\alpha}'_j \mathbf{X}'\mathbf{X} \boldsymbol{\alpha}_j}{1 + \boldsymbol{\alpha}'_j \mathbf{P} \boldsymbol{\alpha}_j} \right]. \tag{5.45}$$

Note that the jth diagonal element of $\mathbf{A}'\mathbf{P}\mathbf{A}$ is only equal to $\boldsymbol{\alpha}'_j \mathbf{P} \boldsymbol{\alpha}_j$ when the former is diagonal, as we are assuming. If (5.45) is now compared with (5.30) it will be seen that each term in the sum is of the same form as in the previous case. L in (5.45) is thus maximised if each term individually is maximised, and this is a problem we have already solved. It is thus clear that the vectors $\boldsymbol{\alpha}_j$ will all be eigenvectors of $\mathbf{P}^{-1}\mathbf{X}'\mathbf{X}$. Let $\boldsymbol{\alpha}_1, \boldsymbol{\alpha}_2, \ldots, \boldsymbol{\alpha}_q$ be any q such eigenvectors; then according to (5.34),

$$L = -\frac{n}{2} \sum_{j=1}^{q} \left(\ln \frac{\delta_j}{n} - \frac{\delta_j}{n} + 1 \right). \tag{5.46}$$

It follows that $\delta_1, \delta_2, \ldots, \delta_j$ should be the largest j eigenvalues and that we may extract factors as long as $\theta_j \geq n$. This restriction imposes a limit on the number of factors that can be extracted.

An interesting property of this solution is that, to the same degree of approximation, the components are uncorrelated. Thus the covariance matrix of $\mathbf{Y}$ is

$$\text{cov}(\mathbf{Y}) = \mathbf{A}' E(\mathbf{x}\mathbf{x}' - E\mathbf{x} E\mathbf{x}')\mathbf{A}.$$

Conditioning on $\mathbf{y}$, we have $E\mathbf{x} = E\boldsymbol{\pi}(\mathbf{y})$ and $E\mathbf{x}\mathbf{x}' = E\pi(\mathbf{y})\pi'(\mathbf{y})$, where

$$\pi'(\mathbf{y}) = (\pi_{i0}(\mathbf{y}), \ldots, \pi_{ic_i-1}(\mathbf{y}), \ldots, \pi_{pc_p-1}(\mathbf{y})).$$

Expanding the (i, r)th element of this vector as far as terms of the first degree in the αs,

$$\begin{aligned} \pi_{ir}(\mathbf{y}) &= \pi_{ir} \left[1 - \sum_{j=1}^{q} y_j \left\{ \alpha_{ij}(r) + \sum_{r=0}^{c_i-1} \pi_{ir} \alpha_{ij}(r) \right\} \right] \\ &= \pi_{ir} \left\{ 1 + \sum_{j=1}^{q} y_j \alpha_{ij}(r) \right\}. \end{aligned}$$

Hence

$$\pi_{ir}(\mathbf{y}) - E\pi_{ir}(\mathbf{y}) = \pi_{ir}\sum_{j=1}^{q} y_j\alpha_{ij}(r)$$

and

$$\text{cov}\{\pi_{ir}(\mathbf{y}), \pi_{ks}(\mathbf{y})\} = \pi_{ir}\pi_{ks}\sum_{j=1}^{q}\alpha_{ij}(r)\alpha_{kj}(s).$$

Expressed in matrix notation,

$$\text{cov}(\mathbf{x}) = \mathbf{PA}(\mathbf{PA})'$$

and therefore,

$$\text{cov}(\mathbf{Y}) = \mathbf{A}'\mathbf{PA}(\mathbf{PA})'\mathbf{A} = (\mathbf{A}'\mathbf{PA})^2. \tag{5.47}$$

It thus follows that if we have constrained $\mathbf{A}'\mathbf{PA}$ to be diagonal then, to a first approximation, the components will be uncorrelated. There are interesting parallels here with results for the normal linear model.

The estimation of the πs requires no new theory and we thus proceed as before, using the estimators of (5.31).

5.7 Binary data as a special case

We now have to justify the results given in Section 4.7, and this is done by putting $c_i = 2$, for all i, in the formulae of this chapter. The vector $\boldsymbol{\alpha}$ is then given by

$$\boldsymbol{\alpha}' = (\alpha_1(0), \alpha_1(1), \alpha_2(0), \alpha_2(1), \ldots, \alpha_p(0), \alpha_p(1)).$$

The elements of the vector have been constrained so that

$$\pi_{i0}\alpha_i(0) + \pi_{i1}\alpha_i(1) = 0 \quad (i = 1, 2, \ldots, p). \tag{5.48}$$

In the binary case we estimated a p-vector of parameters, also called $\boldsymbol{\alpha}$, given by $\boldsymbol{\alpha}' = (\alpha_1, \alpha_2, \ldots, \alpha_p)$; α_i was the score associated with category 1 of each variable – the score for the first category was zero. Since both approaches are essentially estimating the *difference* between the scores for each variable, the relationship between the two sets of αs is that

$$\alpha_i = \alpha_i(1) - \alpha_i(0),$$

or, using (5.48), that

$$\alpha_i(0) = -\pi_{i1}\alpha_i \quad \text{and} \quad \alpha_i(1) = \pi_{i0}\alpha_i \quad (i = 1, 2, \ldots, p).$$

Since, in the binary case, $\pi_{i1} = \pi_i$ and $\pi_{i0} = 1 - \pi_i$ we may write

$$\boldsymbol{\alpha} = \begin{bmatrix} -\pi_1 & 0 & \ldots & 0 \\ 1-\pi_1 & 0 & \ldots & \vdots \\ 0 & -\pi_2 & \ldots & \vdots \\ 0 & 1-\pi_2 & \ldots & \vdots \\ \vdots & 0 & \ldots & \vdots \\ \vdots & \vdots & & -\pi_p \\ 0 & 0 & \ldots & 1-\pi_p \end{bmatrix} \boldsymbol{\alpha}^* = \mathbf{S}\boldsymbol{\alpha}^*, \tag{5.49}$$

say, where, temporarily, $(\boldsymbol{\alpha}^*)' = (\alpha_1, \alpha_2, \ldots, \alpha_p)$. Substituting (5.49) into (5.32) yields

$$\hat{\mathbf{P}}^{-1}\mathbf{X}'\mathbf{X}\mathbf{S}\boldsymbol{\alpha}^* = (n + \mu)\mathbf{S}\boldsymbol{\alpha}^*. \tag{5.50}$$

We convert this into another eigenequation in two steps. First pre-multiply both sides by $\mathbf{S}'\hat{\mathbf{P}}$, giving

$$\mathbf{S}'\mathbf{X}'\mathbf{X}\mathbf{S}\boldsymbol{\alpha}^* = (n + \mu)\mathbf{S}'\hat{\mathbf{P}}\mathbf{S}\boldsymbol{\alpha}^*. \tag{5.51}$$

Next pre-multiply by $(\mathbf{S}'\hat{\mathbf{P}}\mathbf{S})^{-1}$, so that

$$(\mathbf{S}'\hat{\mathbf{P}}\mathbf{S})^{-1}\mathbf{S}'\mathbf{X}'\mathbf{X}\mathbf{S}\boldsymbol{\alpha}^* = (n + \mu)\boldsymbol{\alpha}^*. \tag{5.52}$$

This is the required equation and it remains to simplify the left-hand side. As before, we replace the πs by their estimators given by (5.31). Direct multiplication gives

$$\hat{\mathbf{S}}'\hat{\mathbf{P}}\hat{\mathbf{S}} = \text{diag}\,\{\hat{\pi}_i(1 - \hat{\pi}_i)\} = \hat{\mathbf{Q}},$$

say, and

$$\hat{\mathbf{S}}'\mathbf{X}'\mathbf{X}\hat{\mathbf{S}} = \sum_{h=1}^{n}(\mathbf{x}_h - \hat{\boldsymbol{\pi}})(\mathbf{x}_h - \hat{\boldsymbol{\pi}})' = n\hat{\boldsymbol{\Omega}},$$

say, where $\mathbf{x}_h' = (x_{1h}, x_{2h}, \ldots, x_{ph}) = (x_{1h}(1), x_{2h}(1), \ldots, x_{ph}(1))$. Equation (5.52) may thus be written

$$\hat{\mathbf{Q}}^{-1/2}n\hat{\boldsymbol{\Omega}}\boldsymbol{\alpha}^* = (n + \mu)\hat{\mathbf{Q}}^{1/2}\boldsymbol{\alpha}^*. \tag{5.53}$$

If we make the further transformation $\boldsymbol{\beta} = \mathbf{Q}^{1/2}\boldsymbol{\alpha}^*$, the equation becomes

$$\hat{\mathbf{Q}}^{-1/2}\hat{\boldsymbol{\Omega}}\hat{\mathbf{Q}}^{-1/2}\boldsymbol{\beta} = \left(1 + \frac{\mu}{n}\right)\boldsymbol{\beta} \tag{5.54}$$

and the left-hand matrix will be recognised as the 'phi-coefficient' matrix for the binary variables. It follows that the βs are the principal components derived from the phi-coefficient matrix. The αs are thus not principal component weights because of the weightings $\{\hat{\pi}_i(1 - \hat{\pi}_i)\}^{-1/2}$ arising from multiplication by $\mathbf{Q}^{-1/2}$. Since the αs derived by this means are also the scores that would be derived from a multiple correspondence analysis, this result also shows the connection of that technique with principal components.

5.8 Ordering of categories

Nothing done so far requires or implies any ordering of the categories. Having fitted a model we could, of course, relabel the categories so that their estimated scores were ranked in increasing or decreasing order. The analysis might then be seen as a way of uncovering an underlying metric. Conversely, if we believed the categories to be ordered *a priori* we could impose the order restrictions on the parameters and estimate the αs subject to these constraints. However, the basis for any supposed prior ordering would be likely to rest on the belief that there was some variable underlying the categories. If so it would seem more sensible to make such underlying variables explicit in the model as in the underlying variable binary model. There are, in fact, two such models which we could use. One turns out in some circumstances to be equivalent to the response function model discussed in Sections 5.1 and 5.2 but does not lead to a natural ordering of the categories. The other is the obvious generalisation of the underlying variable model for binary data. This is equivalent to a response function model, but not the one given in Section 5.2. We treat each of them, in turn, taking the latter first, in Sections 5.8.1 and 5.8.4.

5.8.1 A response function model for ordinal variables

A review of the most widely used latent variable models for ordinal variables can be found in van der Linden and Hambleton (1997). There are models that can accommodate ordered categories such as the graded response model generalised from either the probit/normal model or the logit/normal model for binary responses and the partial credit model generalised from the Rasch model or logit/normal model. Samejima (1969) formulated a latent variable model for graded scores as a generalisation of the probit/normal model or logit/normal model. Her unidimensional graded response model was a starting point for later research by Muraki (1990) and Muraki and Carlson (1995) on maximum likelihood estimation for graded response models and multidimensional models. Moustaki (2000) discusses the graded response model within the framework of exponential family distributions.

As before, the c_i ordered categories have, for item i, response category probabilities $\pi_{is}(\mathbf{y})$ $(s = 1, \ldots, c_i)$. To preserve the ordinality property of an item we model the *cumulative response function*

$$\text{link}[\Pi_{is}(\mathbf{y})] = \alpha_{i0(s)} - \sum_{j=1}^{q} \alpha_{ij} y_j \quad (s = 1, \ldots, c_i - 1; i = 1, \ldots, p), \quad (5.55)$$

where $\Pi_{is}(\mathbf{y})$ is the cumulative probability of a response in category s or lower of item x_i, written

$$\Pi_{is}(\mathbf{y}) = \pi_{i1}(\mathbf{y}) + \pi_{i2}(\mathbf{y}) + \cdots + \pi_{is}(\mathbf{y}).$$

To simplify notation we just write Π_{is} in the following.

The link function can be the logit, the complementary log-log function, the inverse normal function, the inverse Cauchy, or the log-log function. All those link functions are monotonically increasing functions that map (0, 1) onto $(-\infty, \infty)$. The parameters $\alpha_{i0(s)}$ are referred to as 'cut-points' on the logistic, probit or other scale, where $\alpha_{i0(1)} \leq \alpha_{i0(2)} \leq \cdots \leq \alpha_{i0,(c_i)} = +\infty$. The negative sign in front of the slope parameter is used to indicate that as y increases the response on the observed item x_i is more likely to fall at the high end of the scale. Note that all category response probabilities have the same discriminating power. The logit link leads to the *proportional odds model*,

$$\ln\left[\frac{\Pi_{is}(\mathbf{y})}{1 - \Pi_{is}(\mathbf{y})}\right] = \alpha_{i0(s)} - \sum_{j=1}^{q} \alpha_{ij} y_j \quad (s = 1, \ldots, c_i - 1). \quad (5.56)$$

From (5.56) we see that the difference between corresponding cumulative logits is independent of the category involved. As in the unordered case of (5.3), we take the conditional probability function of $\mathbf{x}_i$ given $\mathbf{y}$ to be multinomial so that

$$\begin{aligned} g(\mathbf{x}_i \mid \mathbf{y}) &= \prod_{s=1}^{c_i} \pi_{is}(\mathbf{y})^{x_i(s)} \\ &= \prod_{s=1}^{c_i} (\Pi_{is} - \Pi_{i,s-1})^{x_i(s)}, \end{aligned} \quad (5.57)$$

where $x_i(s) = 1$ if a randomly selected individual responds into category s of the ith item and $x_i(s) = 0$ otherwise. We can write (5.57)as

$$g(\mathbf{x}_i \mid \mathbf{y}) = \prod_{s=1}^{c_i - 1} \left(\frac{\Pi_{is}}{\Pi_{i,s+1}}\right)^{x_i^*(s)} \left(\frac{\Pi_{i,s+1} - \Pi_{is}}{\Pi_{i,s+1}}\right)^{x_i^*(s+1) - x_i^*(s)}, \quad (5.58)$$

where $x_i^*(s) = 1$ if a randomly selected individual responds into category s or a lower one of the ith item and $x_i^*(s) = 0$ otherwise. Taking logarithms through (5.58),

$$\begin{aligned}\ln g(\mathbf{x}_i \mid \mathbf{y}) &= \sum_{s=1}^{c_i-1}\left[x_i^*(s)\ln\frac{\Pi_{is}}{\Pi_{i,s+1}-\Pi_{is}} - x_i^*(s+1)\ln\frac{\Pi_{i,s+1}}{\Pi_{i,s+1}-\Pi_{is}}\right]\\ &= \sum_{s=1}^{c_i-1}[x_i^*(s)\theta_{is}(\mathbf{y}) - x_i^*(s+1)b(\theta_{is}(\mathbf{y}))],\end{aligned} \tag{5.59}$$

where

$$\theta_{is}(\mathbf{y}) = \ln\frac{\Pi_{is}}{\Pi_{i,s+1}-\Pi_{is}} \quad (s = 1, \ldots, c_i - 1) \tag{5.60}$$

and

$$b(\theta_{is}(\mathbf{y})) = \ln\frac{\Pi_{i,s+1}}{\Pi_{i,s+1}-\Pi_{is}} = \ln\{1+\exp(\theta_{is}(\mathbf{y}))\} \quad (s = 1, \ldots, c_i - 1). \tag{5.61}$$

5.8.2 Maximum likelihood estimation of the model with ordinal variables

The E-M algorithm used in the binary case in Section 4.5 and for the unordered polytomous case in Section 5.5 can be also applied here. The estimation is presented for a single latent variable, but a generalisation to the multidimensional case is straightforward. For a random sample of size n the complete log-likelihood is written

$$\begin{aligned}L_c &= \sum_{h=1}^{n}\ln f(\mathbf{x}_h, y_h)\\ &= \sum_{h=1}^{n}\left[\sum_{i=1}^{p}\ln g(\mathbf{x}_{ih} \mid y_h) + \ln h(y_h)\right]\end{aligned} \tag{5.62}$$

Substituting (5.59) in (5.62), we have

$$L = \sum_{h=1}^{n}\ln g(\mathbf{x}_h \mid y_h) = \sum_{h=1}^{n}\left[\sum_{i=1}^{p}\sum_{s=1}^{c_i-1}[x_i^*(s)\theta_{is}(y_h) - x_i^*(s+1)b(\theta_{is}(y_h))]\right].$$

The E-step is therefore

$$E[L \mid \mathbf{x}] = \sum_{i=1}^{p}\sum_{s=1}^{c_i-1}\Bigg[\sum_{h=1}^{n}\big(x_i^*(s)E(\theta_{is}(y_h) \mid \mathbf{x}_1, \ldots, \mathbf{x}_n) - x_i^*(s+1)E(b(\theta_{is}(y_h)) \mid \mathbf{x}_1, \ldots, \mathbf{x}_n)\big)\Bigg]$$
$$= \sum_{i=1}^{p}\sum_{s=1}^{c_i-1}\Bigg[\sum_{h=1}^{n}\big(x_i^*(s)E(\theta_{is}(y_h) \mid \mathbf{x}_h) - x_i^*(s+1)E(b(\theta_{is}(y_h)) \mid \mathbf{x}_h)\big)\Bigg].$$

As in the unordered case, for the M-step, one can maximise separately for each item i and each category s the quantity

$$\sum_{h=1}^{n}\big(x_i^*(s)E(\theta_{is}(y_h) \mid \mathbf{x}_h) - x_i^*(s+1)E(b(\theta_{is}(y_h)) \mid \mathbf{x}_h)\big). \tag{5.63}$$

As in Section 5.5, we approximate the averages in (5.63) with Gauss–Hermite quadrature, giving

$$= \sum_{h=1}^{n}\sum_{t=1}^{k} x_{ih}^*(s)\theta_{is}(y^t)h(y^t \mid \mathbf{x}_h) - \sum_{h=1}^{n}\sum_{t=1}^{k} x_{ih}^*(s+1)b(\theta_{is}(y^t))h(y^t \mid \mathbf{x}_h)$$
$$= \sum_{t=1}^{k} r_{it}(s)\theta_{is}(y^t) - \sum_{t=1}^{k} r_{it}(s+1)b(\theta_{is}(y^t)), \tag{5.64}$$

where

$$r_{it}(s) = \sum_{h=1}^{n} x_{ih}^*(s)h(y^t \mid \mathbf{x}_h) \quad (s = 1, 2, \ldots, c_i - 1;\ i = 1, 2, \ldots, p). \tag{5.65}$$

Other approximations to the averages, as discussed in Section 4.5.2, can be used here too. The derivatives required in the M-step for the model in (5.56) are

$$\frac{\partial\theta_{is}(y)}{\partial\alpha_{i0(s)}} = \frac{(1-\Pi_{is})\Pi_{i,s+1}}{(\Pi_{i,s+1}-\Pi_{is})},$$
$$\frac{\partial b(\theta_{is}(y))}{\partial\alpha_{i0(s)}} = \frac{\Pi_{is}(1-\Pi_{is})}{(\Pi_{i,s+1}-\Pi_{is})},$$
$$\frac{\partial\theta_{is}(y)}{\partial\alpha_{i1}} = -y\Pi_{i,s+1} \quad (s = 1, \ldots, c_i - 1)$$

and

$$\frac{\partial b(\theta_{is}(y))}{\partial\alpha_{i1}} = -y\Pi_{is} \quad (s = 1, \ldots, c_i - 1).$$

5.8.3 The partial credit model

A simpler version of the proportional odds model is the partial credit model (Masters 1982) which assumes that the respondent processes the item task in successive steps with probability π_{is} of success in making the step from s to $s+1$, where $\pi_{i1} = 1$. The process terminates when the first failure occurs or when the step s is equal to the last category of the item c_i. The partial credit model is a generalisation of the Rasch model (Rasch 1961) for ordinal responses that assumes one parameter for individuals and one parameter for items. We will outline the main features of the partial credit model here. The main focus is on comparing adjacent categories, and we shall leave out any other probabilities. Therefore any parameters of the model are defined with respect to two adjacent categories rather than taking into account all categories of the item. We let $\pi_{is\,|\,s-1}(y)$ be the conditional probability of selecting category s rather than $s-1$, and that conditional probability is expected to increase monotonically with the latent variable. The conditional probability for a randomly selected individual is modelled as

$$\pi_{is\,|\,s-1}(y) = \frac{\pi_{is}(y)}{\pi_{i,s-1}(y) + \pi_{is}(y)} = \frac{\exp(y - \alpha_{i0,s})}{1 + \exp(y - \alpha_{i0,s})} \quad (s = 1, \ldots, c_i). \tag{5.66}$$

Assuming that each individual is assigned only one score on item i from (5.66), it follows that the response category probability

$$\pi_{is}(y) = \frac{\exp \sum_{l=1}^{s} (y - \alpha_{i0,l})}{\sum_{k=1}^{m_i} \exp\left(\sum_{l=1}^{k} (y - \alpha_{i0,l})\right)}. \tag{5.67}$$

Conditional, joint and marginal maximum likelihood estimation methods can be used for estimating the parameters of the partial credit model (Masters and Wright 1997). A generalised partial credit model that allows for a different slope for each item has been developed by Muraki (1992).

5.8.4 An underlying variable model

For the binary case we imagined that the data had arisen by recording whether or not underlying variables had values above or below fixed thresholds. The obvious generalisation is to suppose that the range of each underlying variable is divided into non-overlapping intervals and that the manifest categorical variables record into which interval the individual falls. Thus we suppose that underlying each set of categories there is a continuous random variable ξ whose variation is explained by the standard linear factor model

$$\boldsymbol{\xi} = \boldsymbol{\mu} + \boldsymbol{\Lambda}\mathbf{y} + \mathbf{e}. \tag{5.68}$$

On the ith dimension there is a sequence of thresholds $\tau_{i1}, \tau_{i2}, \ldots, \tau_{i,c_i-1}$ with associated indicator variables

$$x_i(s) = \begin{cases} 1 & \text{if} \quad \tau_{is} \le \xi_i < \tau_{i,s+1} \quad (s = 0, 1, \ldots, c_i - 1), \\ 0 & \text{otherwise}, \end{cases}$$

where we define $\tau_{i0} = -\infty$, $\tau_{ic_i} = \infty$. The aim is thus to make inferences about the parameters and fit of the model (5.68) on the evidence of a sample of observations on $\mathbf{x}$. We assume that the underlying bivariate distributions are normal and hence we can fit the model provided we can estimate the correlation coefficients. Estimates of these coefficients are known as polychoric correlations. Their computation from ordered contingency tables with more than two categories is more difficult than for 2×2 tables, but a maximum likelihood method is given in Olsson (1979) and an algorithm based on the polychoric series in Martinson and Hadman (1975). Jöreskog (1994) extends this work to include the asymptotic standard errors.

The model just described is not equivalent to the response function model of Section 5.2 unless, of course, we have only two categories on each dimension. This is apparent from the fact that the categories are necessarily ordered in the underlying variable model, whereas in the response function model they are not. However, the response function model of Section 5.8.1 can be made equivalent. By the result of Section 4.4 this is equivalent to a underlying variable model with

$$\alpha_{i0(s)} = \frac{\tau_{i,s-1} - \mu_i}{\psi_i^{1/2}}, \quad \alpha_{ij} = \frac{\lambda_{ij}}{\psi_i^{1/2}} \quad (s = 1, 2, \ldots, c_i - 1;\ i = 1, 2, \ldots, p) \tag{5.69}$$

and the weighted error distribution having a logistic distribution. The threshold parameters $\{\tau_{is}\}$ are thus related to the $\alpha_{i0(s)}$ (or πs) and the factor loadings to the αs.

This model does not arise from the sufficiency principle, and this means that we no longer have the same range of interpretations for the parameters. The parameter π_{is} is still the probability that the median individual falls in category s on variable i, but the αs cannot now be interpreted as category scores or as weights in a component. This is because the posterior density of $\mathbf{y}$ is no longer dependent on a linear function of the manifest variables. To see this we write

$$\begin{aligned} \pi_{is}(\mathbf{y}) &= \Pi_{is}(\mathbf{y}) - \Pi_{i,s+1}(\mathbf{y}) \\ &= \left\{1 + \exp\left(-\alpha_{i0(s)} + \sum_{j=1}^{q} \alpha_{ij} y_j\right)\right\}^{-1} \\ &\quad - \left\{1 + \exp\left(-\alpha_{i0(s+1)} + \sum_{j=1}^{q} \alpha_{ij} y_j\right)\right\}^{-1}. \end{aligned}$$

When this is substituted into (5.4) it does not yield a set of linear components.

Various model specifications under the underlying variable model

The simplest method of fitting this model is to utilise the link with the underlying factor model. If we can estimate the correlation coefficients of the underlying variables from the observed categorical variables, any factor analysis program will do the rest. This method is not fully efficient, and it will not provide asymptotic standard errors for the estimators. Nevertheless, once the correlation matrix has been estimated, the whole analysis is moved onto familiar territory. We will explore three different methods that utilise different parts of the data, ranging from full information maximum likelihood to limited-information maximum likelihood methods.

Since only ordinal information is available about ξ_i, the mean and variance of ξ_i are not identified and are therefore set to zero and one, respectively. Other parameterisations are possible; see, for example, Lee *et al.* (1990a). Utilising the assumptions of the classical factor analysis model, it is further assumed that ys and es are independent and normally distributed with $y_j \sim N(0, 1)$ and $e_i \sim N(0, \psi_i)$. Furthermore, it follows that $\xi_1, \ldots, \xi_p$ has a multivariate normal distribution with zero means, unit variances and correlation matrix $\mathbf{P} = (\rho_{ij})$, where $\rho_{ij} = \sum_{l=1}^{q} \lambda_{il}\lambda_{jl}$.

The parameters of the model are the thresholds τ_{is} $(i = 1, 2, \ldots, p;\ s = 1, 2, \ldots, c_i - 1)$ and the factor loadings λ_{ij} $(i = 1, 2, \ldots, p;\ j = 1, 2, \ldots, q)$. From the multivariate normality of $\xi_1, \ldots, \xi_p$, it follows that the probability of a general p-dimensional response pattern is

$$\Pr(x_1 = s_1, \ldots, x_p = s_p) = \int_{\tau_{i,s_1-1}}^{\tau_{1,s_1}} \cdots \int_{\tau_{p,sp-1}}^{\tau_{p,sp}} \phi_p(u_1, \ldots u_p \mid \mathbf{P}) du_1 \cdots du_p, \quad (5.70)$$

where $\phi_p(u_1, u_2, \ldots u_p \mid \mathbf{P})$ is a p-dimensional normal density function with zero means, unit variances and correlation matrix $\mathbf{P}$. It is apparent that full information maximum likelihood estimation under the multivariate underlying variable approach requires the evaluation of the p-dimensional integral in (5.70) for each response pattern in the sample at several points of the parameter space. It is obvious that this approach is not feasible for more than few variables. Alternative methods are required within the underlying variable approach. Approaches that do not specify a model for the complete p-dimensional response pattern but instead make use of the univariate and bivariate marginals have been developed by Jöreskog (1990, 1994) and in a more general setting by Muthén (1984) and Muthén and Satorra (1995). Those methods have been implemented in LISREL (Jöreskog and Sörbom 2006) and Mplus (Muthén and Muthén 2010). In the first step, the thresholds are estimated from the univariate margins of the observed variables. In the second step, the polychoric correlations are estimated from the bivariate margins of the observed variables for given thresholds. In the third step, the factor model is estimated from the polychoric correlations by weighted least squares using a weight matrix which is the inverse of an estimate of the asymptotic covariance matrix of the polychoric correlations. The asymptotic covariance matrix is often unstable in small samples, particularly if there are zero or small frequencies in the bivariate margins. The link between the parameters of the standard factor model and those of the underlying variable model is given in (5.69).

Jöreskog and Moustaki (2001) explored a composite maximum likelihood approach that estimates the thresholds and the factor loadings in one single step from the univariate and bivariate margins without the use of a weight matrix. This approach only requires bivariate normality of all pairs of underlying response variables and uses only the data in the univariate and bivariate margins to estimate the model. The estimation method has been implemented in LISREL (Jöreskog and Sörbom 2006). From the multivariate normality of $\xi_1, \ldots, \xi_p$, it follows immediately that the probability π_{is} of a response in category s on variable i is

$$\pi_{is}(\mathbf{a}_i) = \int_{\tau_{i,s-1}}^{\tau_{is}} \phi(u)du, \tag{5.71}$$

where $\phi(u)$ is the standard normal density function and $\mathbf{a}_i$ is a vector of model parameters associated with item i, and the probability $\pi_{ij,sh}$ of a response in category s on variable i and a response in category h on variable j is

$$\pi_{ij,sh}(\mathbf{a}_i) = \int_{\tau_{i,s-1}^{(i)}}^{\tau_{is}} \int_{\tau_{j,h-1}}^{\tau_{jh}} \phi_2(u, v \mid \rho_{ij})dudv, \tag{5.72}$$

where $\phi_2(u, v \mid \rho)$ is the density function of the standardised bivariate normal distribution with correlation ρ. Their proposal is to maximise the sum of all univariate and bivariate log-likelihoods

$$L = \sum_{i=1}^{p}\sum_{s=1}^{c_i} p_{is} \ln \pi_{is}(\mathbf{a}_i) + \sum_{i=2}^{p}\sum_{j=1}^{i-1}\sum_{s=1}^{c_i}\sum_{h=1}^{c_j} p_{ij,sh} \ln \pi_{ij,sh}(\mathbf{a}_i), \tag{5.73}$$

where p_{is} is the sample proportion of category s in the univariate margin of variable i, $p_{ij,sh}$ is the sample proportion of category s and h in the bivariate margin of variables i and j, and $\pi_{is}(\mathbf{a}_i)$ and $\pi_{ij,sh}(\mathbf{a}_i)$ are given by (5.71) and (5.72), respectively. The integrals in (5.71) and (5.72) can be computed directly from the univariate and bivariate distribution functions. Only data in the univariate and bivariate margins are used. This approach is quite feasible in that it can handle a large number of variables as well as a large number of factors.

The underlying variable model has fewer parameters than the response function model because the α_{ij} ($j > 0$) do not involve the argument s which indexes the categories. This approach was developed by Lee *et al.* (1990b, 1992), who also give various simplifications based on a two-stage procedure in Lee *et al.* (1995). These authors consider the problem in the context of the broader structural equations model, and they allow mixtures of continuous and categorical variables among the manifest variables. We return to both of these generalisations in Chapters 7 and 8.

As we have already shown, a promising new approach to maximising complicated likelihood functions is provided by the Gibbs sampler algorithm. This was demonstrated for Lee, Poon and Bentler's problem by Shi and Lee (1998).

A third approach is to exploit the connection with the logit/normal model for binary data expressed in (5.56). By reducing each variable to a dichotomy, we can estimate all of the α_{ij} and those $\alpha_{i0}(s)$ corresponding to the point at which the dichotomies have been made. Questions then arise about the best way of making the reduction to two categories, and this requires further investigation, but one would expect that one would aim to get a near 50–50 split. The other values of the $\alpha_{i0}(s)$ could be estimated by making the split at all other possible points, though this would be tedious if there were large numbers of categories. A simpler alternative would be first to estimate the α_{ij}, as described above, and then to equate the observed and expected marginal proportions. This gives the estimating equations

$$E\pi_{is}(\mathbf{y}) = n_{is}/n \quad (i = 1, 2, \ldots, p;\ s = 0, 1, \ldots, c_i - 1),$$

where n_{is} is the number of sample members responding in category s of variable i.

A review of full and limited-information maximum likelihood methods for fitting a latent variable model to ordinal variables can be found in Jöreskog and Moustaki (2001).

5.9 An alternative underlying variable model

Another way of relating categorical responses to underlying variables has been developed for modelling choice behaviour in economics (see, for example, McFadden (1974, 1982)). This approach is of equal relevance whether one is choosing among a set of alternatives in a multiple choice question or among candidates in an election. The model was introduced into latent variable modelling in the first edition of this book (Bartholomew 1987) and developed in a broader context by Arminger and Küsters (1988).

In this model a random variable is associated with *each* category, and the category into which an individual falls is determined by which of these underlying variables turns out to be largest. Thus, suppose that there are variables $\xi_{i0}, \xi_{i1}, \ldots, \xi_{i,c_i-1}$ associated with variable i. The realised values of these variables, for any i, can be thought of as measures of the relative 'attractiveness' of the categories to that individual. The largest ξ thus determines which category 'wins' in the competition for the individual's vote, purchasing decision or whatever. If we postulate that a number of latent variables together with a random 'error' contribute to the final attractiveness, we may suppose that

$$\boldsymbol{\xi}_i = \boldsymbol{\mu}_i + \boldsymbol{\Lambda}_i \mathbf{y} + \mathbf{e}_i \quad (i = 1, 2, \ldots, p), \tag{5.74}$$

with the usual distributional assumptions about the random variables. Here $\boldsymbol{\xi}_i' = (\xi_{i0}, \xi_{i1}, \ldots, \xi_{i,c_i-1})$, $\boldsymbol{\mu}_i' = (\mu_{i0}, \mu_{i1}, \ldots, \mu_{i,c_i-1})$ and $\boldsymbol{\Lambda}_i = \{\lambda_{ij}(s)\}$ is a $c_i \times q$ matrix of factor loadings. We thus have a model in which a common set of latent variables (factors) account for the associations between observed categorical variables.

The parameters $\boldsymbol{\mu}_i$ may be interpreted as measuring the average attractiveness of the categories of variable i. The term $\boldsymbol{\Lambda}_i\mathbf{y}$ shows how the q latent variables influence their attractiveness, and $\mathbf{e}_i$ represents the unexplained variation. All that we can observe is which of the elements in ξ_i is the greatest, and, using this information, we have to devise methods for estimating the parameters of the model. To do this by maximum likelihood we have to specify the form of the distributions of the random variables involved. As in earlier models of the form (5.74) that we have considered, this choice is not likely to be critical since there will be a central limit effect inducing normality in the ξs. We might therefore be content with any reasonable specification which renders the subsequent analysis tractable.

There is one such specification which makes this underlying variable model equivalent to the logit response function model of Section 5.2. This requires us to assume that each residual element in (5.74) has the same Type I extreme-value distribution; that is,

$$P\{e_{is} \leq u\} = \exp(-\exp(-u)), \quad \text{for all } i \text{ and } s. \tag{5.75}$$

The significant part of this assumption is the independence of i and s, implying that the extraneous sources of variation act on the attractiveness of all categories for all variables equally.

We now demonstrate the equivalence referred to by deriving the response function $\pi_{is}(\mathbf{y})$ for this model as follows:

$$\begin{aligned}
\pi_{is}(\mathbf{y}) &= P\{\text{individual falls into category } s \text{ of variable } i \mid \mathbf{y}\} \\
&= P\{\xi_{is} \geq \xi_{ir},\, r = 0, 1, \ldots, c_i - 1 \mid \mathbf{y}\} \\
&= \int_{-\infty}^{+\infty} P\{\xi_{is} \in (u,\, u + du)\, \xi_{ir} \leq u;\, r = 0, 1, \ldots, c_i - 1 \mid \mathbf{y}\}\, du \\
&= \int_{-\infty}^{+\infty} P\{\xi_{is} \in (u,\, u + du) \mid \mathbf{y}\} \prod_{\substack{r=0 \\ r \neq s}}^{c_i - 1} P\{\xi_{ir} \leq u \mid \mathbf{y}\} du.
\end{aligned} \tag{5.76}$$

The last step follows from the independence of the ξs when $\mathbf{y}$ is fixed. Now

$$\begin{aligned}
P\{\xi_{ir} \leq u \mid \mathbf{y}\} &= P\left\{\mu_{ir} + \sum_{j=1}^{q} \lambda_{ij}(r) y_j + e_{ir} \leq u \,\middle|\, \mathbf{y}\right\} \\
&= P\left\{e_{ir} \leq u - \mu_{ir} - \sum_{j=1}^{q} \lambda_{ij}(r) y_j \,\middle|\, \mathbf{y}\right\} \\
&= \exp\left\{-\exp\left(-u + \mu_{ir} + \sum_{j=1}^{q} \lambda_{ij}(r) y_j\right)\right\}
\end{aligned} \tag{5.77}$$

by (5.75). The density function is obtained by differentiation, and when this is substituted into (5.76) we obtain

$$\pi_{is}(\mathbf{y}) = \int_{-\infty}^{+\infty} \exp\left\{-\left(u - \mu_{is} - \sum_{j=1}^{q} \lambda_{ij}(s) y_j\right)\right\}$$
$$\times \exp\left[-\sum_{r=0}^{c_i-1} \exp\left\{-\left(u - \mu_{ir} - \sum_{j=1}^{q} \lambda_{ij}(r) y_j\right)\right\}\right] du$$
$$= \exp\left(\mu_{is} + \sum_{j=1}^{q} \lambda_{ij}(s) y_j\right)$$
$$\times \int_{-\infty}^{+\infty} \exp - \left\{u + e^{-u} \sum_{r=0}^{c_i-1} \exp\left(\mu_{ir} + \sum_{j=1}^{q} \lambda_{ij}(r) y_j\right)\right\} du. \qquad (5.78)$$

The integral may be evaluated by substituting $v = e^{-u}$, giving

$$\pi_{is}(\mathbf{y}) = \frac{\exp\left(\mu_{is} + \sum_{j=1}^{q} \lambda_{ij}(s) y_j\right)}{\sum_{r=0}^{c_i-1} \exp\left(\mu_{ir} + \sum_{j=1}^{q} \lambda_{ij}(r) y_j\right)} \quad (s = 0, 1, \ldots, c_i - 1;\ i = 1, 2, \ldots, p). \qquad (5.79)$$

Comparing this with (5.9), we see that the two models are identical because $\alpha_{i0}(s) \equiv \mu_{is}$ and $\alpha_{ij}(s) \equiv \lambda_{ij}(s)$.

If we wish, therefore, the parameters can be interpreted as required by the 'choice' model. Whether or not this model provides a useful interpretation will depend on what the variables represent in a particular practical situation. If, for example, respondents to a questionnaire are asked which party they would vote for in an election, it seems reasonable to suppose that this would depend partly on their position on a variety of attitude scales (the ys) common to all individuals and partly on their own idiosyncrasies (the es). The same would be true for many other categorical variables arising from the expression of attitudes, preferences or opinions. The argument seems less convincing for biographical variables such as country of birth or level of education attained. However, the fact that the parameters can also be given the more neutral interpretations of Section 5.3 means that nothing vital depends on the choice.

The appearance of the extreme-value distribution in the error term where we might have expected a distribution such as the normal or logistic is unusual and not easy to interpret. The Type I distribution is positively skewed and rather like a log-normal distribution in shape. Fortunately, as noted in Section 5.9 above, its influence on the form of the distribution of ξ is not likely to be large, so it should not be critical for the form of the response function.

The identity between the αs and the λs implies that the origin of the λs is arbitrary for given i and j. This could be deduced directly from the underlying variable model. Here it means that, without loss of generality, we could regard the factors $\mathbf{y}$ as having no effect on the attractiveness of one arbitrarily selected category on each manifest dimension.

An interesting feature emerges if we specialise to the binary case with $c_i = 2$ for all i. We know that our model is then equivalent to the response function model which, in turn, is equivalent to the underlying variable model of Section 4.3. It should then follow that the underlying variable model and the present one are equivalent in the binary case. That this is so is easily demonstrated. From (5.74),

$$\begin{aligned}\xi_{i0} - \xi_{i1} &= (\mu_{i0} - \mu_{i1}) + \{\boldsymbol{\lambda}_i(0) - \boldsymbol{\lambda}_i(1)\}\mathbf{y} \\ &= e_{i0} - e_{i1} \quad (i = 1, 2, \ldots, p),\end{aligned}$$

where $\boldsymbol{\lambda}_i(s) = \{\lambda_{ij}(s)\}$, $s = 0, 1$; the probability of a response in category 1, given $\mathbf{y}$, is then

$$P\{\xi_{i1} - \xi_{i0} \geq 0 \mid \mathbf{y}\} = P\{e_{i1} - e_{i0} \geq \mu_{i0} - \mu_{i1} + \{\boldsymbol{\lambda}_i(0) - \boldsymbol{\lambda}_i(1)\}\mathbf{y} \mid \mathbf{y}\}. \quad (5.80)$$

For the underlying variable model of Chapter 4 the corresponding probability is

$$P\{\xi_i \geq \tau_i \mid \mathbf{y}\} = P\left\{\frac{e_i}{\psi_i^{\frac{1}{2}}} \geq \frac{\tau_i - \mu_i - \boldsymbol{\Lambda}_i\mathbf{y}}{\psi_i^{1/2}} \middle| \mathbf{y}\right\}, \quad (5.81)$$

where $\boldsymbol{\Lambda}_i$ is the ith row of $\boldsymbol{\Lambda}$. For the two to be the same the parameter values must be chosen so that

$$\mu_{i1} - \mu_{i2} = (\tau_i - \mu_i)\psi_i^{-1/2} \quad \text{and} \quad \boldsymbol{\lambda}_i(1) - \boldsymbol{\lambda}_i(0) = \boldsymbol{\Lambda}_i\psi_i^{-1/2},$$

and the distribution of $e_i\psi_i^{-1/2}$ must be the same as that of $e_{i1} - e_{i0}$. This last requirement is easily verified, since if e_{i0} and e_{i1} have independent Type I extreme-value distributions their difference has a logistic distribution with variance $\pi^2/3$. This is precisely the form which the distribution of $e_i\psi_i^{-1/2}$ must have if it is to be equivalent to the response function model. The two underlying variable models thus coincide for binary data.

5.10 Posterior analysis

There are no essentially new issues arising in the posterior analysis of polytomous models. Those models which are based on the sufficiency principle yield linear components which can be used to rank individuals on the latent scale as before. Some examples are given below. The posterior expectations and variances can be computed from the posterior distribution as given in (5.4).

In the case of the underlying variable model, methods based on the E-M algorithm have been provided by Shi and Lee (1997a,b). The underlying variable model is basically an NLFM and hence the posterior distribution, when expressed in terms of the underlying variables, is multivariate normal. For such a distribution the mean and the mode coincide and Shi and Lee's method is to find the mode of that distribution.

5.11 Further observations

There is now a fairly complete theory of latent trait models available for polytomous data, both nominal and ordinal, but caution must be exercised in applying it. Much less practical experience of these methods is available than for those for binary data. The very large number of parameters involved, especially in the nominal model, magnifies the problems we have already noted with binary models. For example, there may be problems with the convergence of the estimators, and their estimated asymptotic standard errors may be very large and, hence, unreliable. Again, the problems of testing goodness of fit arising from the sparseness of the contingency table are potentially more serious. The number of cells is now $\prod_{i=1}^{p} c_i$, which may be substantially larger than 2^p. In the binary case we saw that this problem could be alleviated by pooling response patterns to avoid very small expected frequencies. With polytomous data there is a second option. This is to combine categories. In the social sciences, especially, this can often be done in a substantively meaningful way. Categories of the Likert variety often consist of five or seven possible responses, but one finds that the bulk of observations fall into only two or three of them. Little information is then lost by amalgamating these little used categories with larger ones. Of course, even if this procedure is taken to the limit in which all variables become dichotomies, sparseness may remain. In that case, resort must be had to pooling response patterns and/or using the Monte Carlo test recommended for use in these circumstances in the case of binary data. Section 4.9 discusses various methods for model testing that can be applied to the polytomous case. In the polytomous case as in the binary case, tests that are based on residuals calculated from the marginal frequencies of various orders developed for multidimensional contingency tables (Maydeu-Olivares and Joe 2006) can be used.

An important practical question concerns the choice of model. Should we use the logit or probit response function model (divide-by-total) or the underlying variable (difference) model in any particular case? The underlying variable model has received most attention in the literature primarily, one suspects, because it fits easily into the familiar framework of factor analysis. However, it is clear that there are circumstances in which it is not appropriate. If there were an underlying variable, the categories formed from it would necessarily be ordered. Without some *a priori* ordering it would therefore be pointless to consider such a model. Even where ordering occurs, as when respondents in a social survey are asked to express a strength of opinion, there is likely to be a 'don't know' category which does not fit into the ordering of the remaining categories. In such cases, the response function model may be more appropriate.

The response function model can always be used and it provides a means of checking any supposed ordering. We saw that the $\alpha_{ij}(s)$ could be viewed as category scores and their ordering should correspond to the *a priori* ordering of the categories. A more sophisticated way of accommodating ordered categories into the response function model mentioned in Section 5.8 would be to maximise the likelihood subject to order restrictions on the category scores using the techniques of monotonic regression. This would be particularly useful if only some of the manifest variables had ordered categories. This avenue remains to be explored.

The choice between the two kinds of model in practice is likely to depend on a blend of substantive plausibility and goodness of fit. A good fit is a necessary attribute of a good model, but it may not be sufficient.

One of the most intriguing things revealed by our analysis of the response function model is the connection with multiple correspondence analysis (see also Chapter 9). Both techniques can be viewed as methods for assigning scores to the categories of a multi-way contingency table. The link between the sets of scores obtained from the two approaches was established by our demonstration that the correspondence analysis scores were first-order approximations to the $\alpha_{ij}(s)$ of the response function model. In Example 5.12.1 below, the correlation coefficient between the two sets of scores was 0.97, and our limited experience in other cases suggests that this may not be untypical. Multiple correspondence analysis is very easy to carry out as it only involves the eigenanalysis of a matrix containing the two-way margins.

The link with correspondence analysis suggests a further way of using the latent trait model. Correspondence analysis has traditionally been concerned with the two-way contingency tables, sometimes with many categories on each dimension. Provided that there are enough cells, it would be possible to fit a latent trait model as a way of explaining or measuring the association between only two variables. This too is a matter for further research.

5.12 Examples of the analysis of polytomous data using the logit model

Example 5.12.1 This first example comes from the results of a staff assessment exercise in a large public organisation. A five-point ordered scale was used to assess 397 managers on 13 aspects of their work. Three of the variables are used in this illustration. The higher categories were rarely used, so some amalgamation was carried out. On variable 1, categories 4 and 5 were amalgamated, and on variables 2 and 3, categories 4, 5 and 6. The resulting data constitute a $4 \times 3 \times 3$ contingency table. The frequencies are set out in Table 5.1, where they are listed in order of posterior mean. We shall fit a one-factor response function model to see whether the three categorical variables can be replaced by a single underlying latent variable and, if so, how each of the observed variables contributes to the latent scale.

The calculations were done with the IRTPRO program, (Cai *et al.* 2011) and all entries have been rounded for ease of presentation. In order to economise on space,

Table 5.1 The fit and scores obtained by fitting the logit/normal model to staff assessment data.

Response pattern	Frequency	Expected frequency	$E(y \mid \mathbf{x})$	Standard deviation
111	1	2.3	−2.17	0.61
211	7	6.4	−1.92	0.59
311	1	1.2	−1.64	0.59
112	2	2.8	−1.61	0.59
212	13	11.6	−1.37	0.59
121	3	2.0	−1.34	0.60
131	1	0.7	−1.14	0.60
221	10	9.9	−1.09	0.61
312	5	3.4	−1.09	0.61
231	5	3.9	−0.88	0.62
321	3	3.6	−0.79	0.62
122	12	9.1	−0.76	0.62
331	1	1.7	−0.57	0.63
132	4	4.3	−0.54	0.63
222	64	68.6	−0.48	0.63
232	36	38.1	−0.26	0.63
322	38	40.9	−0.17	0.64
332	31	26.8	0.06	0.64
123	1	2.1	0.13	0.64
422	4	3.3	0.21	0.64
133	1	1.6	0.36	0.65
223	37	29.0	0.42	0.65
432	3	2.7	0,44	0.65
233	23	26.8	0.66	0.66
323	34	35.6	0.76	0.67
333	41	40.1	1.02	0.69
423	5	6.8	1.20	0.71
433	11	9.9	1.49	0.74

$\chi^2 = 11.44$ on 21 degrees of freedom, $P = 0.9537$.

results are presented only for response patterns which have a non-zero observed frequency.

It is clear that the one-factor model provides an excellent fit, so the observed response patterns are entirely consistent with a single underlying dimension which one might describe as success in the job.

One can further elucidate the relationship between the latent variable and the observed variables by looking at the parameter estimates given in Table 5.2. We notice first that the standard errors of the αs are very large, so if we interpret the αs as category scores they must be regarded as very imprecisely determined on a

Table 5.2 Maximum likelihood estimates and standard errors for the logit/normal model fitted to staff assessment data.

Variable(i)	Category (s)	$\hat{\alpha}_i(s)$	s.e.$\hat{\alpha}_i(s)$	Approx $\alpha_i(s)$	$\hat{\pi}_{is}$	s.e.$(\hat{\pi}_{is})$
1	0	0	—	0	0.05	—
	1	0.70	0.43	0.76	0.52	0.03
	2	1.49	0.51	1.33	0.40	0.03
	3	2.41	0.80	1.79	0.03	0.02
2	0	0	—	0	0.02	—
	1	2.37	1.51	1.22	0.59	0.03
	2	2.93	1.53	1.68	0.40	0.03
3	0	0	—	0	0.02	—
	1	1.57	0.76	1.00	0.67	0.10
	2	3.77	1.58	1.68	0.31	0.09

sample of this size. The categories were ordered, and it is reassuring to note that for each variable the αs form a monotonic sequence. In Section 5.3 we noted several ways in which the αs could be interpreted. Viewing them as weights in the formation of the components, we see that variable 3 carries most weight, and this accounts for the preponderance of response patterns in the highest category of this variable at the upper end of the scale of Table 5.1. Similarly, variable 2 is more important than variable 1. Interpreted as discrimination parameters, it appears that variable 3, with the largest spread of scores, is the best single variable for distinguishing the overall quality of individuals. Looking at the αs as category scores, the categories of variables 1 and 3 are roughly equally spaced, whereas for variable 2, categories 2 and 3 are very close together and might be amalgamated with little loss of information.

The πs are much more precisely estimated. Recall that π_{is} is the probability of the 'median individual' falling into category s on variable i. As one might have anticipated, such an individual will most likely be placed in the middle category.

Table 5.2 also gives the approximate maximum likelihood estimates for the αs obtained by the method described in Section 5.6; they are also used as starting values for the iterative procedure used to obtain the estimates. They are equivalent to the scores obtained by multiple correspondence analysis and therefore of independent interest. At first sight the approximation is very poor; the true estimates exceed the starting values by a factor of roughly 2, and this seems to be a typical result. However, as we observed earlier, it is the profile of the scores which is important for interpretation. In every case the approximate scores place the categories in the same order and at roughly the same spacing. (It is this fact which makes them good starting values for iteration, because it appears to be important to have the right ordering to get rapid convergence.) One way of judging the accuracy of the approximate estimates is to compute the approximate components and to see how the rankings assigned to response patterns compare. If this is done for the data of this example, we noted above that the ranking assigned to response patterns, using the final and initial estimates in constructing the components, has correlation coefficient 0.97. In this sense therefore the approximate estimates give results which differ little from the exact estimates.

Example 5.12.2 It often happens that a category such as 'don't know' is included among a set of possible answers to a question in a survey. For example, in a simple case there might be three answers: yes, no and don't know.

If we were to use an underlying variable model, we would have to decide on an ordering of the responses and, in particular, whether 'don't know' should be the middle category. One advantage of the response function model is that the data are allowed to speak for themselves by determining the category scores. Our second example illustrates this using data from Leimu (1983).

The study was of a sample of 469 employees, obtained in 1975, from a small industry in Finland. There are three variables defined by the following questions.

1. Was there any alternative choice of job when coming to your present job? (0 = no, 1 = don't know, 2 = yes)
2. Is the job permanent? (0 = very unsure or quite unsure, 1 = don't know, 2 = quite sure or very sure)
3. Were you unemployed in the last three years? (0 = no, 1 = yes)

The results of fitting the one-factor response function model are given in Tables 5.3 and 5.4 in the usual format.

Although the standard errors of the αs are large, their ordering across the categories strongly suggests that the 'don't know' response is correctly placed in the middle position on each variable. The πs are more precisely estimated. Had the 'yes' response on variable 3 been coded 0, the $\hat{\alpha}$s would have had the same signs on all variables and, in retrospect, this would have been more natural. It is clear from the goodness-of-fit test that the response patterns are satisfactorily explained by a single latent dimension. The component score may thus be thought of as an index which ranks employees according to the ease with which they can find secure employment. A person who ranks high will be someone who has been unemployed, had little choice of job in the past and no secure future in the present job. A person with the complementary attributes will be at the opposite end of the scale.

The foregoing examples involved only three manifest variables and one latent variable. This was sufficient to illustrate the main features of the model, but it falls short of what is often required in practice. Further examples with two latent variables and many more manifest variables can be found in Leung (1992). One of these relates to a survey of secondary school pupils in Hong Kong and uses 29 items on democratic belief. The responses were on a five-point scale: strongly agree, agree, marginally agree, disagree and strongly disagree. It is common practice to use equally spaced scores for such categories and then to analyse the data as though those scores were continuous variables. Leung treats them as unordered categories and uses the model

Table 5.3 Observed and expected frequencies and component scores for Leimu's (1983) employment data.

Response pattern	Frequency	Expected frequency	Component score
220	145	145.3	−1.78
120	54	57.4	−1.46
020	72	67.5	−1.17
210	33	33.6	−1.06
110	22	15.7	−0.74
221	17	16.6	−0.65
200	24	23.1	−0.62
010	14	21.6	−0.44
121	9	8.5	−0.33
100	9	11.9	−0.30
021	11	12.7	−0.04
000	21	17.9	0.00
211	7	6.9	0.07
111	6	4.1	0.39
201	6	6.6	0.51
011	7	7.1	0.69
101	2	4.4	0.83
001	10	8.2	1.13
Total	469	469.10	

$\chi^2 = 7.55$ on 6 degrees of freedom.

Table 5.4 Parameter estimates of the response function model for Leimu's (1983) employment data.

Variable (i)	Category (s)	$\hat{\alpha}_i(s)$	s.e.	$\hat{\pi}_{is}$	s.e.
1	0	0	—	.28	—
	1	−0.30	0.30	0.22	0.02
	2	−0.62	0.27	0.50	0.03
2	0	0	—	0.12	—
	1	−0.44	0.46	0.19	0.02
	2	−1.67	0.61	0.69	0.03
3	0	0	—	0.88	—
	1	1.13	0.56	0.12	0.04

of Section 5.2 to estimate the category scores. This allows the data to speak for themselves. The way is now open to use models which are designed for categorical data rather than, as in the past, to adapt continuous models purely because of the computational advantages which they offer.

Example 5.12.3 We now turn to an example based on a data set that has been frequently analysed with factor models. It is a US sample of the political action survey. The data was made available by the Zentralarchiv für Empirische Sozialforschung, University of Cologne (Barnes and Kaase 1979). The sample consists of 1554 respondents (after listwise deletion) who were asked their opinions on six variables with four response categories each ranging from 'agree strongly' to 'disagree strongly'. These six items have been used in many surveys as indicators of the theoretical construct *political efficacy*. The variables are:

1. People like me have no say in what the government does.
2. Voting is the only way that people like me can have any say about how the government runs things.
3. Sometimes politics and government seem so complicated that a person like me cannot really understand what is going on.
4. I don't think that public officials care much about what people like me think.
5. Generally speaking, those we elect to Parliament lose touch with the people pretty quickly.
6. Parties are only interested in people's votes but not in their opinions.

The six variables have been thoroughly analysed in Jöreskog and Moustaki (2001) using different factor analysis models for ordinal responses. Using different measures of fit and goodness of fit statistics, they found that the one-factor solution was not adequate, and we therefore show here the results for the two-factor model under three different model specifications – the proportional odds model (5.56), the probit/normal model obtained from (5.55) with a probit link (inverse of the distribution function of the normal) and the parameter estimates obtained when univariate and bivariate information is used (5.73). The first two methods are full maximum likelihood, while the third is a limited-information estimation method. LISREL (Jöreskog and Sörbom 2006) provides estimates from the maximisation of the sum of the univariate and bivariate likelihoods. Constraints are imposed to allow for comparisons among the three model solutions. We fix the second loading of the first observed variable to zero. Table 5.5 gives the estimated standardised parameters. The three methods show similar patterns. The loadings for the proportional odds model tend to be higher than the loadings obtained under the other two methods.

Example 5.12.4 We conclude this chapter with an example based on a survey of 301 university students (204 women and 97 men) on body satisfaction using seven items

Table 5.5 Standardised parameter estimates for the logit/normal, the probit/normal and the underlying bivariate model to political efficacy data.

	Logit/normal		Probit/normal		Underlying bivariate	
Item	$\hat{\alpha}_{i1}$	$\hat{\alpha}_{i2}$	$\hat{\alpha}_{i1}$	$\hat{\alpha}_{i2}$	$\hat{\lambda}_{i1}$	$\hat{\lambda}_{i2}$
1	0.88	0.00	0.75	0.00	0.72	0.00
2	0.70	−0.13	0.44	−0.03	0.46	−0.07
3	0.75	0.14	0.65	0.05	0.52	0.12
4	0.85	0.43	0.81	0.38	0.75	0.38
5	0.68	0.63	0.54	0.62	0.57	0.58
6	0.72	0.61	0.60	0.65	0.62	0.58

selected from the *Body Esteem Scale* (Franzoi and Shield 1984). Students were asked how satisfied they were with certain parts of their bodies – in particular, thighs, body build, buttocks, hips, legs, figure and weight. Responses were given on a five-point scale (1 = very dissatisfied; 2 = moderately dissatisfied; 3 = slightly satisfied; 4 = moderately satisfied; and 5 = very satisfied). The data are analysed again in Chapter 6 using a latent class model. The proportional odds model of (5.56) is fitted to the seven items. The analysis is mainly exploratory. The AIC and BIC model selection criteria favoured the two-factor solution (see Table 5.6). The upper triangle of Table 5.7 gives a standardised χ^2 local dependence (LD) index (Chen and Thissen 1997) for the one-factor solution, while the lower triangle gives the residuals from the two-factor solution. For each pair of variables a sum of χ^2 residuals across all response categories is calculated in order to produce the LD index. The standardised version of LD produced by the IRTPRO program (Cai *et al.* 2011) is $(\chi^2 - df)/\sqrt{2df}$, where $df = (r-1)(c-1)$ and r and c are the number of categories for the row and column variables respectively. There are still some positive dependencies that have not been well explained by the introduction of the second factor. Table 5.8 gives the unstandardised and standardised loadings. The two-factor solution suggests two groups of items. Thighs, buttocks, hips and legs seem to be highly correlated with factor 2, while body build, figure and weight are highly correlated with factor 1. Although one might have expected the seven items to form a one-dimensional scale this is not the case. Analysing both the females and males together without accounting for gender in the analysis is another possible reason for expecting complications with

Table 5.6 AIC and BIC for the body satisfaction data.

	One-factor	Two-factor
AIC	4734.70	4608.58
BIC	4864.45	4764.25

Table 5.7 Standardised χ^2 statistic for one-factor model (upper triangle) and two-factor model (lower triangle) for the body satisfaction data.

Item	1	2	3	4	5	6	7
1		6.2	4.6	6.2	2.0	0.4	1.4
2	2.4		4.2	8.7	6.7	13.9	4.1
3	5.2	4.7		6.6	3.8	3.0	1.1
4	2.6	7.0	6.6		9.3	2.5	3.5
5	1.6	4.9	3.8	9.5		1.4	1.0
6	−0.4	3.2	2.6	2.9	1.6		4.0
7	1.2	4.3	0.7	3.3	0.8	2.8	

Table 5.8 Unstandardised and standardised parameter estimates with standard errors in brackets for the proportional odds model for the body satisfaction data.

Item	$\hat{\alpha}_{i1}$	$\hat{\alpha}_{i2}$	$\hat{\lambda}_{i1}$	$\hat{\lambda}_{i2}$
1	2.60 (0.62)	5.34 (0.50)	0.42 (0.15)	0.86 (0.07)
2	3.52 (0.56)	1.40 (0.37)	0.85 (0.08)	0.34 (0.15)
3	1.62 (0.27)	2.22 (0.26)	0.50 (0.12)	0.69 (0.09)
4	2.19 (0.34)	2.90 (0.35)	0.55 (0.12)	0.72 (0.09)
5	1.83 (0.31)	2.54 (0.30)	0.51 (0.12)	0.71 (0.09)
6	6.91 (0.39)	2.44 (0.74)	0.92 (0.06)	0.32 (0.16)
7	1.75 (0.21)	1.14 (0.21)	0.65 (0.09)	0.42 (0.12)

the analysis. Gender could be included in the analysis as a covariate directly affecting the seven items together with the latent variables. In order to accommodate covariate effects we need to add the covariates in the right-hand side of (5.56). The latent class analysis performed in Chapter 6 indicates the need for five classes to explain the dependencies for the female sample only.

6

Latent class models

6.1 Introduction

In Chapter 1 we used a model with two latent classes to illustrate our general approach. In this chapter we shall extend that model in three principal directions by allowing:

(a) more than two latent classes;

(b) the manifest variables to be polytomous, ordered or unordered;

(c) the manifest variables to be metrical.

The first systematic treatment of latent class models was by Lazarsfeld and Henry (1968). They tended to emphasise the differences between factor analysis and latent structure analysis, remarking (p. 6) that 'Latent structure analysis is in some ways similar and in many very basic ways different from factor analysis'. This judgement, which perhaps had greater justification 40 years ago, may have inhibited the development of the common approach on which this book is based. Ten years later Goodman (1978, Part Four) gave many interesting developments, though the connections with other latent variable methods tend to be obscured by his choice of notation. Everitt (1984) gave a good elementary introduction to the topic and since then there has been a rapid growth in applications. The two volumes of conference proceedings (Langeheine and Rost 1988; Rost and Langeheine 1997) provide a focal point for much of this work. Heinen (1996) links latent class models with latent trait models and with log-linear models (see also Haberman (1979) and Formann (1992)).

It is worth pausing to consider the circumstances under which it might be substantively plausible to treat the latent space as categorical. We might distinguish those cases where there is prior evidence or theory which leads us to expect there to be latent categories, from those where it would be practically useful if such a representation proved to be possible.

Latent Variable Models and Factor Analysis: A Unified Approach, Third Edition.
David Bartholomew, Martin Knott and Irini Moustaki.

An example of the former is provided by criterion-referenced testing in education. Here the focus is on whether subjects have acquired the mastery of a concept or skill. If it were possible to design test items which infallibly discriminated between 'masters' and 'non-masters' there would be no problem. But if masters sometimes make mistakes, and non-masters sometimes get items correct by accident or guessing, it will not be possible to allocate an individual with certainty to either class on the basis of their responses. To meet this situation, Macready and Dayton (1977) and others have proposed that allocation be based on a latent class model. Here, theory predicts that there will be two classes, and it seems reasonable to suppose that members of any one class will have the same fixed probabilities of getting each item correct. An example of such a data set is given later in Table 6.3.

An example of the second kind is provided by a study of nearly 50 000 Scottish births carried out by Pickering and Forbes (1984). Eleven categorical variables were recorded for each baby: eight were binary, concerning for example whether the baby had convulsions or jaundice, and three, including birthweight, had more than two categories. Here, the aim was to identify classes of abnormal birth which, it was hoped, might show some geographical variation. If this proved to be the case it would then be possible to target resources more effectively. The expectation was that there would be one (large) normal class and an unknown number of abnormal classes. The analysis here was exploratory, in the sense that it was not known in advance whether classes existed or how many there might be. The 'mastery' model, on the other hand, involves a confirmatory approach, though if we failed to find two classes, we might go on to see if there were more.

6.2 The latent class model with binary manifest variables

The extension of the two-class model to a K-class model is almost immediate. If the model is used in exploratory mode, K will be a parameter to be determined, otherwise it will be given.

Let π_{ij} be the probability of a positive response on variable i for a person in category j $(i = 1, 2, \ldots, p;\ j = 0, 1, \ldots, K-1)$ and let η_j be the prior probability that a randomly chosen individual is in class j $(\sum_{j=0}^{K-1} \eta_j = 1)$. For the case of K classes (1.3) becomes

$$f(\mathbf{x}) = \sum_{j=0}^{K-1} \eta_j \prod_{i=1}^{p} \pi_{ij}^{x_i}(1-\pi_{ij})^{1-x_i}. \tag{6.1}$$

The posterior probability that an individual with response vector $\mathbf{x}$ belongs to category j is thus

$$h(j \mid \mathbf{x}) = \eta_j \prod_{i=1}^{p} \pi_{ij}^{x_i}(1-\pi_{ij})^{1-x_i} / f(\mathbf{x}) \quad (j = 0, 1, \ldots, K-1). \tag{6.2}$$

We can use (6.2) to construct an allocation rule according to which an individual is placed in the class for which the posterior probability is greatest. The principal statistical problem is thus the estimation of the parameters and testing goodness of fit. On the substantive side the main problem is to identify the latent classes – that is, to interpret them in terms which make practical sense. In this connection it is worth noting that our model includes the case where there are two or more cross-classified latent categorical variables. For example, if there were two binary cross-classified latent categorical variables, their joint distribution would involve four classes representing the possible combinations of two factors at two levels. If, in such a case, we wished to require the latent variables to be independent, this would impose a constraint on the ηs.

6.3 The latent class model for binary data as a latent trait model

The latent class model is a special case of the latent trait model in which the prior distribution consists of discrete probability masses. The conditional distribution of x_i given $\mathbf{y}$ is identical for both models. Given an individual's latent location, x_i is assumed to be a Bernoulli random variable. All of the general results which do not depend on the prior distribution will therefore apply equally to the latent trait and the latent class model. In particular, the form of the sufficient statistic will be a linear combination of the xs, as we shall show in Section 6.4 below. Since the two models differ only in the way they treat the prior distribution, and since the prior has relatively little influence on the fit of the model, especially when p is large, we may anticipate that it may not be easy to distinguish empirically between the two families. We can get further insight into this relationship by considering the correspondence in more detail, beginning with the case $K = 2$.

If we start with the logit/normal model with a single latent variable we have

$$\operatorname{logit} \pi_i(y) = \alpha_{i0} + \alpha_{i1} y \quad (i = 1, 2, \ldots, p), \tag{6.3}$$

where y has zero mean and unit standard deviation. Suppose that y takes the two values $\sqrt{(1-\eta)/\eta}$ and $-\sqrt{\eta/(1-\eta)}$ with probabilities η and $1-\eta$, respectively. It follows that $E(y) = 0$ and $\operatorname{var}(y) = 1$ as required. We then have a latent class model with

$$\operatorname{logit} \pi_{i0} = \alpha_{i0} - \alpha_{i1} \sqrt{\frac{\eta}{1-\eta}}. \tag{6.4}$$

and

$$\operatorname{logit} \pi_{i1} = \alpha_{i0} + \alpha_{i1} \sqrt{\frac{1-\eta}{\eta}}. \tag{6.5}$$

This differs from the two-class model defined above only in the way we designate the classes and in the parameterisation. For class 1, corresponding to $y = \sqrt{(1-\eta)/\eta}$,

$$\pi_{i1} = \frac{1}{1 + \exp\left\{-\alpha_{i0} - \alpha_{i1}\sqrt{\frac{\eta}{1-\eta}}\right\}},$$

and for class 0,

$$\pi_{i0} = \frac{1}{1 + \exp\left\{-\alpha_{i0} + \alpha_{i1}\sqrt{\frac{1-\eta}{\eta}}\right\}}.$$

These equations relate the class-specific response probabilities of the latent class model to the intercept and slope parameters of the latent trait model. If, for example, we had estimated π_{i0}, π_{i1} and η using the theory for the latent class model, we could express them as latent trait parameters. They will not, of course, be equal to the estimates of α_{i0} and α_{i1} obtained from IRTPRO (Cai *et al.* 2011) because that program assumes a normal prior distribution. How close they turn out to be will give some indication of how sensitive the estimates are to the choice of prior distribution.

We have used this method for the Law School Admission Test data given in Table 4.2. The results are set out in Table 6.1.

The two sets of estimates are remarkably close, especially in the case of the α_{i0}. The first set of estimates uses the normal prior and the second the two-point distribution appropriate for the latent class model. We can interpret this result either by saying that the prior distribution has little effect or that it is difficult to distinguish empirically between the latent trait and latent class models. The reader can confirm the latter interpretation by using the Latent GOLD program (Vermunt and Magidson 2000) to fit a two-class model to the Law School Admission Test data. The predicted

Table 6.1 Parameter estimates for the two-class model fitted to the Law School Admission Test VI data together with a comparison of the estimates of $\{\alpha_{i0}\}$ and $\{\alpha_{i1}\}$ obtained from (6.4) and (6.6) and from the logit/normal model (from Table 4.1).

			Logit/normal estimates		Latent class estimates	
i	$\hat{\pi}_{i0}$	$\hat{\pi}_{i1}$	$\hat{\alpha}_{i0}$	$\hat{\alpha}_{i1}$	$\hat{\alpha}_{i0}$	$\hat{\alpha}_{i1}$
1	0.852	0.965	2.77	0.83	2.75	0.74
2	0.528	0.812	0.99	0.72	0.97	0.64
3	0.301	0.693	0.25	0.89	0.22	0.79
4	0.611	0.849	1.28	0.69	1.27	0.61
5	0.776	0.924	2.05	0.66	2.03	0.59

$\hat{\eta} = 0.638$

frequencies are as close to the observed, as in the case of the latent trait model. The $\chi^2 = 19.42$ and the log-likelihood ratio $G^2 = 22.76$ on 20 degrees of freedom, with $P = 0.49$ and 0.30 respectively, indicate a good fit.

There is a second way in which a two-class model can be expressed as a latent trait model. In this case we retain the standard normal prior but choose the item response function to be

$$\pi_i(y) = \begin{cases} \pi_{i0} & \text{if } y \leq y_0, \\ \pi_{i1} & \text{if } y > y_0, \end{cases}$$

where $\pi_{i1} > \pi_{i0}$ $(i = 1, 2, \ldots, p)$. All individuals with $y \leq y_0$ will thus have the same response probability and the proportion of the population in this class will be $\Phi(y_0) = \eta$, say. The difference between the latent class model formulated in this way and the logit/normal model is in the form of the *response function*. This example illustrates very clearly the intimate relationship between the prior and the response function. We have two substantively different hypotheses which give rise to the same model. In one the prior is normal with a step-function as its item response curve. In the other the prior is a two-point distribution with an arbitrary response function constrained only by the values it takes at two values of y.

The step-function approach can easily be extended to models with more than two latent classes by increasing the number of segments into which the latent scale is divided.

6.4 *K* latent classes within the GLLVM

We now return to the representation in (6.4) and (6.5) of the two-class model as a latent trait model and extend it to the general case of K classes. It has to be shown that the model as defined in Section 6.2 can be written in the form (4.3) for some q. This was easy to do when $K = 2$ because we could immediately see how to construct a single binary variable with zero mean and unit standard deviation so that the response probabilities coincided. In the general case we proceed as follows. Let $\mathbf{y}$ be an indicator vector showing into which class an individual falls. Thus, for $j = 0, \ldots, K-1$, $y_j = 1$ if the individual is in class j and zero otherwise. It follows that $\sum y_j = 1$. So far as the formal treatment is concerned, we can treat $\mathbf{y}$ just like any vector of latent variables because the general theory places no restrictions whatsoever on the form of their distribution. This would lead to a GLLVM of the form

$$\text{logit}\,\pi_i(\mathbf{y}) = \sum_{j=0}^{K-1} \pi_{ij} y_j \quad (i = 1, 2, \ldots p). \tag{6.6}$$

The prior distribution $h(\mathbf{y})$ is highly degenerate. For $j = 0, \ldots, K-1$ it puts probability η_j at the point where $y_j = 1$ and all other ys are 0.

This form defines a parameterisation of the latent class model which is essentially the log-linear representation discussed by Haberman (1979). A discussion of various parameterisations of this kind is given by Heinen (1996, Chapter 2).

6.5 Maximum likelihood estimation

The log-likelihood function from (6.1) is complicated, but it can be maximised using standard optimisation routines. This might be done as in McHugh (1956, 1958) using the standard Newton–Raphson technique. However, as with many other latent variable models, an easier method which enables larger problems to be tackled is offered by the E-M algorithm. The standard reference is Dempster *et al.* (1977), supplemented by Wu (1983), but a version suitable for the latent class model was also given by Goodman (1978). Here we shall develop the method from first principles in a manner which gives some insight into the nature of the estimators.

From (6.1) we find the log-likelihood for a random sample of size n to be

$$L = \sum_{h=1}^{n} \ln \left\{ \sum_{j=0}^{K-1} \eta_j \prod_{i=1}^{p} \pi_{ij}^{x_{ih}} (1 - \pi_{ij})^{1-x_{ih}} \right\}. \tag{6.7}$$

This has to be maximised subject to $\sum \eta_j = 1$, so we find the unrestrained maximum of

$$\phi = L + \theta \sum_{j=0}^{K-1} \eta_j,$$

where θ is an undetermined multiplier. Finding partial derivatives, we have

$$\begin{aligned} \frac{\partial \phi}{\partial \eta_j} &= \sum_{h=1}^{n} \left\{ \prod_{i=1}^{p} \pi_{ij}^{x_{ih}} (1 - \pi_{ij})^{1-x_{ih}} / f(\mathbf{x}_h) \right\} + \theta \quad (j = 0, 1, \ldots, K-1) \\ &= \sum_{h=1}^{n} \{ g(\mathbf{x}_h \mid j) / f(\mathbf{x}_h) \} + \theta, \end{aligned} \tag{6.8}$$

where $g(\mathbf{x}_h \mid j)$ is the joint probability of $\mathbf{x}_h$ for an individual in class j.

Also,

$$\frac{\partial \phi}{\partial \pi_{ij}} = \sum_{h=1}^{n} \eta_j \frac{\partial}{\partial \pi_{ij}} g(\mathbf{x}_h \mid j) / f(\mathbf{x}_h) \quad (i = 1, 2, \ldots, p; j = 0, 1, \ldots, K-1).$$

Now

$$\frac{\partial g(\mathbf{x}_h \mid j)}{\partial \pi_{ij}} = \frac{\partial}{\partial \pi_{ij}} \exp \sum_{i=1}^{p} \{x_{ih} \ln \pi_{ij} + (1 - x_{ih}) \ln(1 - \pi_{ij})\}$$
$$= g(\mathbf{x}_h \mid j) \left\{ \frac{x_{ih}}{\pi_{ij}} - \frac{(1 - x_{ih})}{(1 - \pi_{ij})} \right\}$$
$$= (x_{ih} - \pi_{ij}) g(\mathbf{x}_h \mid j) / \pi_{ij}(1 - \pi_{ij}). \tag{6.9}$$

Therefore

$$\frac{\partial \phi}{\partial \pi_{ij}} = \{\eta_j / \pi_{ij}(1 - \pi_{ij})\} \sum_{h=1}^{n} (x_{ih} - \pi_{ij}) g(\mathbf{x}_h \mid j) / f(\mathbf{x}_h). \tag{6.10}$$

The resulting equations can be simplified by expressing (6.8) and (6.10) in terms of the posterior probabilities $\{h(j \mid \mathbf{x})\}$. By Bayes' theorem,

$$h(j \mid \mathbf{x}_h) = \eta_j g(\mathbf{x}_h \mid j) / f(\mathbf{x}_h). \tag{6.11}$$

Substituting in (6.8) and setting equal to zero, we find

$$\sum_{h=1}^{n} h(j \mid \mathbf{x}_h) = -\theta \eta_j.$$

Summing both sides over j and using $\sum \eta_j = 1$ gives $\theta = -n$, and hence the first estimating equation is

$$\hat{\eta}_j = \sum_{h=1}^{n} h(j \mid \mathbf{x}_h) / n \quad (j = 0, 1, \ldots, K - 1). \tag{6.12}$$

The second is

$$\sum_{h=1}^{n} (x_{ih} - \pi_{ij}) h(j \mid \mathbf{x}_h) / \pi_{ij}(1 - \pi_{ij}) = 0,$$

whence

$$\hat{\pi}_{ij} = \frac{\sum_{h=1}^{n} x_{ih} h(j \mid \mathbf{x}_h)}{\sum_{h=1}^{n} h(j \mid \mathbf{x}_h)}$$
$$= \sum_{h=1}^{n} x_{ih} h(j \mid \mathbf{x}_h) / n \hat{\eta}_j \quad (i = 1, 2, \ldots, p;\ j = 0, 1, \ldots, K - 1). \tag{6.13}$$

Although these equations have a simple form it must be remembered that $h(j \mid \mathbf{x}_h)$ is a complicated function of $\{\eta_j\}$ and $\{\pi_{ij}\}$ given by

$$h(j \mid \mathbf{x}_h) = \frac{\eta_j \prod_{i=1}^{p} \pi_{ij}^{x_{ih}}(1-\pi_{ij})^{1-x_{ih}}}{\sum_{k=0}^{K-1} \eta_j \prod_{i=1}^{p} \pi_{ik}^{x_{ih}}(1-\pi_{ik})^{1-x_{ih}}}. \tag{6.14}$$

However, if $h(j \mid \mathbf{x}_h)$ were known it would be easy to solve (6.12) and (6.13) for $\{\eta_j\}$ and $\{\pi_{ij}\}$. The E-M algorithm takes advantage of this fact, proceeding in a 'zig-zag' fashion as follows:

1. Choose an initial set of posterior probabilities $\{h(j \mid \mathbf{x}_h)\}$.
2. Use (6.12) and (6.13) to obtain a first approximation to $\{\hat{\eta}_j\}$ and $\{\hat{\pi}_{ij}\}$.
3. Substitute these estimates into (6.14) to obtain improved estimates of $\{h(j \mid \mathbf{x}_h)\}$.
4. Return to step 2 to obtain second approximations to the parameters and continue the cycle until convergence is attained.

The solution reached will be a local maximum (or saddle point). It is known that models of this kind may have multiple maxima and the risk of this appears to increase as K, the number of classes, increases and to decrease with increasing sample size. Aitkin *et al.* (1981) provided an illustration of multiple maxima arising with only three latent classes. By using different starting values one can guard against the risk of mistaking a local for a global maximum, but if such multiple maxima do occur it is not clear what interpretation should be placed on the different sets of latent classes implied by the various local maxima.

A reasonable way of starting the iteration is to allocate individuals, arbitrarily, to latent classes on the basis of their total score $\left(\sum_{i=1}^{p} x_i\right)$ – that is, to take $h(j \mid \mathbf{x}_h) = 1$ if $\mathbf{x}_h$ is allocated to class j and $h(j \mid \mathbf{x}_h) = 0$ otherwise. Although the method may take a very large number of iterations to converge, the steps are simple and fast, so the total computing time is unlikely to be excessive. As well as providing parameter estimates, the method provides the posterior probabilities that each individual belongs to a given latent class. It does not provide the second derivatives needed for the calculation of standard errors, but these can easily be found and evaluated at the solution point, as we now see. The program Latent GOLD (Vermunt and Magidson 2000) implements the theory given above.

6.6 Standard errors

Since the ηs are constrained to sum to one, we eliminate η_0 using $\eta_0 = 1 - \sum_{j=1}^{K-1} \eta_j$, and then second derivatives and cross-derivatives of L can be expressed in terms of the posterior distribution as follows:

$$\frac{\partial^2 L}{\partial \eta_j \partial \eta_k} = -\sum_{h=1}^{n} \{h(j \mid \mathbf{x}_h) - h(0 \mid \mathbf{x}_h)\}\{h(k \mid \mathbf{x}_h) - h(0 \mid \mathbf{x}_h)\}/\eta_j \eta_k, \tag{6.15}$$

$$\frac{\partial^2 L}{\partial \pi_{ij} \partial \pi_{lk}} = \sum_{h=1}^{n} (x_{ih} - \pi_{ij})(x_{lh} - \pi_{lk}) h(j \mid \mathbf{x}_h) \times \{\delta_{jk}(1 - \delta_{il}) - h(k \mid \mathbf{x}_h)\}/\pi_{ij}(1 - \pi_{ij})\pi_{lk}(1 - \pi_{lk}), \tag{6.16}$$

where

$$\delta_{rs} = \begin{cases} 1 & \text{if } r = s, \\ 0 & \text{otherwise,} \end{cases}$$

$$\frac{\partial^2 L}{\partial \eta_j \partial \pi_{ik}} = \sum_{h=1}^{n} (x_{ih} - \pi_{ij}) h(j \mid \mathbf{x}_h)\{\delta_{jk} - h(k \mid \mathbf{x}_h)\} \quad (j, k = 0, 1, 2, \ldots, K - 1;\ i, l = 1, 2, \ldots, p). \tag{6.17}$$

The asymptotic variance–covariance matrix of the estimates is then the inverse of the expectation of the $(K(p + 1) - 1) \times (K(p + 1) - 1)$ matrix of the negatives of the derivatives set out above. The exact computation of the expected values involves summation over the 2^p possible score patterns $\mathbf{x}$. This is feasible if p is small, but with large p the number of terms becomes extremely large and the magnitude of each term so small that accurate calculation becomes impossible. In this case the expectation can be approximated by taking the inverse of the observed second derivative matrix.

It was shown by de Menezes (1999) that the asymptotic standard errors may be very poor approximations for sample sizes which are commonly encountered. This arises especially, but not exclusively, when the parameter estimates are close to the boundary values of 0 and 1. For this reason commercial software for latent class analysis usually provides the facility for estimating standard errors by the parametric bootstrap method. An example is given in Table 6.2 for the Law School Admission Test data where the asymptotic standard errors are compared with estimates based on 5000 bootstrap samples.

The asymptotic standard error of $\hat{\eta}$ is 0.141 and the bootstrap estimate is 0.129. In this example the asymptotic values underestimate the true values for most of the parameters as estimated by the bootstrap. This discrepancy may be partly due to the skewness of the sampling distribution which gives rise to a few extreme estimates but, as de Menezes showed, the position is far from clear and caution needs to be exercised.

Table 6.2 Comparison of asymptotic standard errors and bootstrap estimates for the Law School Admission Test data (5000 replications).

	$\hat{\pi}_{i0}$		$\hat{\pi}_{i1}$	
i	Asymptotic	Bootstrap	Asymptotic	Bootstrap
1	0.039	0.043	0.015	0.016
2	0.068	0.084	0.043	0.041
3	0.092	0.093	0.053	0.058
4	0.063	0.071	0.035	0.035
5	0.044	0.051	0.024	0.023

6.7 Posterior analysis of the latent class model with binary manifest variables

Posterior analysis in the case of a latent class model is concerned with what we can say about an individual's class membership after their response pattern has been observed. All the information relevant to this question is contained in the posterior probability distribution $h(j \mid \mathbf{x})$. The link with the latent trait model is made more explicit by noting that

$$E(\mathbf{y} \mid \mathbf{x}) = (h(0 \mid \mathbf{x}), h(1 \mid \mathbf{x}), \ldots, h(K-1 \mid \mathbf{x})).$$

In practice the class to which an individual is most likely to belong could be discovered by inspection of this set of conditional posterior probabilities. Equivalently we can do it by reference to the components. Thus, class j is more likely than class k if $h(j \mid \mathbf{x})/h(k \mid \mathbf{x}) > 1$, which is true if

$$(\eta_j/\eta_k) \exp \sum_{i=1}^{p} [\{x_i \ln \pi_{ij} + (1-x_i) \ln(1-\pi_{ij})\} \\ - \{x_i \ln \pi_{ik} + (1-x_i) \ln(1-\pi_{ik})\}] > 1.$$

This inequality is equivalent to

$$\sum_{i=1}^{p} x_i \operatorname{logit} \pi_{ij} + \sum_{i=1}^{p} \ln(1-\pi_{ij}) + \ln \eta_j > \sum_{i=1}^{p} x_i \operatorname{logit} \pi_{ik} \\ + \sum_{i=1}^{p} \ln(1-\pi_{ik}) + \ln \eta_k. \tag{6.18}$$

When $K = 2$ this reduces to the inequality of (1.5). The first term on each side of the inequality is simply a component, but Theorem 2.15.1 is not applicable to comparisons across latent variables.

6.8 Goodness of fit

There is very little to be added to what was said in Section 4.9 for the latent trait model. All of the methods described there apply equally to the latent class model. In fact some of the relevant research quoted there was directed towards the latent class model. The calculation of the number of degrees of freedom for a model with K classes is

$$\begin{aligned} df &= 2^p - 1 - pK(\text{response probabilities}) - (K - 1) \\ &= 2^p - K(p + 1). \end{aligned}$$

6.9 Examples for binary data

Example 6.9.1 Having fitted a model we will normally wish to consider what can be learned from the classification at which we have arrived. In the case of Macready and Dayton's mastery model we would expect to find two classes with one having the characteristics we would expect of 'masters' and the other of 'non-masters'. The results of the analysis for the example from Macready and Dayton (1977) are given in Tables 6.3 and 6.4.

It is immediately clear from Table 6.3 that the fit is extremely good. The $\chi^2 = 9.79$ and the $G^2 = 9.08$ on 6 degrees of freedom, with $P = 0.13$ and 0.17 respectively, also indicate a very good fit. The posterior analysis shown in the last column reveals an interesting distribution of probabilities. For many response patterns, for example 1111, 1011 and 0110, we would be extremely confident in allocating individuals to the 'master' category. Similarly, we would have little hesitation in classifying someone responding 0000 as a non-master, and scarcely more if it was 1000. There are several response patterns, however, which are ambiguous. These are 0100, 0010 and 0001, where the posterior probability is close to a $\frac{1}{2}$. There are only 11 individuals in this situation out of 142, so the method has been successful in allocating the great majority of individuals to their respective groups. A little further light can be shed on these conclusions by inspecting the parameter estimates in Table 6.4.

Members of class 0 have a small estimated probability of responding correctly to any item; these are clearly the 'non-masters'. Those in class 1 have, in each case, a much higher probability of answering correctly; these are the 'masters'. The standard errors are quite large, but not sufficiently so to put this interpretation at risk. The uncertainty about those who get only one item correct arises from the need to balance the very small probability of a non-master getting that item correct, against the high probability that a non-master would get the other three wrong. In other words, such a response pattern shows a somewhat inconsistent performance, and it is therefore

Table 6.3 Macready and Dayton's (1977) data with posterior probabilities of belonging to the mastery state.

Response pattern	Frequency	Expected frequency	Pr{master | **x**}
1111	15	15.0	1.00
1110	7	6.1	1.00
1101	23	19.7	1.00
1100	7	8.9	0.90
1011	1	4.1	1.00
1010	3	1.9	0.85
1001	6	6.1	0.87
1000	13	13.1	0.16
0111	4	4.8	1.00
0110	2	2.0	0.95
0101	5	6.5	0.96
0100	6	5.9	0.43
0011	4	1.4	0.95
0010	1	1.6	0.33
0001	4	4.4	0.38
0000	41	41.3	0.02
	142	142	

reassuring that there are so few responses of that kind. The response pattern 1000 is less surprising because, with a probability of 0.22, it is much more likely that a non-master will get item 1 correct.

Example 6.9.2 For our second example we turn to a case, discussed in de Menezes and Bartholomew (1996), where there was no prior theory to suggest whether there might be any latent classes at all or, if there were, how many. Fuller details are given in the above paper, including a discussion of how the data were reduced to binary form. The data are taken from the British Social Attitudes survey 1990 and relate to

Table 6.4 Parameter estimates for the fit of Macready and Dayton's (1977) model to the data of Table 6.3 (standard errors in brackets).

j	$\hat{\pi}_{1j}$	$\hat{\pi}_{2j}$	$\hat{\pi}_{3j}$	$\hat{\pi}_{4j}$	$\hat{\eta}$
0	0.22	0.08	0.03	0.06	0.43
	(0.07)	(0.06)	(0.04)	(0.05)	(0.06)
1	0.76	0.79	0.43	0.71	0.57
	(0.06)	(0.06)	(0.06)	(0.06)	(0.06)

1077 respondents who provided views on the following ten issues relating to sexual attitudes:

1. Should divorce be easier here?
2. Do you support the law against sexual discrimination?
3. View on pre-marital sex: not at all wrong, . . . , always wrong.
4. View on extra-marital sex: not at all wrong, . . . , always wrong.
5. View on sexual relationship between individuals of the same sex: not at all wrong, . . . , always wrong.
6. Should gays teach in schools?
7. Should gays teach in higher education?
8. Should gays hold public positions?
9. Should a female homosexual couple be allowed to adopt children?
10. Should a male homosexual couple be allowed to adopt children?

The parameter estimates are given in Table 6.5 for a two-, three- and four-class model.

The fits in the first two cases were poor, but they are included to show how the latent class structure evolves as further classes are added. The two-class model produces a pattern not dissimilar in some respects to that in the 'mastery' model. Members of the second class have a higher probability of responding positively to every item than those of the first class. This suggests a permissive/non-permissive dichotomy in the population into groups of roughly equal size.

Table 6.5 Parameter estimates $\hat{\pi}_{ij}$ for the British Social Attitudes 1990 survey.

Item	Two classes		Three classes			Four classes			
1	0.14	0.18	0.13	0.09	0.21	0.14	0.07	0.10	0.21
2	0.76	0.88	0.76	0.87	0.92	0.77	0.60	0.87	0.93
3	0.64	0.88	0.64	0.86	0.96	0.63	0.87	0.87	0.96
4	0.09	0.17	0.09	0.13	0.30	0.08	0.27	0.13	0.31
5	0.08	0.49	0.08	0.38	0.82	0.07	0.59	0.38	0.83
6	0.01	0.92	0.01	0.87	0.97	0.01	0.00	0.87	1.00
7	0.08	0.99	0.06	0.98	1.00	0.06	0.18	0.98	1.00
8	0.23	0.94	0.21	0.91	0.99	0.21	0.30	0.91	1.00
9	0.07	0.30	0.07	0.10	0.98	0.04	0.99	0.10	0.98
10	0.02	0.19	0.03	0.00	0.83	0.00	0.91	0.00	0.84
$\hat{\eta}_j$	0.49	0.51	0.47	0.41	0.11	0.46	0.02	0.42	0.10

Table 6.6 Two-class model fitted to items 5–10 in the British Social Attitudes 1990 survey.

Item	$\hat{\pi}_{i1}$	$\hat{\pi}_{i2}$
5	0.08	0.49
6	0.01	0.92
7	0.09	0.99
8	0.23	0.94
9	0.07	0.30
10	0.02	0.19
$\hat{\eta}_j$	0.49	0.51

When we add a third class, the new first class is almost the same in size and response probabilities as before, but the second has split into two parts. These differ most strikingly on the last two items concerning adoption. The third group, comprising 11% of the population, are permissive on all issues. The middle group are only marginally less permissive on the first eight items, but are clearly distinguished by their negative views on adoption.

The final and best-fitting model has four latent classes. The non-permissive class remains almost unchanged, but more subtle differences emerge among the permissives. The third class is almost identical to the second class of the previous model; its members are permissive on everything except adoption. The new element is a very small group, comprising 1% of the population, whose distinguishing feature is very positive support for homosexual adoption, combined with a strong negative attitude to items 6, 7 and 8, concerning teaching and holding public positions.

This provides a coherent and plausible interpretation of the situation but the increasing uncertainty of the parameter estimates as the number of classes increases should be borne in mind.

In a further attempt to confirm these conclusions, de Menezes and Bartholomew (1996), repeated the analysis using items 5–10 only: the results are given in Table 6.6. These appeared to be the items on which views were most sharply divided. This produced a result which was identical to the relevant part of the two-class analysis for the full set of items, though the fit, in this case, was much better.

6.10 Latent class models with unordered polytomous manifest variables

The extension of the foregoing theory to polytomous data is straightforward, though applications have been less common. Some theory of estimation was given by Goodman (1978), and we shall describe applications by Pickering and Forbes (1984) and Clogg (1979), the latter reanalysed by Masters (1985).

When there are more than two categories the indicator variable x_i used previously is replaced by a vector $\mathbf{x}_i$ with c_i elements $x_i(s)$ defined by

$$x_i(s) = \begin{cases} 1 & \text{if the response is in category } s \text{ of variable } i, \\ 0 & \text{otherwise } (s = 0, 1, \ldots, c_i - 1). \end{cases}$$

Note that $\sum_s x_i(s) = 1$. The complete response vector for an individual is then written

$$\mathbf{x}' = (\mathbf{x}_1', \mathbf{x}_2', \ldots, \mathbf{x}_p').$$

The conditional response probabilities are defined by

$$\pi_{ij}(s) = P\{\text{Response of an individual in class } j \text{ is in category } s \text{ on variable } i\}.$$

The joint probability function of $\mathbf{x}$ is then

$$f(\mathbf{x}) = \sum_{j=0}^{K-1} \eta_j \prod_{i=1}^{p} \prod_{s=0}^{c_i-1} \{\pi_{ij}(s)\}^{x_i(s)}. \tag{6.19}$$

The posterior distribution is

$$h(j \mid \mathbf{x}) = \eta_j \prod_{i=1}^{p} \prod_{s=0}^{c_i-1} \{\pi_{ij}(s)\}^{x_i(s)} / f(\mathbf{x}). \tag{6.20}$$

The model for polytomous manifest variables can be reparameterised in the log-linear form in exactly the same way as the model for binary data. In place of the logit relationship given in (6.6) we need the multivariate logits given in (5.10) and (5.15). On the right-hand sides of the latter equations we replace the latent vector $\mathbf{y}$ by the indicator vector $\mathbf{z}$ showing into which class the individual falls.

6.11 Latent class models with ordered polytomous manifest variables

We have seen that a latent class model is, essentially, a latent variable model whose prior distribution consists of probability masses on a finite set of points. From this it follows that methods for dealing with ordered categories already discussed in Chapter 5 for that model can be used here also. As already defined in Section 6.10, $\pi_{ij}(s)$ is the conditional probability of a response of an individual in category s on variable i given the latent class j. To preserve the ordinality property of the variable x_i, we model the cumulative probability of responding in category s or above on variable i for a member of class j,

$$\Pi_{ij}(s) = \pi_{ij}(s) + \pi_{ij}(s+1) + \cdots + \pi_{ij}(c_i - 1) \quad (i = 1, \ldots, p; j = 1, \ldots, K).$$

One would then estimate the $\Pi_{ij}(s)$ from which the $\pi_{ij}(s)$ could be obtained by differencing. The joint probability function of $\mathbf{x}$ is then

$$f(\mathbf{x}) = \sum_{j=0}^{K-1} \eta_j \prod_{s=0}^{c_i-1} \pi_{ij}(s)^{x_i(s)} = \sum_{j=0}^{K-1} \eta_j \prod_{s=0}^{c_i-1} \{\Pi_{ij}(s) - \Pi_{ij}(s+1)\}^{x_i(s)}, \qquad (6.21)$$

where $x_i(s) = 1$ if a randomly selected object belongs into category s of the ith variable and $x_i(s) = 0$ otherwise.

6.12 Maximum likelihood estimation

As with binary data, the log-likelihood may be written

$$L = \sum_{h=1}^{n} \ln f(\mathbf{x}_h),$$

but the maximisation must now be effected under two sets of constraints. The original constraint $\sum \eta_j = 1$ remains, but we must now impose

$$\sum_{s=0}^{c_i-1} \pi_{ij}(s) = 1 \quad (i = 1, 2, \ldots, p). \qquad (6.22)$$

This did not arise in the binary case because we used (6.22) to eliminate one of the two probabilities for each dimension. The function now to be maximised is thus

$$\phi = L + \theta \sum_{j=0}^{K-1} \eta_j + \sum_{j=0}^{K-1} \sum_{i=1}^{p} \beta_{ij} \sum_{s=0}^{c_i-1} \pi_{ij}(s), \qquad (6.23)$$

where θ and $\{\beta_{ij}\}$ are undetermined multipliers. The partial derivatives with respect to $\{\eta_j\}$ are

$$\frac{\partial \phi}{\partial \eta_j} = \sum_{h=1}^{n} g(\mathbf{x}_h \mid j)/f(\mathbf{x}_h) + \theta \qquad (6.24)$$

leading, as before, to

$$\hat{\eta}_j = \frac{1}{n} \sum_{h=1}^{n} h(j \mid \mathbf{x}_h). \qquad (6.25)$$

Similarly,

$$\frac{\partial L}{\partial \pi_{ij}(s)} = \sum_{h=1}^{n} \eta_j \frac{\partial g(\mathbf{x}_h \mid j)}{\partial \pi_{ij}(s)} \bigg/ f(\mathbf{x}_h).$$

Now

$$\begin{aligned}\frac{\partial g(\mathbf{x}_h \mid j)}{\partial \pi_{ij}(s)} &= \frac{\partial}{\partial \pi_{ij}(s)} \exp \sum_{i=1}^{p} \sum_{s=0}^{c_i-1} x_{ih}(s) \ln \pi_{ij}(s) \\ &= g(\mathbf{x}_h \mid j) x_{ih}(s)/\pi_{ij}(s), \end{aligned} \tag{6.26}$$

giving

$$\begin{aligned}\frac{\partial \phi}{\partial \pi_{ij}(s)} &= \eta_j \sum_{h=1}^{n} \frac{g(\mathbf{x}_h \mid j)}{f(\mathbf{x}_h)} \frac{x_{ih}(s)}{\pi_{ij}(s)} + \beta_{ij} \\ &= \sum_{h=1}^{n} h(j \mid \mathbf{x}_h) x_{ih}(s)/\pi_{ij}(s) + \beta_{ij}. \end{aligned} \tag{6.27}$$

Setting the right-hand side equal to zero yields

$$0 = \sum_{h=1}^{n} h(j \mid \mathbf{x}_h) x_{ih}(s) + \pi_{ij}(s) \beta_{ij}. \tag{6.28}$$

Summing both sides over s,

$$0 = \sum_{h=1}^{n} h(j \mid \mathbf{x}_h) + \beta_{ij} \quad \text{or} \quad \beta_{ij} = -\sum_{h=1}^{n} h(j \mid \mathbf{x}_h).$$

Substitution into (6.28) finally gives the estimating equations

$$\begin{aligned}\hat{\pi}_{ij}(s) &= \frac{\sum_{h=1}^{n} h(j \mid \mathbf{x}_h) x_{ih}(s)}{\sum_{h=1}^{n} h(j \mid \mathbf{x}_h)} \\ &= \frac{\sum_{h=1}^{n} h(j \mid \mathbf{x}_h) x_{ih}(s)}{n \hat{\eta}_j} \quad (i = 1, 2, \ldots, p; j = 0, 1, \ldots, K-1; s = 0, \ldots, c_i - 1). \end{aligned} \tag{6.29}$$

In the ordered polytomous case (6.29) can be also written

$$\hat{\Pi}_{ij(s)} = \hat{\Pi}_{ij(s+1)} + \frac{\sum_{h=1}^{n} x_{ih(s)} h(j \mid \mathbf{x}_h)}{n \hat{\eta}_j}$$
$$(i = 1, 2, \ldots, p; j = 0, 1, \ldots, K-1; s = 0, \ldots, c_i - 1). \tag{6.30}$$

Equations (6.25) and (6.29) or (6.30) may then be solved as in the binary E-M algorithm. We first choose starting values for $\{\hat{h}(j \mid \mathbf{x}_h)\}$ and then obtain first approximations to $\{\hat{\eta}_j\}$ and $\{\hat{\pi}_{ij}(s)\}$. These are then used to improve the estimates of $\{\hat{h}(j \mid \mathbf{x}_h)\}$ and so on.

6.12.1 Allocation of individuals to latent classes

Reference to (6.20) shows that

$$h(j \mid \mathbf{x}_h)/h(k \mid \mathbf{x}_h) = (\eta_j/\eta_k) \exp \sum_{i=1}^{p} \sum_{s=0}^{c_i-1} x_i(s) \ln \pi_{ij}(s)/\pi_{ik}(s).$$

Discrimination is thus based on comparing the linear functions

$$X_j = \sum_{i=1}^{p} \sum_{s=0}^{c_i-1} x_i(s) \ln \pi_{ij}(s) \quad (j = 0, 1, \ldots, K-1).$$

Thus class j is preferred to class k if

$$X_j - X_k > \ln(\eta_k/\eta_j). \tag{6.31}$$

This implies that we allocate to the class for which

$$X_j + \ln \eta_j$$

is greatest.

6.13 Examples for unordered polytomous data

Example 6.13.1 We return to the classification of Scottish infants investigated by Pickering and Forbes (1984). This involved an unusually large sample size (45 426 usable cases relating to the year 1980) which permitted some interesting additions to the usual analysis. There were 11 variables, of which 8 were binary; they are listed in Table 6.7, which also gives the parameter estimates obtained using the E-M algorithm.

It is only necessary to give the estimates for $c_i - 1$ categories of each variable because the response probabilities for each variable sum to 1. Although the value of χ^2 continues to decrease as the number of classes fitted increases, multiple maxima began to appear when four classes were fitted. The decision to present results up to the four-class model was based on the interpretability of the results rather than on statistical criteria. However, with such a large sample it was possible to test the stability of the results by dividing the sample in half at random and refitting the model to each sub-sample. The authors showed that the parameter estimates obtained

Table 6.7 Parameter estimates ($\hat{\pi}$ s) for the Pickering and Forbes (1984) data.

Variable	c_i	Levels	2 Classes		3 Classes			4 Classes			
Birthweight	4	2001–2500 g	0.01	0.48	0.00	0.18	0.79	0.00	0.20	0.78	0.03
		1501–2000 g	0.00	0.15	0.00	0.26	0.09	0.00	0.32	0.09	0.00
		≤1500 g	0.00	0.08	0.00	0.21	0.01	0.00	0.25	0.01	0.00
Birthweight for Gestation age	2	< 10 th percentile	0.07	0.43	0.07	0.19	0.62	0.07	0.21	0.62	0.10
Apgar at 5 min	2	< 7	0.01	0.12	0.01	0.26	0.01	0.00	0.21	0.01	0.32
Resuscitation	3	Intermediate	0.08	0.19	0.08	0.26	0.13	0.07	0.25	0.13	0.31
		By intubation	0.02	0.17	0.02	0.33	0.04	0.01	0.29	0.03	0.52
Assisted ventilation after 30 mins	2	Present	0.00	0.10	0.00	0.25	0.00	0.00	0.29	0.00	0.01
Recurrent apnoea	2	present	0.00	0.06	0.00	0.17	0.00	0.00	0.20	0.00	0.00
Jaundice	2	Present	0.28	0.58	0.28	0.67	0.49	0.28	0.71	0.49	0.32
Convulsions	2	Present	0.00	0.03	0.00	0.07	0.00	0.00	0.07	0.00	0.01
In tube feeding	2	Present	0.01	0.30	0.01	0.60	0.10	0.01	0.67	0.10	0.05
Dead at discharge	2	Present	0.00	0.05	0.00	0.13	0.00	0.00	0.15	0.00	0.00
Age at discharge	3	4–10 days	0.84	0.34	0.83	0.09	0.53	0.83	0.04	0.50	0.84
		>11 days	0.03	0.61	0.03	0.79	0.45	0.03	0.82	0.46	0.14
$\hat{\eta}_j$			0.92	0.08	0.92	0.03	0.05	0.89	0.03	0.05	0.04

were virtually the same. The four-class solution distinguishes between a large class of normal healthy infants (class I) and three additional classes of infants whose births are abnormal in some way. Two classes (III and IV) represent moderately ill infants requiring special neonatal care and the third (II) represents severely ill infants with low birthweight.

Using these results one can predict the prevalence of each class and study their geographical distribution.

Example 6.13.2 The second example is taken from Clogg (1979) who fitted a three-class latent variable model to data from the 1975 US General Household Survey with a sample size of 1472. The data were reanalysed by Masters (1985) using a latent trait model.

The data concern three questions about degree of satisfaction with family (F), hobbies (H) and residence (R). Answers to each question were classified as low, medium or high. It is reasonable to suppose that if individuals tend to give the same reply to each question then the distribution might be explained by a model with three classes, and this is what Clogg fitted. The method used was maximum likelihood using Goodman's E-M routine. The parameter estimates are given in Table 6.8 and the observed and expected frequencies in Table 6.9.

The goodness of fit may be judged by the value of $\chi^2 = 2.36$ which, with $27 - 18 - 2 - 1 = 6$ degrees of freedom, indicates an extremely good fit.

We note from the $\hat{\eta}$s in Table 6.8 that the first class is very small, but the pattern of the πs confirms our expectation that there might be a tendency to give similar answers to all questions. Thus those in class I tend to have low satisfaction, those in class II somewhat higher satisfaction and those in class III highest of all. Indeed, the choice of questions with three alternatives may well have divided the respondents into three classes by requiring them to think about life satisfaction in this way. To

Table 6.8 Parameter estimates ($\hat{\pi}$s) of three-class latent variable models for life satisfaction data.

Class		I	II	III
Family	low	0.48	0.06	0.03
	medium	0.28	0.36	0.02
	high	0.24	0.58	0.95
Hobbies	low	0.85	0.18	0.07
	medium	0.00	0.50	0.14
	high	0.15	0.32	0.79
Residence	low	0.53	0.21	0.06
	medium	0.31	0.53	0.22
	high	0.16	0.26	0.72
	$\hat{\eta}_j$	0.04	0.41	0.55

Table 6.9 Observed and expected frequencies for life satisfaction data ($n = 1472$).

Response pattern			Observed frequency	Expected frequency	Response pattern			Observed frequency	Expected frequency
F	H	R			F	H	R		
0	0	0	15	14.1	2	1	0	45	43.3
0	0	1	11	11.3	2	1	1	117	116.4
0	0	2	7	6.8	2	1	2	126	123.0
1	0	0	16	15.6	0	2	0	5	5.7
1	0	1	26	25.3	0	2	1	14	11.7
1	0	2	12	13.2	0	2	2	16	17.5
2	0	0	23	22.8	1	2	0	18	16.6
2	0	1	49	48.9	1	2	1	38	40.4
2	0	2	54	57.0	1	2	2	27	27.8
0	1	0	3	4.0	2	2	0	64	61.2
0	1	1	12	10.3	2	2	1	191	193.6
0	1	2	5	7.2	2	2	2	466	466.9
1	1	0	23	23.0					
1	1	1	58	58.1	Total			1472	1472.0
1	1	2	31	29.9					

that extent the excellent fit may be, in part, an artefact, even though responses to all questions reflect a basic degree of satisfaction.

Individuals may be allocated to classes using the estimated parameters and the allocation rule of Section 6.12.1. The result is that those with response patterns 000, 001 and 002 go into class I, those with response patterns 222, 221, 220, 212, 202 and 022 into class III, and the remainder into class II. Membership of class I is thus determined by low satisfaction with family and hobbies; the make-up of class III is less transparent, but high satisfaction in at least two areas is the main qualification.

Example 6.13.3 In this third example we return to the survey of 301 university students (204 female and 97 men) on body satisfaction using seven items selected from the Body Esteem Scale (Franzoi and Shield 1984) discussed in Example 5.12.4. Students were asked how satisfied they were with certain parts of their bodies – in particular, thighs, body build, buttocks, hips, legs, figure and weight. Responses were given on a five-point scale (1 = very dissatisfied; 2 = moderately dissatisfied; 3 = slightly satisfied; 4 = moderately satisfied; and 5 = very satisfied). We present here the analysis only for the female part of the sample. Table 6.10 gives the BIC for five different models. According to the BIC, the five-class model is marginally preferred to the six-class model.

Table 6.11 gives the estimated category probabilities for the five-class solution obtained with Latent GOLD (Vermunt and Magidson 2000). Nearly 9% of the female students are dissatisfied with all seven parts of their body (class 4), followed by 19.6%

Table 6.10 Body satisfaction: log-likelihood, BIC and number of parameters.

Model	Log-likelihood	BIC	Number of parameters
2-Class	−1747.5965	3686.6454	36
3-Class	−1643.7688	3521.5349	44
4-Class	−1593.8066	3464.1554	52
5-Class	−1555.1662	3429.4195	60
6-Class	−1534.3650	3430.3622	68

of students in class 3 who are mainly dissatisfied but some slightly satisfied. Only 5.8% are allocated in class 5, which shows high satisfaction with all parts of the body, followed by 33% in class 1 with students who are at least slightly satisfied. Finally, 33% are students in the middle range of the satisfaction scale (class 2). Attention should be drawn to the extreme sparseness of the data due to the large number of response categories and the relatively small sample size. Model selection criteria favoured the five-class solution over less classes, but a more thorough investigation of residuals is needed to identify possible outliers. The seven items have been also analysed using marginal models that look not at the correlations among the seven items but on the marginal proportions (Bergsma *et al.* 2009).

6.14 Identifiability

As with all models involving many parameters, there is the possibility that the latent class model may not be identifiable. That is, there may be more than one point in the parameter space which yields the same likelihood. We can approach this question by thinking of the data set out in a $c_1 \times c_2 \times \cdots \times c_p$ contingency table. The probability for the cell designated by $\mathbf{x}$ is then given by (6.19). There are only $\prod_{i=1}^{p} c_i - 1$ independent probabilities because of the requirement that they sum to 1. These cell probabilities are functions of the model parameters which appear on the right-hand side of (6.19). If there are more model parameters than there are independent cell probabilities then, clearly, there will be many sets of model parameters leading to the same set of $f(\mathbf{x})$s and hence to the same likelihood. The number of parameters in the model is

$$\underbrace{K\sum_{i=1}^{p}(c_i - 1)}_{\pi s} + \underbrace{K - 1}_{\eta s}.$$

The model will, therefore, certainly be unidentifiable if

$$\prod_{i=1}^{p} c_i - 1 < K\sum_{i=1}^{p}(c_i - 1) + K - 1. \tag{6.32}$$

However, this condition is not sufficient because the parameters are probabilities and thus are subject to constraints on their ranges.

Table 6.11 Body satisfaction: estimated category and class probabilities.

Item	Class 1	Class 2	Class 3	Class 4	Class 5
Thighs					
1	0.0000	0.0088	0.1422	0.8251	0.0000
2	0.0157	0.3815	0.7355	0.1737	0.0000
3	0.2883	0.5191	0.1197	0.0012	0.0066
4	0.6769	0.0904	0.0025	0.0000	0.5135
5	0.0191	0.0002	0.0000	0.0000	0.4799
Body build					
1	0.0000	0.0001	0.0101	0.3690	0.0000
2	0.0095	0.0546	0.5189	0.6154	0.0000
3	0.1704	0.3627	0.4005	0.0155	0.0005
4	0.6902	0.5446	0.0698	0.0001	0.1011
5	0.1299	0.0380	0.0006	0.0000	0.8984
Buttocks					
1	0.0003	0.0128	0.1331	0.7734	0.0000
2	0.0370	0.2846	0.6143	0.2213	0.0014
3	0.3341	0.5264	0.2366	0.0053	0.0624
4	0.5240	0.1692	0.0158	0.0000	0.4755
5	0.1045	0.0069	0.0001	0.0000	0.4606
Hips					
1	0.0000	0.0017	0.1049	0.8794	0.0000
2	0.0166	0.1654	0.7184	0.1200	0.0001
3	0.2805	0.5702	0.1714	0.0006	0.0118
4	0.6162	0.2554	0.0053	0.0000	0.3439
5	0.0868	0.0073	0.0000	0.0000	0.6443
Legs					
1	0.0001	0.0090	0.0391	0.8309	0.0000
2	0.0253	0.3410	0.5891	0.1678	0.0010
3	0.2396	0.4700	0.3233	0.0012	0.0429
4	0.6131	0.1750	0.0479	0.0000	0.5013
5	0.1219	0.0051	0.0006	0.0000	0.4547
Figure					
1	0.0000	0.0012	0.0538	0.5476	0.0000
2	0.0043	0.0815	0.5169	0.4272	0.0000
3	0.1251	0.4241	0.3706	0.0249	0.0000
4	0.8102	0.4869	0.0586	0.0003	0.0321
5	0.0604	0.0064	0.0001	0.0000	0.9679
Weight					
1	0.0011	0.0236	0.1682	0.4275	0.0001
2	0.0239	0.1600	0.3864	0.4055	0.0033
3	0.2021	0.4211	0.3445	0.1493	0.0687
4	0.5372	0.3479	0.0964	0.0173	0.4472
5	0.2358	0.0475	0.0045	0.0003	0.4808
$\hat{\eta}_j$	0.3316	0.3269	0.1959	0.0875	0.0581

There appear to be no known global conditions for identifiability, but Goodman (1974) was able to make progress with local identifiability. This means that in a small neighbourhood the maximum of the likelihood is unique. Goodman's condition is based on the fact that the maximum is unique if the transformation from the cell probabilities to the parameters is one-to-one in a small neighbourhood. This can be tested by finding the Jacobian of the transformation. The matrix of the partial derivatives of the $f(\mathbf{x})$s with respect to the model parameters has $\sum_{i=1}^{p} c_i - 1$ rows and $K\{\sum_{i=1}^{p}(c_i - 1) + 1\}$ columns. The model is locally identifiable if the rank of the Jacobian is equal to the number of columns.

In practice the questions of identifiability and precision become intertwined. Roughly speaking, unidentifiability is a limiting case of decreasing precision of estimation. High asymptotic errors imply that the likelihood is fairly flat in the neighbourhood of the maximum – unidentifiability that it is precisely flat. Unidentifiability is therefore unlikely to be a problem if we restrict ourselves to models which can be estimated with reasonable precision. In our experience this means models with no more than three or four classes.

6.15 Starting values

The initial allocation of objects into classes can be done both randomly and by using hierarchical cluster analysis or k-means cluster analysis. Many different starting values need to be used before choosing a final model to avoid the problem of likelihood convergence into a local maximum. That phenomenon often occurs in latent class models (finite mixture models) especially when the number of variables is large and the sample size is small (Wedel and DeSarbo 1995). Searching for a global maximum could increase the computational time needed to estimate a latent class model significantly. Methodology that allows the estimation of the number of classes together with the model parameters is desirable to avoid refitting the model too many times.

6.16 Latent class models with metrical manifest variables

At the beginning of Section 1.3.2 we gave an example of a latent class model with metrical data and we noted that the use of such a model would require the choice of a suitable form for the conditional distribution of $\{x_i\}$. When dealing with binary or polytomous data the binomial and multinomial distributions were the obvious choice, but with metrical data there may be some difficulty in knowing what to assume. Some information can be gleaned from the marginal distributions, as we note below. In this section we are concerned with a model where the joint distribution of the xs has the form

$$f(\mathbf{x}) = \sum_{j=0}^{K-1} \eta_j \prod_{i=1}^{p} g_i(x_i \mid j) \tag{6.33}$$

for some K, where $g_i(x_i \mid j)$ is the conditional distribution of x_i for members of class j. Such models have been termed *latent profile models*.

In the absence of information on the form of $g_i(x_i \mid j)$, some guidance may be obtained from an inspection of the marginal distributions given by

$$f(x_i) = \sum_{j=0}^{K-1} \eta_j g_i(x_i \mid j) \quad (i = 1, 2, \ldots, p). \tag{6.34}$$

A bimodal distribution would suggest a two-class normal mixture, whereas an extremely skew distribution would argue against a normal mixture of any number of components. At best we are only likely to obtain very crude information in this way, but it should be possible to avoid grossly inappropriate assumptions. In favourable circumstances it may be possible to use the marginal distributions by methods given in Everitt and Hand (1981), but this would not solve our present problem unless the ηs were constrained to be the same for each margin. We shall describe two approaches to fitting the model of (6.33), the second of which throws some interesting light on the interpretation of covariance structures.

6.16.1 Maximum likelihood estimation

The same approach can be used as with categorical manifest variables, though the details depend on our choice of $\{g_i(x_i \mid j)\}$. The equations obtained by setting the derivatives of L with respect to η equal to zero are the same for all choices including the distributions used for categorical data and thus are

$$\hat{\eta}_j = \sum_{h=1}^{n} h(j \mid \mathbf{x}_h)/n \quad (j = 0, 1, \ldots, K-1). \tag{6.35}$$

Suppose we now take

$$g_i(x_i \mid j) = g(x_i \mid \theta_{ij});$$

then

$$L = \sum_{h=1}^{n} \ln \sum_{j=0}^{K-1} \eta_j \prod_{i=1}^{p} g(x_{ih} \mid \theta_{ij}) \tag{6.36}$$

and

$$\frac{\partial L}{\partial \theta_{ij}} = \sum_{h=1}^{n} \frac{\eta_j}{f(\mathbf{x}_h)} \prod_{r=1}^{p} g(x_{rh} \mid \theta_{rj}) \frac{\partial \log g(x_{ih} \mid \theta_{ij})}{\partial \theta_{ij}}. \tag{6.37}$$

Setting these derivatives equal to zero and solving for $\{\theta_{ij}\}$, we shall have equations of the form

$$\theta_{ij} = \psi(\mathbf{x}_h, \eta_j) \quad (i = 1, 2, \ldots, p; j = 0, 1, \ldots, K-1). \tag{6.38}$$

The E-M procedure can then be used, as before, by alternating between (6.35) and (6.38). The second derivatives can be used to obtain asymptotic standard errors.

Equations (6.38) take a particularly simple form if $g(x_i \mid \theta_{ij})$ is a member of the exponential family and especially if it is normal with mean θ_{ij} and unit variance. We then find

$$\frac{\partial \log g(x_{ih} \mid \theta_{ij})}{\partial \theta_{ij}} = x_{ih} - \theta_{ij}$$

and hence that

$$\sum_{h=1}^{n} h(j \mid \mathbf{x}_h)(x_{ih} - \theta_{ij}) = 0$$

or

$$\hat{\theta}_{ij} = \frac{\sum_{h=1}^{n} x_{ih} h(j \mid \mathbf{x}_h)}{\sum_{h=1}^{n} h(j \mid \mathbf{x}_h)}. \tag{6.39}$$

Equations (6.39) and (6.35) then lend themselves to a straightforward application of the E-M algorithm.

Furthermore, if one assumes that $g(x_i \mid \theta_{ij}, \sigma_i^2)$ is normal with mean θ_{ij} and variance σ_i^2 the same in all classes then

$$\hat{\sigma}_i^2 = \sum_{h=1}^{n} \sum_{j=0}^{K-1} (x_{ih} - \hat{\theta}_{ij})^2 h(j \mid \mathbf{x}_h)/n.$$

6.16.2 Other methods

The original method of estimation proposed for the latent profile model by Lazarsfeld and Henry (1968) involved fitting by the method of moments. In this case one writes down sufficient moment, and cross-moment, equations to determine the unknown parameters. As a method of estimation this has been superseded by the maximum likelihood approach, but it brings out an important link with the factor analysis model which is not otherwise evident.

Let $\mu_i(j)$ be the mean of x_i for members of latent class j and let $\sigma_i^2(j)$ be its variance. Then

$$E(x_i) = \sum_{j=0}^{K-1} \eta_j \int \cdots \int x_i \prod_{i=1}^{p} g_i(x_i \mid j)\, d\mathbf{x} = \sum_{j=0}^{K-1} \eta_j \mu_i(j),$$

$$E(x_i^2) = \sum_{j=0}^{K-1} \eta_j \{\sigma_i^2(j) + \mu_i^2(j)\},$$

$$E(x_i x_k) = \sum_{j=0}^{K-1} \eta_j \mu_i(j)\mu_k(j) \quad (i, k = 1, 2, \ldots, p; i \neq j). \tag{6.40}$$

We then have

$$\begin{aligned}
\text{var}(x_i) &= \sum_{j=0}^{K-1} \eta_j \sigma_i^2(j) + \sum_{j=0}^{K-1} \eta_j \{\mu_i(j) - \bar{\mu}_i\}^2 \quad (i = 1, 2, \ldots, p), \\
\text{cov}(x_i, x_k) &= \sum_{j=0}^{K-1} \eta_j \mu_i(j)\mu_k(j) - \left\{ \sum_{j=0}^{K-1} \eta_j \mu_i(j) \right\}^2 \\
&= \sum_{j=0}^{K-1} \eta_j \{\mu_i(j) - \bar{\mu}_i\}\{\mu_k(j) - \bar{\mu}_k\} \quad (i, k = 1, 2, \ldots, p; i \neq k),
\end{aligned} \tag{6.41}$$

where $\bar{\mu}_i = \sum_{j=0}^{K-1} \eta_j \mu_i(j)$. The covariance matrix may thus be written

$$\text{cov}(\mathbf{x}) = \mathbf{L}\mathbf{L}' + \boldsymbol{\Psi}, \tag{6.42}$$

where $\boldsymbol{\Psi}$ is now a diagonal matrix with (i, i)th element $\sum_{j=0}^{K-1} \eta_j \sigma_i^2(j)$ and the element l_{ij} of $\mathbf{L}$ is given by

$$l_{ij} = \sqrt{\eta_j}\{\mu_i(j) - \bar{\mu}_i\}. \tag{6.43}$$

Thus cov($\mathbf{x}$) is of exactly the same form as the covariance matrix for the linear factor model with a normally distributed latent variable, but there is one important difference. The columns of $\mathbf{L}$ are linearly dependent because

$$\sum_{j=0}^{K-1} \sqrt{\eta_j}\, l_{ij} = 0 \quad (i = 1, 2, \ldots, p)$$

from the definition of $\bar{\mu}_i$. For this reason any attempt to estimate $\mathbf{L}$ and $\boldsymbol{\Psi}$ by fitting the sample covariance matrix to (6.42) by standard methods (for which see Chapter 3) would fail. However, there exists a $p \times (K-1)$ matrix $\mathbf{\Lambda}$ with linearly independent columns such that $\mathbf{LL}' = \mathbf{\Lambda\Lambda}'$. We can then write

$$\text{cov}(\mathbf{x}) = \mathbf{\Lambda\Lambda}' + \mathbf{\Psi}, \tag{6.44}$$

which makes the correspondence complete.

This result is very important because it shows that, at the level of second moments, the latent class model with K classes is indistinguishable from a factor model with $K-1$ factors. Thus if we have successfully fitted a linear factor model to a given covariance matrix there is no guarantee that the latent space is continuous. It could equally well be categorical. This was one reason why we emphasised the desirability of examining the marginal distributions since multimodality there would be one way of distinguishing the two models. Unfortunately mixtures of distributions do not always exhibit distinct modes, so an inspection of the margins may leave the matter unresolved.

The analysis given above has been taken further by Molenaar and von Eye (1994). They show that the converse is also true, namely that if a covariance matrix has arisen from a linear factor model it is always possible to find a latent class model with the same covariance structure. They have also shown that the indeterminacy associated with the factor model that allows rotations in the factor space also carries over to the latent class model, and have provided the means to find rotated solutions. This means that any set of classes we identify by a latent class analysis is not unique.

The result of (6.44) suggests that we might be able to use software for fitting factor models to fit latent profile models. However, knowledge of $\mathbf{\Lambda}$ and $\mathbf{\Psi}$ is not sufficient to determine the parameters of the latent profile model, as (6.43) shows. Molenaar and von Eye (1994) have, however, shown that it is possible to construct constrained factor models, having the same covariance structure as a latent profile model, for which the prior distribution $\{\eta_j\}$ and the conditional means $\{\mu_i(j)\}$ (but not the variances) can be identified. For practical purposes, maximum likelihood is to be preferred as a method of estimation. The real importance of the moment results is that they illustrate the intrinsic difficulty of obtaining precise knowledge about latent variables.

A full analysis of the relationship between the moment structure of the latent profile and factor models would take us too far afield, but it is instructive and informative to look at the link between $\mathbf{\Lambda}$ and $\mathbf{L}$ when $K = 2$. We note first that

$$\begin{aligned} \mu_i(1) - \bar{\mu}_i &= (1-\eta_1)\{\mu_i(1) - \mu_i(2)\}, \\ \mu_i(2) - \bar{\mu}_i &= -\eta_1\{\mu_i(1) - \mu_i(2)\}. \end{aligned}$$

$\mathbf{L}$ is thus a matrix with ith row equal to

$$[\sqrt{\eta_1}(1-\eta_1)\{\mu_i(1) - \mu_i(2)\},\ -\sqrt{\eta_2}\eta_1\{\mu_i(1) - \mu_i(2)\}].$$

The (i, k)th element of $\mathbf{LL}'$ is thus

$$\eta_1(1-\eta_1)\{\mu_i(1)-\mu_i(2)\}\{\mu_k(1)-\mu_k(2)\},$$

and this is identical to the corresponding element in $\mathbf{\Lambda\Lambda}'$ when $\mathbf{\Lambda}$ is a $p \times 1$ matrix with ith element

$$\sqrt{\eta_1(1-\eta_1)}\{\mu_i(1)-\mu_i(2)\}. \tag{6.45}$$

The same result could have been obtained by starting with the linear model of (1.11), setting $q = 1$ and allowing y to have the Bernoulli distribution with $P\{y = 1\} = \eta_1$. Looking at the problem this way shows why the case $K > 2$ is less simple since then we need a vector-valued indicator for $\mathbf{y}$.

6.16.3 Allocation to categories

If $g_i(x_i \mid j)$ is a member of the exponential family for all i then it follows from the general theory that the posterior distribution, and hence the allocations rule, will be based on a linear function of the xs. In particular, if x_i given j is normal with mean $\mu_i(j)$ and unit standard deviation then

$$g(\mathbf{x} \mid j) = (2\pi)^{-\frac{1}{2}p} \exp\left\{-\frac{1}{2}\sum_{i=1}^{p}(x_i - \mu_i(j))^2\right\}.$$

Category j is more probable than category k if

$$\eta_j\, g(\mathbf{x} \mid j) > \eta_k\, g(\mathbf{x} \mid k),$$

which holds if

$$\sum_{i=1}^{p}\mu_i(j)x_i - \frac{1}{2}\sum_{i=1}^{p}\mu_i^2(j) + \ln\eta_j > \sum_{i=1}^{p}\mu_i(k) - \frac{1}{2}\sum_{i=1}^{p}\mu_i^2(k) + \ln\eta_k.$$

6.17 Models with ordered latent classes

In some circumstances we may expect the latent classes to be ordered. If, for example, they represent political beliefs we might suppose that they are ordered on a left/right scale. In that case the probability of giving a positive response to a binary item might be expected to change monotonically as we move along the sequence of classes. A model with ordered classes represents an intermediate stage between the latent trait model, in which the latent space is a continuum, and the latent class model. Models of this kind were proposed by Croon (1990) who gave an algorithm for fitting and a numerical example. Croon also considers the more general case in which the manifest variables are polytomous and ordered. There is then a choice of ways for specifying the ordering. Croon's method involves requiring that the monotonicity property holds for all possible ways of dichotomising the ordered manifest variables.

Given the difficulty of distinguishing a latent class model from a latent trait model, which we have already noted, one would expect it to be even more difficult to distinguish either from an intermediate form. This point should be borne in mind when using the model with ordered latent categories.

6.18 Hybrid models

Table 1.3 can be extended to allow for mixed categorical and metrical latent variables. The purpose of this might be to identify latent sub-populations and to measure latent constructs within each such latent class. To illustrate the value of such models we shall describe a typical example. A two-parameter latent trait model assumes that the item difficulty and item discrimination parameters are the same for all population members. This is often a plausible assumption, but not always. For example, the item parameters may take different values in different subgroups of the population. In particular, a majority of respondents might give responses according to their ability whereas others might use another strategy such as random guessing. These different strategies could not be identified from the respondents considered as a whole but only within the latent classes. In some cases it might be possible to identify classes in advance using covariates such as gender or age; otherwise the classes could be latent. A hybrid model allows us to fit latent trait models within the unobserved latent classes. This is a useful way of extending the traditional approach which has many new applications (see, for example, Mislevy and Verhelst (1990); Muthén (2008); Muthén and Asparouhov (2006): Rost (1990)). Such models can take into account various types of population heterogeneity arising from different response strategies and so resolve the lack of local (or conditional) independence in latent classes by imposing further structure.

6.18.1 Hybrid model with binary manifest variables

We discuss in this section a latent class model with K classes with a two-parameter item response theory model fitted in each class. Other models can be also fitted in a class, or a class can be chosen to be without any further structure, capturing for example aberrant response patterns or outliers. Let $\pi_{ij}(y)$ be the conditional probability of a positive response on variable i for a person in category j ($i = 1, 2, \ldots, p$; $j = 0, 1, \ldots, K-1$) with latent variable value y. Let η_j be the prior probability that a randomly chosen individual is in class j $\left(\sum_{j=0}^{K-1} \eta_j = 1\right)$ and $h(y)$ be the prior distribution of the latent variable. For the case of K classes and one latent variable y,

$$f(\mathbf{x}) = \int_{R_y} h(y) \sum_{j=0}^{K-1} \eta_j \prod_{i=1}^{p} \{\pi_{ij}(y)\}^{x_i} \{1 - \pi_{ij}(y)\}^{1-x_i} dy. \tag{6.46}$$

The posterior probability that an individual with response vector $\mathbf{x}$ belongs to category j and is at the latent position y is thus

$$h(j, y \mid \mathbf{x}) = h(y)\eta_j \prod_{i=1}^{p} \{\pi_{ij}(y)\}^{x_i} \{1 - \pi_{ij}(y)\}^{1-x_i} / f(\mathbf{x}) \quad (j = 0, 1, \ldots, K-1). \tag{6.47}$$

Let us assume that in each class the logit/normal model with a single latent variable is fitted,

$$\operatorname{logit} \pi_{ij}(y) = \alpha_{i0,j} + \alpha_{i1,j} y \quad (i = 1, 2, \ldots, p; j = 0, 1, \ldots, K-1), \tag{6.48}$$

where y has standard normal distribution.

6.18.2 Maximum likelihood estimation

The log-likelihood for a random sample of size n is

$$L = \sum_{h=1}^{n} \ln \left\{ \int_{R_y} h(y) \sum_{j=0}^{K-1} \eta_j \prod_{i=1}^{p} \{\pi_{ij}(y)\}^{x_{ih}} \{1 - \pi_{ij}(y)\}^{1-x_{ih}} dy \right\}. \tag{6.49}$$

The integral in (6.49) can be approximated using the Gauss–Hermite method already discussed in Section 4.5. Suppose that the Gauss–Hermite quadrature uses the discrete distribution at values $y^1, \ldots, y^T$ with probabilities $h(y^1), \ldots, h(y^T)$. Then

$$L = \sum_{h=1}^{n} \ln \left\{ \sum_{t=1}^{T} h(y^t) \sum_{j=0}^{K-1} \eta_j \prod_{i=1}^{p} \{\pi_{ij}(y^t)\}^{x_{ih}} \{1 - \pi_{ij}(y^t)\}^{1-x_{ih}} \right\}. \tag{6.50}$$

As before, this has to be maximised subject to $\sum \eta_j = 1$, so we find the unrestrained maximum of

$$\phi = L + \theta \sum_{j=0}^{K-1} \eta_j,$$

where θ is an undetermined multiplier. Finding partial derivatives, we have

$$\begin{aligned} \frac{\partial \phi}{\partial \eta_j} &= \sum_{h=1}^{n} \sum_{t=1}^{T} h(y^t) \left\{ \prod_{i=1}^{p} \{\pi_{ij}(y^t)\}^{x_{ih}} \{1 - \pi_{ij}(y^t)\}^{1-x_{ih}} / f(\mathbf{x}_h) \right\} + \theta \\ &\quad (j = 0, 1, \ldots, K-1) \\ &= \sum_{h=1}^{n} \sum_{t=1}^{T} \{h(y^t) g(\mathbf{x}_h \mid j, y^t) / f(\mathbf{x}_h)\} + \theta, \end{aligned} \tag{6.51}$$

where $g(\mathbf{x}_h \mid j, y^t)$ is the joint probability of $\mathbf{x}_h$ for an individual in class j with latent ability y^t.

Also,

$$\frac{\partial \phi}{\partial \alpha_{il,j}} = \eta_j \sum_{h=1}^{n} \sum_{t=1}^{T} h(y^t) \frac{\partial g(\mathbf{x}_h \mid j, y^t)}{\partial \alpha_{il,j}} \Bigg/ f(\mathbf{x}_h)$$

$$(i = 1, 2, \ldots, p; j = 0, 1, \ldots, K - 1; l = 0, 1),$$

giving

$$\frac{\partial \phi}{\partial \alpha_{il,j}} = \eta_j \sum_{t=1}^{T} y_l^t h(y^t) \sum_{h=1}^{n} (x_{ih} - \pi_{ij}(y^t)) g(\mathbf{x}_h \mid j, y^t)/f(\mathbf{x}_h) \quad (l = 0, 1). \quad (6.52)$$

The resulting equations can be simplified by expressing (6.51) and (6.52) in terms of the posterior probabilities $\{h(j, y^t \mid \mathbf{x})\}$. Substituting in (6.51) and equating to zero, we find

$$\sum_{h=1}^{n} \sum_{t=1}^{T} h(j, y^t \mid \mathbf{x}_h) = -\theta \eta_j.$$

Summing both sides over j and y^t and using $\sum \eta_j = 1$ gives $\theta = -n$, and hence the first estimating equation (6.51) is

$$\hat{\eta}_j = \sum_{h=1}^{n} \sum_{t=1}^{T} h(j, y^t \mid \mathbf{x}_h)/n \quad (j = 0, 1, \ldots, K - 1). \quad (6.53)$$

The second estimating equation (6.52) becomes

$$\sum_{t=1}^{T} \sum_{h=1}^{n} y_l^t \{x_{ih} - \pi_{ij}(y^t)\} h(j, y^t \mid \mathbf{x}_h) = 0 \quad (l = 0, 1) \quad (6.54)$$

or

$$\sum_{t=1}^{T} y_l^t \{r_{i,jt} - n_{jt} \pi_{ij}(y^t)\} = 0 \quad (l = 0, 1),$$

where

$$r_{i,jt} = \sum_{h=1}^{n} x_{ih} h(j, y^t \mid \mathbf{x}_h),$$

$$n_{jt} = \sum_{h=1}^{n} h(j, y^t \mid \mathbf{x}_h)$$

and

$$h(j, y^t \mid \mathbf{x}_h) = \frac{\eta_j h(y^t) g(\mathbf{x}_h \mid j, y^t)}{\sum_{t=1}^{T} \sum_{k=0}^{K-1} \eta_j h(y^t) g(\mathbf{x}_h \mid j, y^t)}. \tag{6.55}$$

As in Section 4.5, it may be helpful to give an interpretation of $r_{i,jt}$ and n_{jt}. The quantity $h(j, y^t \mid \mathbf{x}_h)$ is the probability that an individual with response vector $\mathbf{x}_h$ is located in class j and at y^t; $\sum_{h=1}^{n} h(j, y^t \mid \mathbf{x}_h)$ is thus the expected number of individuals in class j and y^t. By a similar argument, $r_{i,jt}$ is the expected number of those predicted to be in class j and y^t who will respond positively.

An E-M algorithm is used again here to obtain maximum likelihood estimates using the following steps

1. Choose starting values for all model parameters.
2. Compute the values $r_{i,jt}$, n_{jt} and $h(j, y^t \mid \mathbf{x}_h)$.
3. Obtain improved estimates of the αs by solving (6.54) for $l = 0, 1$ and $i = 1, \ldots, p$, treating $r_{i,jt}$ and n_{jt} as given numbers, and for the ηs by solving (6.53).
4. Obtain improved estimates of $\{h(j, y^t \mid \mathbf{x}_h)\}$ from (6.55).
5. Return to step 2 and continue until convergence is attained.

7

Models and methods for manifest variables of mixed type

7.1 Introduction

So far we have supposed that all manifest variables were of the same type – that is, were all continuous or all categorical with the same distribution. In the former case their conditional distributions were assumed to be normal; and in the latter Bernoulli for binary responses, or multinomial for polytomous responses. In many fields of application, especially with data arising from sample surveys, both types of variables often occur together. The purpose of this chapter is to show how such problems can easily be handled within the framework of the GLLVM, as set out in Chapter 2. There is nothing in the general treatment given there which requires the xs to have distributions of the same form, provided that they are all members of the exponential family. As a by-product we shall obtain new results for cases where all the variables are of the same type but have distributions other than those already covered.

The traditional way of handling problems with manifest variables of mixed type has been to convert the problem into one where the variables are of the same type. One way of doing this is by categorising the continuous variables. At the price of some loss of information the problem can then be handled by established methods for categorical variables. The more usual course has been to resort to the device of introducing underlying variables. All categorical variables are thereby regarded as incompletely observed continuous variables which are supposed to conform to the normal linear factor model. The correlations required for fitting such models are then estimated as tetrachoric, polychoric or polyserial correlations according to the type of variable involved. This approach was pioneered by Muthén (1984) and has been

Latent Variable Models and Factor Analysis: A Unified Approach, Third Edition.
David Bartholomew, Martin Knott and Irini Moustaki.

implemented in the Mplus program (Muthén and Muthén 2010). A similar approach is available in the LISREL package (Jöreskog and Sörbom 2006) within the broader framework of linear structural relations modelling (see Chapter 8). A comprehensive theoretical treatment of such models, where the manifest variables are of mixed type, is given by Arminger and Küsters (1988). We shall return to this approach in Section 7.9 in order to compare it with our treatment based on the GLLVM. The latter has been developed by Moustaki (1996) for mixtures of continuous and binary variables and, more generally, in Moustaki and Knott (2000a). A very similar treatment has been given by Sammel *et al.* (1997).

7.2 Principal results

Suppose the manifest variables are of k types, with p_1 of the first type, p_2 of the second and so on ($p_1 + p_2 + \cdots + p_k = p$). The vector of manifest variables can then be partitioned, according to type, as $\mathbf{x}' = (\mathbf{x}_1', \mathbf{x}_2', \ldots, \mathbf{x}_k')$, where $\mathbf{x}_i' = (x_{i1}, x_{i2}, \ldots, x_{ip_i})$ ($i = 1, 2, \ldots, k$). All elements of $\mathbf{x}_i$ belong to the same member of the exponential family. In the general formulation of the model in (2.1) and (2.2) we allowed each manifest variable to have a different distribution, as indicated by the suffix of g in (2.1). In the present case, the first p_1 elements of $\mathbf{x}$ share the same form of distribution and hence the functions F_i, G_i and u_i in (2.1) will be common to all members of that subset. The parameter θ_i will, however, be influenced differently by each member. The same is true for the remaining subsets of $\mathbf{x}$. For our present purpose it is therefore convenient to rewrite the model of (2.1) and (2.2) as

$$g_i(x_{il} \mid \theta_{il}) = F_i(x_{il})G_i(\theta_{il}) \exp(\theta_{il} u_i(x_{il})) \tag{7.1}$$

and

$$\theta_{il} = \alpha_{il0} + \alpha_{il1} y_1 + \alpha_{il2} y_2 + \cdots + \alpha_{il} y_q, \tag{7.2}$$

where, in both cases, $i = 1, 2, \ldots, k$ and $l = 1, 2, \ldots, p_i$. When expressed in this form (2.4) becomes

$$h(\mathbf{y} \mid \mathbf{x}) \propto h(\mathbf{y}) \prod_{i=1}^{k} \prod_{l=1}^{p_i} G_i(\theta_{i\ell}) \exp \sum_{j=1}^{q} y_j X_j, \tag{7.3}$$

where

$$X_j = \sum_{i=1}^{k} \sum_{l=1}^{p_i} \alpha_{ilj} u_i(x_{il}). \tag{7.4}$$

$\mathbf{X}' = (X_1, X_2, \ldots, X_q)$ is again the Bayesian minimal sufficient statistic or component. For those distributions for which u_i is a linear function the components will be linear functions of the manifest variables. They may, therefore, be interpreted in essentially the same way as when the xs were all of the same type.

Although the foregoing argument has been conducted as though the latent variables were continuous, the same result follows if some or all of the latent variables are categorical, as we saw in Chapter 6. In principle, therefore, no new theory is required for manifest variables of mixed type. It emerges as a special case of the GLLVM. There are, however, new technical difficulties concerned with fitting the models and with judging their goodness of fit. Many of these problems remain unsolved, but methods are available for some of the commonly occurring problems and these will be reviewed in the remainder of the chapter. We shall conclude by setting our results in the rather broader context mentioned in Section 7.1.

7.3 Other members of the exponential family

Hitherto we have worked only with normal, Bernoulli and multinomial distributions for the conditional distributions of manifest variables. Before proceeding we shall therefore extend the repertoire to include the binomial, Poisson and gamma distributions which are all easily derived from the GLLVM. The results for the gamma distribution are needed for Example 7.8.2.

7.3.1 The binomial distribution

This can be dealt with using the results for the Bernoulli distribution because a binomial random variable is the sum of independent Bernoulli random variables. However, a direct approach starts with

$$g_i(x_i \mid \mathbf{y}) = \binom{n}{x_i} \{\pi_i(\mathbf{y})\}^{x_i} \{1 - \pi_i(\mathbf{y})\}^{n-x_i} \quad (x_i = 0, 1, 2, \ldots, n), \tag{7.5}$$

where n is the numbers of 'trials'. Writing this in the form

$$g_i(x_i \mid \mathbf{y}) = \binom{n}{x_i} \{1 - \pi_i(\mathbf{y})\}^n \exp\{x_i \operatorname{logit} \pi_i(\mathbf{y})\}, \tag{7.6}$$

the linear model becomes

$$\operatorname{logit} \pi_i(\mathbf{y}) = \alpha_{i0} + \sum_{j=1}^{q} \alpha_{ij} y_j \tag{7.7}$$

exactly as for the Bernoulli case. The components are then

$$X_j = \sum_{i=1}^{p} \alpha_{ij} x_i. \tag{7.8}$$

7.3.2 The Poisson distribution

If

$$g_i(x_i \mid \mathbf{y}) = \frac{(\mu_i(\mathbf{y}))^{x_i}}{x_i!} e^{-\mu_i(\mathbf{y})} \quad (x_i = 0, 1, 2, \ldots), \tag{7.9}$$

the standard approach yields

$$\ln \mu_i(\mathbf{y}) = \alpha_{i0} + \sum_{j=1}^{q} \alpha_{ij} y_j \tag{7.10}$$

and

$$X_j = \sum_{j=1}^{q} \alpha_{ij} x_i, \tag{7.11}$$

a result which we might have anticipated from the results for the binomial distribution.

7.3.3 The gamma distribution

Like the normal distribution, the gamma distribution depends on two parameters, and different models arise according to how we parameterise the model and which parameter is made to depend on the latent variables. In the case of the normal distribution we assumed that only the location of the conditional distribution depended on the latent variables. If we are to consider mixtures of normal and gamma variates, say, it is natural to treat each type of variable in the same way, but it must be remembered that other options are possible. Here, therefore, we shall parameterise the distribution so that the mean is one of the parameters and it will be the mean which depends on the latent variables. Thus suppose

$$g_i(x_i \mid \mathbf{y}) = \frac{1}{\Gamma(\nu_i)} \left(\frac{\nu_i}{\mu_i(\mathbf{y})} \right)^{\nu_i} x_i^{\nu_i - 1} e^{-\nu_i x_i / \mu_i(\mathbf{y})}, \tag{7.12}$$

where $\mu_i(\mathbf{y})$ is the mean and ν_i is a nuisance parameter determining the shape of the distribution. For this distribution

$$\theta_i(\mathbf{y}) = \frac{1}{\mu_i(\mathbf{y})} = \alpha_{i0} + \sum_{j=1}^{q} \alpha_{ij} y_j \tag{7.13}$$

and the components are therefore

$$X_j = \sum_{i=1}^{p} \alpha_{ij} \nu_i x_i. \tag{7.14}$$

It should be noted that, since a gamma variate is non-negative, $\mu_i(\mathbf{y})$ must also be non-negative. If we are considering several types of manifest variables we shall not wish to impose constraints on the form or range of the prior distribution since we shall want it to serve for all such variables. The model proposed will therefore only be reasonable if α_{i0} is sufficiently large to make $\mu_i(\mathbf{y})$ positive with high probability.

Returning to (7.4) in Section 7.2, we can see immediately that if we have p_1 variables of one type, p_2 of another and so on, the components will simply be the sum of the components for the constituent parts – one for each type. The maximum likelihood estimators will follow by constructing the likelihood from the appropriate number of constituents of each type, as illustrated in the examples which follow.

7.4 Maximum likelihood estimation

We have provided several methods for finding maximum likelihood estimators when all manifest variables are of the same type. Now we shall develop a general approach for the GLLVM when the individual xs may all have different distributions within the exponential family. This will cover, as a special case, the problem when the manifest variables are grouped according to type. The basic idea is the same as that which was used for the latent trait and latent class models in Chapters 4–6. The method lends itself to implementation using the E-M approach but can also be the starting point for other approaches.

As in Chapter 2, we now suppose that

$$g_i(x_i \mid \theta_i) = F_i(x_i)G_i(\theta_i)\exp\{\theta_i u_i(x_i)\}, \tag{7.15}$$

where

$$\theta_i = \alpha_{i0}y_0 + \alpha_{i1}y_1 + \alpha_{i2}y_2 + \cdots + \alpha_{iq}y_q \quad (i = 1, 2, \ldots, p) \tag{7.16}$$

and $y_0 \equiv 1$.

We consider the problem of estimating the αs, any nuisance parameters being assumed known at this stage. If we have a random sample of size n, the log-likelihood may be written

$$\begin{aligned} L &= \sum_{\ell=1}^{n} \ln f(\mathbf{x}_\ell) \\ &= \sum_{\ell=1}^{n} \ln \int h(\mathbf{y}) \prod_{i=1}^{p} F_i(x_{i\ell})G_i(\theta_i)\exp\left\{\sum_{i=1}^{p} \theta_i u_i(x_{i\ell})\right\} d\mathbf{y}, \end{aligned} \tag{7.17}$$

the integral here and below being over the range space of $\mathbf{y}$.

The partial derivatives of L with respect to the αs are then given by

$$\frac{\partial L}{\partial \alpha_{ij}} = \sum_{\ell=1}^{n} \frac{1}{f(\mathbf{x}_\ell)} \int h(\mathbf{y}) \prod_{i=1}^{p} F_i(x_{i\ell}) \frac{\partial}{\partial \alpha_{ij}} \left\{ \exp\left(\sum_{i=1}^{p} \theta_i u_i(x_{i\ell}) + \ln G_i(\theta_i) \right) \right\} d\mathbf{y}$$
$$(i = 1, 2, \ldots, p;\ j = 1, 2, \ldots, q).$$

The partial derivatives, within the integral, may be evaluated by first differentiating with respect to θ_i to give

$$\left\{ u_i(x_{i\ell}) + \frac{d \ln G_i(\theta_i)}{d\theta_i} \right\} \exp\left\{ \sum_{i=1}^{p} (\theta_i u_i(x_{i\ell}) + \ln G_i(\theta_i)) \right\} \frac{\partial \theta_i}{\partial \alpha_{ij}}.$$

Hence

$$\frac{\partial L}{\partial \alpha_{ij}} = \sum_{\ell=1}^{n} \int h(\mathbf{y} \mid \mathbf{x}_\ell) \left\{ u_i(x_{i\ell}) + \frac{d \ln G_i(\theta_i)}{d\theta_i} \right\} y_j \, d\mathbf{y} \tag{7.18}$$
$$(i = 1, 2, \ldots, p;\ j = 0, 1, 2, \ldots, q),$$

where $h(\mathbf{y} \mid \mathbf{x}_\ell)$ is the posterior distribution of $\mathbf{y}$ given $\mathbf{x}_\ell$, $y_0 = 1$.

Interchanging the order of summation and integration and setting the partial derivatives equal to zero, we obtain the basic estimating equations

$$\int y_j \sum_{\ell=1}^{n} u_i(x_{i\ell}) h(\mathbf{y} \mid \mathbf{x}_\ell) d\mathbf{y} = \int y_j \frac{d \ln G_i(\theta_i)}{d\theta_i} \sum_{\ell=1}^{n} h(\mathbf{y} \mid \mathbf{x}_\ell) d\mathbf{y} \tag{7.19}$$
$$(i = 1, 2, \ldots, p;\ j = 0, 1, \ldots, q).$$

If a maximum of the likelihood exists, it will satisfy these equations. Estimation for all models covered in this book, belonging to the exponential family, can be tackled using the appropriate special case of these equations. If there happen to be additional unknown (nuisance) parameters, each of these will give rise to a further estimating equation.

7.4.1 Bernoulli manifest variables

In order to establish the link between the general case and the special cases we have already met, we look briefly at the cases of Bernoulli and normal manifest variables.

When the xs are all binary random variables with a conditional Bernoulli distribution, we saw in Section 2.4 that

$$\theta_i = \operatorname{logit} \pi_i(\mathbf{y}), \quad G_i(\theta_i) = 1 - \pi_i(\mathbf{y}) = 1/(1 + e^{\theta_i})$$

and $u_i(x_i) = x_i$. Consequently,

$$-\frac{d \ln G_i(\theta_i)}{d\theta_i} = \frac{1}{1+e^{-\theta_i}} = \pi_i(\mathbf{y}). \tag{7.20}$$

Substitution into (7.19) gives

$$\sum_{\ell=1}^{n} x_{i\ell} E(\mathbf{y} \mid \mathbf{x}_\ell) = \sum_{\ell=1}^{n} E(y_j \pi_i(\mathbf{y}) \mid \mathbf{x}_\ell), \tag{7.21}$$

which is essentially the same as (4.20) after it has been differentiated with respect to model parameters and in which the integrals for the expectations have been approximated by sums.

7.4.2 Normal manifest variables

If the conditional distributions of the xs are normal with means $\{\mu_i\}$ and standard deviations $\{\sigma_i\}$, we showed in Section 2.6 that $\theta_i = \mu_i/\sigma_i, G_i(\theta_i) = \exp\left\{-\frac{1}{2}\theta_i^2\right\}$ and $u_i(x_i) = x_i/\sigma_i$, $(i = 1, 2, \ldots, p)$. In this case

$$\frac{d \ln G_i(\theta_i)}{d\theta_i} = -\theta_i \quad (i = 1, 2, \ldots, p). \tag{7.22}$$

To facilitate comparison with our earlier results for the normal linear model we write (7.19) in matrix notation after substituting $-\theta_i$ for $d \ln G_i(\theta_i)/d\theta_i$, writing $\psi_i = \sigma_i^2$ and putting $u_i(x_i) = x_i/\sigma_i$. This gives

$$\begin{aligned} \sum_{\ell=1}^{n} E(\mathbf{y} \mid \mathbf{x}_\ell)\mathbf{x}_\ell' \boldsymbol{\psi}^{-1/2} &= \sum_{\ell=1}^{n} E(\mathbf{y}\boldsymbol{\theta}' \mid \mathbf{x}_\ell) \\ &= \sum_{\ell=1}^{n} E(\mathbf{y}\mathbf{y}'\mathbf{A}_0' \mid \mathbf{x}_\ell) \\ &= \sum_{\ell=1}^{n} \{\operatorname{var}(\mathbf{y} \mid \mathbf{x}_\ell) + E(\mathbf{y} \mid \mathbf{x}_\ell)E(\mathbf{y}' \mid \mathbf{x}_\ell)\}\mathbf{A}_0', \end{aligned} \tag{7.23}$$

where $\mathbf{A}_0 = (\boldsymbol{\alpha}_0, \ \mathbf{A})$ with

$$\boldsymbol{\alpha}_0' = (\alpha_{01}, \alpha_{02}, \ldots, \alpha_{0p}) \quad \text{and} \quad \mathbf{A} = \{\alpha_{ij}\} \quad (i = 1, 2, \ldots; \ j = 1, 2, \ldots, q).$$

For this model, the conditional expectations in (7.23) have already been obtained in (1.13) and (3.6). On the left-hand side, we have that

$$E(\mathbf{y} \mid \mathbf{x}_\ell) = \left[\begin{array}{c} 1 \\ \boldsymbol{\Lambda}'\boldsymbol{\Sigma}^{-1}(\mathbf{x}_\ell - \bar{\mathbf{x}}) \end{array}\right], \tag{7.24}$$

where we have replaced the $\boldsymbol{\mu}$ of (3.6) by its maximum likelihood estimator, $\bar{\mathbf{x}} = \frac{1}{n}\sum_{\ell=1}^{n} \mathbf{x}_\ell$, and

$$\text{var}(\mathbf{y} \mid \mathbf{x}_\ell) = \begin{bmatrix} 0 & \mathbf{0}' \\ \mathbf{0} & (\mathbf{I}+\boldsymbol{\Gamma})^{-1} \end{bmatrix}. \tag{7.25}$$

Substituting (7.24) and (7.25) into (7.23) and carrying out the summation over l, we find that the left-hand side becomes

$$n \begin{bmatrix} \bar{\mathbf{x}}' \\ \boldsymbol{\Lambda}'\boldsymbol{\Sigma}^{-1}\mathbf{S}_\mathbf{x} \end{bmatrix} \psi^{-1/2},$$

where $\mathbf{S}_\mathbf{x}$ is the sample covariance matrix of the xs, given by

$$\mathbf{S}_\mathbf{x} = \frac{1}{n}\sum_{\ell=1}^{n}(\mathbf{x}_\ell - \bar{\mathbf{x}})(\mathbf{x}_\ell - \bar{\mathbf{x}})'.$$

On the right-hand side we obtain

$$n\left[\begin{pmatrix} 0 & \mathbf{0}' \\ \mathbf{0} & (\mathbf{I}+\boldsymbol{\Gamma}) \end{pmatrix} + \begin{pmatrix} 1 & \mathbf{0}' \\ \mathbf{0} & \boldsymbol{\Lambda}'\boldsymbol{\Sigma}^{-1}\mathbf{S}_\mathbf{x}\boldsymbol{\Sigma}^{-1}\boldsymbol{\Lambda} \end{pmatrix}\right]\mathbf{A}_0'.$$

The resulting estimation equations may therefore be written

$$\bar{\mathbf{x}}\boldsymbol{\Psi}^{-1/2} = \boldsymbol{\alpha}_0 \tag{7.26}$$

and

$$\boldsymbol{\Lambda}'\boldsymbol{\Sigma}^{-1}\mathbf{S}_\mathbf{x}\boldsymbol{\Psi}^{-1/2} = (\mathbf{I}+\boldsymbol{\Gamma})^{-1}\mathbf{A}' + \boldsymbol{\Lambda}'\boldsymbol{\Sigma}^{-1}\mathbf{S}_\mathbf{x}\boldsymbol{\Sigma}^{-1}\boldsymbol{\Lambda}\mathbf{A}'. \tag{7.27}$$

To complete the derivation we replace $\mathbf{A}$ in (7.27) by the equivalent expression in $\boldsymbol{\Lambda}$ and $\boldsymbol{\Psi}$. In the standard normal model $E(x_i \mid \mathbf{y}) = \mu_i + \sum_{j=1}^{q} \lambda_{ij} y_j$ and the GLLVM formulation is

$$E(x_i \mid \mathbf{y}) = \sigma_i \left\{ \alpha_{i0} + \sum_{j=1}^{q} \alpha_{ij} y_j \right\}. \tag{7.28}$$

Therefore, $\mathbf{A} \equiv \boldsymbol{\Psi}^{-1/2}\boldsymbol{\Lambda}$. Substituting into (7.27) and rearranging gives

$$(\mathbf{I}+\boldsymbol{\Gamma})\boldsymbol{\Lambda}'\boldsymbol{\Sigma}^{-1}\mathbf{S}_\mathbf{x}\boldsymbol{\Psi}^{-1/2} = \boldsymbol{\Lambda}'\boldsymbol{\Psi}^{-1/2} + (\mathbf{I}+\boldsymbol{\Gamma})\boldsymbol{\Lambda}'\boldsymbol{\Sigma}^{-1}\mathbf{S}_\mathbf{x}\boldsymbol{\Sigma}^{-1}\boldsymbol{\Lambda}\boldsymbol{\Lambda}'\boldsymbol{\Psi}^{-1/2}. \tag{7.29}$$

We showed in (3.8) that

$$\boldsymbol{\Lambda}'\boldsymbol{\Sigma}^{-1} = (\mathbf{I}+\boldsymbol{\Gamma})^{-1}\boldsymbol{\Lambda}'\boldsymbol{\Psi}^{-1}$$

and, therefore, the estimating equations become

$$\mathbf{\Lambda}'\mathbf{\Psi}^{-1}\mathbf{S_x}\mathbf{\Psi}^{-1/2} = \mathbf{\Lambda}'\mathbf{\Psi}^{-1/2} + \mathbf{\Lambda}'\mathbf{\Psi}^{-1}\mathbf{S_x}\mathbf{\Psi}^{-1}\mathbf{\Lambda}(\mathbf{I}+\mathbf{\Gamma})^{-1}\mathbf{\Lambda}'\mathbf{\Psi}^{-\frac{1}{2}} \tag{7.30}$$

or

$$\mathbf{A}'\mathbf{V} = \mathbf{A}' + \mathbf{A}'\mathbf{V}\mathbf{A}(\mathbf{I}+\mathbf{\Gamma})^{-1}\mathbf{A}', \tag{7.31}$$

where $\mathbf{V} = \mathbf{\Psi}^{-1/2}\mathbf{S_x}\mathbf{\Psi}^{-1/2}$. Transposing and rearranging, we find

$$(\mathbf{I} - \mathbf{A}(\mathbf{I}+\mathbf{\Gamma})^{-1}\mathbf{A}')\mathbf{V}\mathbf{A} = \mathbf{A}. \tag{7.32}$$

This may be simplified further as follows. Pre-multiplying both sides by $\mathbf{A}'$ and noting that $\mathbf{A}'\mathbf{A} = \mathbf{\Gamma}$,

$$\mathbf{A}'\mathbf{V}\mathbf{A} + \mathbf{\Gamma}(\mathbf{I}+\mathbf{\Gamma})^{-1}(\mathbf{A}'\mathbf{V}\mathbf{A}) = \mathbf{\Gamma} \tag{7.33}$$

or

$$\mathbf{A}'\mathbf{V}\mathbf{A} = \mathbf{\Gamma}(\mathbf{I}+\mathbf{\Gamma}). \tag{7.34}$$

Substituting back into the transpose of (7.31), we have

$$\mathbf{V}\mathbf{A} = \mathbf{A}\{\mathbf{I} + (\mathbf{I}+\mathbf{\Gamma})^{-1}\mathbf{\Gamma}(\mathbf{I}+\mathbf{\Gamma})\} = \mathbf{A}(\mathbf{I}+\mathbf{\Gamma}) \tag{7.35}$$

or

$$(\mathbf{\Psi}^{-1/2}\mathbf{S_x}\mathbf{\Psi}^{-1/2})(\mathbf{\Psi}^{-1/2}\mathbf{\Lambda}) = (\mathbf{\Psi}^{-1/2}\mathbf{\Lambda})(\mathbf{I}+\mathbf{\Gamma}). \tag{7.36}$$

This is the equation arrived at in standard treatments of the normal linear model. Here we have derived it as a special case of the general exponential family model. An interesting by-product of this result is that we can use the general E-M approach described below as an alternative estimation method for the normal linear model.

7.4.3 A general E-M approach to solving the likelihood equations

We can rewrite (7.19) in a form which makes the equations amenable to solution by the E-M method and which also gives some insight into their interpretation. Let

$$n_{\mathbf{y}} = \sum_{\ell=1}^{n} h(\mathbf{y} \mid \mathbf{x}_\ell) \tag{7.37}$$

and

$$r_{i\mathbf{y}} = \sum_{\ell=1}^{n} u_i(x_{i\ell}) h(\mathbf{y} \mid \mathbf{x}_\ell). \tag{7.38}$$

These are generalisations of (4.23) and (4.22). If $\mathbf{y}$ has a discrete distribution, $n_{\mathbf{y}}$ may be interpreted as the expected number of sample members located at $\mathbf{y}$, given the data. If $\mathbf{y}$ is continuous, it is the density function of the expectation. In either event we may think of it, loosely, as specifying the distribution of individuals in the latent space.

If $u_i(x_{i\ell})$ is regarded as a score attached to individual ℓ on variable i then $\sum_{\ell=1}^{n} u_i(x_{i\ell})$ is the total score associated with variable i. The quantity $r_{i\mathbf{y}}$ then specifies how that total score is distributed across the latent space. In this notation the estimation equations may be written

$$\int y_j r_{i\mathbf{y}} d\mathbf{y} = -\int y_j n_{\mathbf{y}} \frac{d \ln G_i(\theta_i)}{d\theta_i} d\mathbf{y} \quad (i = 1, 2, \ldots, p : j = 1, 2, \ldots, q). \tag{7.39}$$

The notation $r_{i\mathbf{y}}$ and $n_{\mathbf{y}}$ has been chosen to bring out the similarity between the general equations and those in the special case of the latent trait model discussed in Chapter 4. The E-M algorithm used there applies here also. Its essential steps are as follows:

1. Choose initial values of $\{r_{i\mathbf{y}}\}$ and $n_{\mathbf{y}}$.
2. Solve (7.39) for $\{\alpha_{ij}\}$.
3. Use these estimates to give new values of $r_{i\mathbf{y}}$ and $n_{\mathbf{y}}$.
4. Repeat the cycle until convergence is achieved.

In practice the integrals in (7.39) are approximated by sums over a suitable grid of points in the y-space. The choice of starting values for $\{r_{i\mathbf{y}}\}$ and $n_{\mathbf{y}}$ is made by specifying a discrete version of $h(\mathbf{y} \mid \mathbf{x}_\ell)$.

The algorithm takes an especially simple form when all of the manifest variables are conditionally normal. For, in that case, (7.23) is linear in the αs and so step 2 involves the solution of a set of linear simultaneous equations. In Section 3.5 we gave a different E-M algorithm for fitting the NLFM.

7.4.4 Interpretation of latent variables

In Chapter 2 we described three ways in which the question of interpretation – or naming – of the latent variables could be approached. All of these led to comparisons among the αs and, in particular, on their relative values. When the manifest variables are of different types some care is needed because the interpretation depends on the scaling of the xs. A Bernoulli variable takes only values 0 and 1 whereas a normal variable varies continuously with an unknown standard deviation which has to be estimated. In order to interpret the latent variables we shall therefore have to ensure that the αs are calibrated so that they may be meaningfully compared across variable types. This may be done in a variety of ways, but our aim will be to adopt a parameterisation which keeps the interpretation as close as possible to the familiar methods of traditional factor analysis. Since these questions are best addressed in relation to particular applications, we shall take them up again in Examples 7.8.1 and 7.8.2.

7.5 Sampling properties and goodness of fit

The sampling properties of the parameter estimates can be found in the mixed case by exactly the same method as when all of the manifest variables have the same distribution. Thus if $\boldsymbol{\beta}$ is the vector of all parameters to be estimated in a particular model, the asymptotic variances and covariances are given by the elements of the inverse of the information matrix. As before, this may be approximated by

$$I(\hat{\beta}) = \left\{ \sum_{h=1}^{n} \frac{1}{\{f(\mathbf{x}_h)\}^2} \frac{\partial f(\mathbf{x}_h)}{\partial \beta_j} \frac{\partial f(\mathbf{x}_h)}{\partial \beta_h} \right\}^{-1}_{\beta=\beta^*} . \tag{7.40}$$

Similar problems to those encountered before concerning the accuracy of the approximation will arise if some of the parameters take extreme values. It is therefore preferable to supplement or replace the asymptotic sampling theory by bootstrap calculations.

No global methods exist for testing the goodness of fit of the mixed model. When all manifest variables were discrete, we based such a test on the observed and expected values of the frequencies of the score patterns – that is, we compared the observed with the expected joint frequency distributions. In the case of normal, continuous variables, the closeness of the observed and expected distributions was judged by the respective covariances. With other types of continuous variable we would have to look for similar relevant parameters of their joint distribution. If the manifest variables are of mixed type, it is not immediately clear how the overall fit should be judged, and further research is needed. Nevertheless, it is possible to make partial judgements of fit by utilising the notion of residuals used for this purpose in earlier chapters. Thus, for example, we can compare the observed and expected frequencies for two- or three-way margins of the categorical variables. Similarly, we can compare observed and expected correlation matrices for any normal continuous variables.

If our main concern is with model selection, as when we wish to choose the appropriate number of latent variables to include in the model, a goodness-of-fit test is not the most appropriate instrument. For that we need a model selection criterion such as the AIC proposed by Akaike (1983, 1987) or the BIC proposed by Schwarz (1978). The AIC takes account of the maximised value of the likelihood and the number of parameters estimated, while the BIC also takes account of the sample size. Those criteria, first given in Section 3.7.1, can be applied regardless of the type of variable and are given by

$$\text{AIC} = -2(\text{max log-likelihood}) + 2(\text{number of parameters}).$$

and

$$\text{BIC} = -2(\text{max log-likelihood}) + \ln(n)(\text{number of parameters}).$$

The model with the smallest value of AIC or BIC is judged to be the most appropriate. Other variants of this approach are available. They are reviewed and compared in Sclove (1987).

7.6 Mixed latent class models

As noted in Section 7.1, the treatment which we have given in the foregoing sections is sufficiently general to cover latent class models. In the case of just two latent classes, all that is necessary is to have a single binary latent variable. More latent classes can be accommodated by the device described in Section 6.4. Here we shall illustrate the method when the manifest variables are a mixture of normal and binary variables.

In the case of two types of variable the joint distribution may be written

$$f(\mathbf{x}) = \sum_{j=0}^{K-1} \eta_j g(\mathbf{x}_1 \mid j)\, g(\mathbf{x}_2 \mid j), \tag{7.41}$$

where $\mathbf{x}_1$ is the vector of normal variables, say, $\mathbf{x}_2$ is that of the binary variables and K is the number of classes. Since the normal distribution depends on two parameters, the mean and the variance, we must specify how these parameters depend on the latent class. In this example we shall suppose that the mean depends on the latent class but that the variance is constant across classes. Thus we suppose that

$$x_{1\ell j} \sim N(\mu_{\ell j}, \sigma_\ell^2) \quad (j = 0, 1, \ldots, K-1;\ \ell = 1, 2, \ldots, p_1). \tag{7.42}$$

For the binary variables we assume

$$P(x_{2\ell j} = 1 \mid j) = \pi_{\ell j} \quad (j = 0, 1, \ldots, K-1;\ \ell = 1, 2, \ldots, p_2). \tag{7.43}$$

The log-likelihood is then

$$\begin{aligned} L &= \sum_{h=1}^{n} \ln f(\mathbf{x}_h) \\ &= \sum_{h=1}^{n} \ln \sum_{j=0}^{K-1} \eta_j \left[\prod_{\ell=1}^{p_i} (2\pi)^{-1/2} \sigma_\ell^{-1/2} \exp\left\{ -\frac{1}{2\sigma_\ell^2} (x_{1\ell h} - \mu_{\ell j})^2 \right\} \right] \\ &\quad \times \left[\prod_{\ell=1}^{p_2} \left\{ \pi_{\ell j}^{x_{2\ell h}} (1 - \pi_{\ell j})^{1 - x_{2\ell h}} \right\} \right]. \end{aligned} \tag{7.44}$$

Finding the partial derivatives with respect to the unknown parameters then gives the following estimating equations:

$$\hat{\eta}_j = \sum_{h=1}^{n} h(j \mid \mathbf{x}_{1h}, \mathbf{x}_{2h})/n, \tag{7.45}$$

$$\hat{\pi}_{\ell j} = \sum_{h=1}^{n} x_{2\ell h} h(j \mid \mathbf{x}_{1h}, \mathbf{x}_{2h})/n\hat{\eta}_j, \tag{7.46}$$

$$\hat{\mu}_{\ell j} = \sum_{h=1}^{n} x_{1\ell h} h(j \mid \mathbf{x}_{1h}, \mathbf{x}_{2h})/n\hat{\eta}_j, \tag{7.47}$$

$$\hat{\sigma}_\ell^2 = \frac{\sum_{h=1}^{n}\sum_{j=0}^{K-1}(x_{1\ell h} - \mu_{\ell j})^2 h(j \mid \mathbf{x}_{1h}, \mathbf{x}_{2h})}{\sum_{h=1}^{n}\sum_{j=0}^{K-1} h(j \mid \mathbf{x}_{1h}, \mathbf{x}_{2h})}, \tag{7.48}$$

where

$$h(j \mid \mathbf{x}_{1h}, \mathbf{x}_{2h}) \propto \eta_j g(\mathbf{x}_{1h} \mid j) g(\mathbf{x}_{2h} \mid j)$$

is the posterior probability that an individual with response pattern $\mathbf{x}_h = (\mathbf{x}_{1h}, \mathbf{x}_{2h})$ is in class j ($j = 0, 1, \ldots, K-1$). These equations can be solved in the usual way as follows:

1. Choose initial values for the posterior probabilities.
2. Obtain first approximations for $\hat{\eta}_j$, $\hat{\pi}_{\ell j}$, $\hat{\mu}_{\ell j}$ and $\hat{\sigma}_\ell^2$.
3. Use these to re-estimate the posterior probabilities.
4. Repeat the cycle until convergence is achieved.

See Example 7.8.3 below.

7.7 Posterior analysis

Posterior analysis flows from the conditional distribution of $\mathbf{y}$ given $\mathbf{x}$ which, for the case of variables of mixed type, is given by (7.3) and (7.4). It is clear from (7.4) that the jth component X_j is simply the sum of the components for the constituent subsets of variables of each type. Thus, for example, if we have normal and binary manifest variables the components will be

$$X_j = \sum_{\ell=1}^{p_1} \frac{\lambda_{\ell j}}{\psi_\ell} x_{1\ell} + \sum_{\ell=1}^{p_2} \alpha_{\ell j} x_{2\ell}. \tag{7.49}$$

In the case of a single latent variable, Theorem 2.15.1 applies. This shows that the component ranks individuals in the same order as does the conditional expectation

$E(y \mid \mathbf{x})$. For most practical purposes the components will serve. However, if we need the expectations or the posterior variances it will be necessary to compute them by numerical integration starting with the posterior distribution of (7.3).

For the latent class model, posterior analysis involves the determination of the posterior probabilities of falling in the latent classes. These, too, will be functions of the components.

7.8 Examples

Several examples where the manifest variables are mixtures of binary and continuous (treated as normal) variables are given in Moustaki (1996) and Moustaki and Papageorgiou (2004). Further examples covering a wider range of types of variables, including multinomial and gamma variates, will be found in Moustaki and Knott (2000a). Here we use two of these examples to illustrate the use and interpretation of the models of this chapter. For our purposes it is not necessary to enquire into questions such as the number and choice of items or the justification of the distributional assumptions. More detail will be found in the original papers. Here we shall concentrate on the fitting and interpretation of the models.

Example 7.8.1 The origin of these data come from a study of the phenomenon of 'flashbulb' memories reported in Wright *et al.* (1994). In the present example 485 individuals were asked about what they could remember of the Hillsborough football disaster on 15 April 1989. There were five questions, the first two of which were as follows.

Q.1 Taking your answer from this list, how clear is your recollection of the event?

Cannot remember it. (1)
Vague. (2)
Fairly clear. (3)
Clear. (4)
Completely clear. (5)

Q.2 Thinking back to when you first heard about the disaster, can you remember – just answer yes or no.

Where you were.
Who you were with.
How you heard about it.
What you were doing.

Answers to the four items in Question 2 were coded 1 for yes and 0 for no. Answers to Question 1 were scored on an equally spaced scale as indicated above. The distribution

Table 7.1 Hillsborough disaster: estimates for the single latent trait model.

Variable(x_1)	ℓ	$\hat{\mu}_\ell$	$\hat{\lambda}_{\ell 1}$	$\hat{\psi}_\ell$	$\hat{\lambda}^*_{\ell 1}$
Vividness of recollection	1	3.19 (0.05)	0.56 (0.05)	0.75 (0.06)	0.54
Variable($\mathbf{x}_2$)		$\hat{\alpha}_{\ell 0}$	$\hat{\alpha}_{\ell 1}$	$\hat{\pi}_\ell$	$\hat{\alpha}^*_{\ell 1}$
Where you were	1	4.41 (1.54)	6.45 (2.33)	0.99	0.988
Who you were with	2	2.91 (0.73)	5.29 (1.31)	0.95	0.983
How you heard about it	3	5.55 (0.83)	3.24 (0.61)	0.99	0.955
What you were doing	4	1.39 (0.19)	2.03 (0.28)	0.80	0.897

was roughly normal, so the problem was treated as one with four binary variables and one normal variable.

In fitting a latent variable model we are asking whether the variation between individuals can be explained by the supposition that they are located in a small number of latent dimensions or classes. Moustaki (1996) investigated the fit of a latent trait model with one and two dimensions and a latent class model with two classes. The two-dimensional model was marginally poorer than the one-dimensional model, and we only report the latter. The parameter estimates are given in Tables 7.1 and 7.2. Asymptotic standard errors are given in brackets in Table 7.1.

The interpretation of the latent class model fitted in Table 7.2 is straightforward. In the first class individuals have a high probability of being able to remember the circumstances in which they heard the news, whereas those in the second have, with the partial exception of x_3, a much lower probability. Those in the first class score higher on the x_1 scale (mean 3.44) than those in the second (mean 2.53). This is consistent with the interpretation suggested by the x_2-variables, though one might perhaps have expected a bigger difference in the means given the very clear separation in the πs. A difference of barely one unit on a scale from 1 to 5 and a standard deviation

Table 7.2 Hillsborough disaster: estimates for a latent class model with two classes.

Variable(x_1)	ℓ	$\hat{\mu}_{10}$	$\hat{\mu}_{11}$	$\hat{\sigma}^2_1$
Vividness of recollection	1	3.44	2.53	0.90
Variable ($\mathbf{x}_2$)		$\hat{\pi}_{\ell 0}$	$\hat{\pi}_{\ell 1}$	
Where you were	1	0.979	0.141	
Who you were with	2	0.933	0.090	
How you heard about it	3	0.995	0.774	
What you were doing	4	0.863	0.274	
$\hat{\eta}_j$		0.724	0.276	

of 0.95 ($\sqrt{0.9}$) suggests that the reported vividness of the recollection correlates rather weakly with the class to which the individual belongs.

The interpretation of the latent trait model needs some care. If we look first at the part concerning the variables $\mathbf{x}_2$ we find the coefficients $\{\alpha_{\ell 1}\}$ are typical of what we would expect where there is a dominant latent variable with which each item is strongly correlated. Further, the πs show that the median individual on the latent scale is very likely to remember all the circumstances. The complications arise when we attempt to link the results for the continuous variable with those for the remainder. The question is how to calibrate the two types of loading. In particular, does $\hat{\lambda}_{\ell 1} = 0.56$ represent a bigger 'loading' than, say, $\hat{\alpha}_{21} = 5.29$? To answer such questions we turn back to the approaches to interpretation given in Chapter 2 (especially Section 2.12). One is via the components. For this example the component is

$$X = 0.75x_1 + 6.45x_{21} + 5.29x_{22} + 3.24x_{23} + 2.03x_{24}.$$

The importance of variables is judged by the impact which they have on the value of X. For example, the presence or absence of x_{22} (which is a binary variable) changes X by an amount 5.29. If x_1 changes from 1, at one end of its range, to 5, X will change by an amount (4 × 0.56)/0.75=2.99. A change of 1 unit in the value of x_1 produces a change of 0.75 in X. It seems, therefore, that x_1 is certainly less important than x_{22} and, by the same token, than any of the other binary variables. This is consistent with the results of the latent class analysis. A second approach to interpretation, used previously for the normal model, is to look at the correlations between the manifest and latent variables. For the continuous part of this model the correlation between x_1 and y is

$$\hat{\lambda}^*_{\ell 1} = \frac{\hat{\lambda}_{11}}{\sqrt{\hat{\lambda}^2_{11} + \hat{\psi}_1}} = 0.54, \tag{7.50}$$

as given in the last column of Table 7.1. One could calculate the correlation between each of the binary variables and the latent variable, but this is not strictly comparable with a product-moment correlation between two continuous variables. A better method is to use the equivalence of the binary model and an underlying variable model established in equation (4.31). The correlation between the underlying variable and the latent variable is then obtained by replacing λ_{ij} by $\alpha_{ij}\psi_i^{1/2}$. The standardised version of α_{ij} thus becomes

$$\frac{\alpha_{ij}\psi_i^{1/2}}{\sqrt{\alpha^2_{ij}\psi_i + \psi_i}} = \frac{\alpha_{ij}}{\sqrt{\alpha^2_{ij} + 1}} = \alpha^*_{ij}, \tag{7.51}$$

say. In the general case, when there are q latent variables,

$$\alpha^*_{ij} = \frac{\alpha_{ij}}{\sqrt{\sum_{j=1}^{q} \alpha^2_{ij} + 1}}. \tag{7.52}$$

The estimated standardised αs are given in the last column of Table 7.1. They show a very strong correlation with the latent variables – substantially greater than that for the variable x_1 represented by the loading $\hat{\alpha}_{11}^* = 0.54$.

Taking the two analyses together it is clear that the variation in the ability of individuals to recall salient facts about the circumstances in which they heard of the Hillsborough disaster is satisfactorily explained by the hypothesis of a single latent variable. It is equally clear that the binary variables are much better indicators of that latent dimension than the subjectively assessed vividness of that memory.

Example 7.8.2 This example is taken from Moustaki and Knott (1997) and is chosen to illustrate the use of the model to construct a measurement scale from variables of mixed type. The items are from the Enquête sur la Consommation 1990 Survey of the Office Fédéral de la Statistique. Information was collected from a survey of 1963 households. After removing cases and variables judged to be inappropriate, the sample used consisted of 923 households with five variables as follows:

1. Presence of a dishwasher (1/0) [DishWasher]
2. Presence of a video recorder (1/0) [Video]
3. Presence of a car (1/0) [Car]
4. Equivalent food expenditures [FoodExp]
5. Equivalent expenditures for clothing [ClothExp]

The first three items are binary and items 4, 5 are metrical. Judged by their frequency distributions it seemed reasonable to treat item 4 as normally distributed and item 5 as gamma distributed. Both of the metrical variables were standardised so that item 4 had mean zero and variance one and item 5 variance one. The one-factor model gave a good fit to the binary variables taken alone, judging from the two- and three-way margins. The AIC was 16 546 for the one-factor model and 16 544 for the two-factor model.

The parameter estimates for the full model are given in Table 7.3. From the last column of the table we see that the median individual is very likely to have a car ($\hat{\pi}_3 = 0.91$) and less likely to have a dish washer ($\hat{\pi}_3 = 0.40$) and video recorder ($\hat{\pi}_3 = 0.38$).

As an index of welfare we might then use the component given by

$$X = 1.49x_1 + 0.74x_2 + 1.72x_3 + 0.67x_4 + 0.46x_5$$

where the xs are listed in the same order as in Table 7.3.

Example 7.8.3 We conclude this section with an example taken from Moustaki and Papageorgiou (2004). It is an application of latent class analysis for mixed binary and continuous data in archaeometry. The 73 sample terracotta objects are majolica vases and floor tiles, manufactured between the sixteenth and eighteenth centuries. Most

Table 7.3 Measurement of welfare: estimates and standard errors for the one-factor latent trait model.

Binary	$\hat{\alpha}_{i0}$	$\hat{\alpha}_{i1}$	$\hat{\pi}_i$
DishWasher	−0.43 (0.07)	1.49 (0.15)	0.40
Video	−0.51 (0.05)	0.74 (0.08)	0.38
Car	2.30 (0.15)	1.72 (0.18)	0.91
Normal	$\hat{\alpha}_{i0}$	$\hat{\alpha}_{i1}$	$\hat{\Psi}_i$
FoodExp	0.00 (0.02)	0.50 (0.03)	0.75 (0.03)
Gamma	$\hat{\alpha}_{i0}$	$\hat{\alpha}_{i1}$	$\hat{\nu}_i^{-1}$
ClothExp	−0.74 (0.02)	0.18 (0.01)	0.39 (0.01)

of them can be found in the Museo Regionale della Ceramica of Caltagirone. There are 21 metric variables that provide chemical information and 12 binary variables from the petrological analysis. A latent class model is fitted to the 33 variables. The investigation of univariate, bivariate and trivariate chi-squared residuals suggested that a three-class solution is adequate: 33% of the objects fall in class I, 37% in class II and 30% in class III. The estimates for the conditional class probabilities and the parameters of the metric variables, with standard errors in brackets, are given in Tables 7.4 and 7.5 respectively. More specifically, Table 7.4 shows that variables 4 and 8 do not discriminate between the three classes whereas variables 3 and 7 do not discriminate between classes 1 and 3.

The estimated posterior class-membership probability was greater than 0.99 for each object. Therefore, the allocation into classes is almost deterministic. The classification obtained was the same as the grouping with respect to the origin of the objects (class I contains ceramics from Napoli, class II contains ceramics from Caltagirone and class III from Palermo).

7.9 Ordered categorical variables and other generalisations

All of the models considered in this chapter emerge as special cases of the sufficiency principle. However, we have already noted in Chapters 5 and 6 that there is an alternative way of modelling ordered categorical variables which does not fall within this framework. In that approach we adopted a logit model for the cumulative probability that the response occurred in a given category or one which ranked higher. This was shown to be equivalent to an underlying variable model in which the categories were formed by grouping a continuous distribution. This, in turn, offered a simple, but

Table 7.4 Terracotta data: estimated conditional probabilities, $\hat{\pi}_{ij}$, for the binary variables and class probabilities, $\hat{\eta}_j$, with standard errors in brackets, for the three-class model.

Variables	Class I $\hat{\pi}_{i1}$	Class II $\hat{\pi}_{i2}$	Class III $\hat{\pi}_{i3}$
1	0.04 (0.04)	0.11 (0.06)	0.18 (0.08)
2	0.66 (0.19)	0.59 (0.09)	0.00 (0.02)
3	1.00 (0.01)	0.44 (0.09)	1.00 (0.01)
4	1.00 (0.06)	0.85 (0.07)	1.00 (0.00)
5	0.50 (0.10)	0.08 (0.05)	0.00 (0.01)
6	0.08 (0.06)	0.11 (0.06)	0.27 (0.09)
7	1.00 (0.01)	0.19 (0.08)	1.00 (0.01)
8	0.92 (0.06)	0.93 (0.05)	1.00 (0.00)
9	0.17 (0.08)	0.07 (0.05)	0.82 (0.08)
10	0.12 (0.07)	0.18 (0.07)	0.05 (0.05)
11	0.46 (0.10)	0.59 (0.09)	0.64 (0.11)
12	0.50 (0.10)	0.81 (0.07)	0.77 (0.09)
$\hat{\eta}_j$	0.33 (0.05)	0.37 (0.06)	0.30 (0.05)

not fully efficient, method of estimating the parameters by estimating first the correlations between the underlying variables. If ordered categorical variables are mixed with normal variables, the same approach can be used except that different kinds of correlations will be involved. There will be polychoric correlations for pairs of categorical variables, polyserial correlations where one member is categorical and the other continuous, and product-moment correlations for pairs of continuous variables. For full maximum likelihood, the form of the likelihood will be exactly the same as for the unordered model derived from the sufficiency principle, but the πs will now be given by the expression in (5.79).

All of these models, together with those treated earlier in the chapter, are included within the general framework provided by Arminger and Küsters (1988). Their treatment starts, like this book and earlier related work, from the basic idea that all variables, latent and manifest, are random variables. Whether or not such variables are continuous, categorical or whatever, is then a secondary matter which is irrelevant to the essential structure of the model which is expressed in the various conditional distributions. Once this framework is accepted, there is no problem about having manifest variables of mixed type. Arminger and Küsters (1988) treat the estimation problem at a sufficient level of generality to cover all the problems of this chapter and many more. The lack of software needed to implement this approach may have inhibited the wider recognition of its generality.

Arminger and Küsters's model goes beyond the subject of this chapter in several respects. It provides for relationships between the latent variables to which we come in the next chapter. It also allows for other exogenous manifest variables which are assumed to act causally on the latent variables. Of more relevance to the content

Table 7.5 Terracotta data: estimated within-class means and variances for the metric variables, with standard errors in brackets, for the three-class model.

Variables	Class I $\hat{\mu}_{i1}$	Class II $\hat{\mu}_{i2}$	Class III $\hat{\mu}_{i3}$	$\hat{\sigma}_i^2$
1	0.67 (0.20)	0.05 (0.13)	−0.80 (0.17)	0.64
2	−0.72 (0.06)	1.04 (0.13)	−0.49 (0.15)	0.34
3	−0.29 (0.12)	−0.41 (0.17)	0.82 (0.21)	0.70
4	−0.74 (0.09)	−0.05 (0.14)	0.87 (0.22)	0.58
5	0.59 (0.15)	−0.25 (0.21)	−0.34 (0.18)	0.81
6	−0.36 (0.12)	1.06 (0.11)	−0.90 (0.08)	0.29
7	0.31 (0.34)	−0.14 (0.02)	−0.16 (0.03)	0.94
8	0.54 (0.17)	0.26 (0.12)	−0.90 (0.19)	0.63
9	0.24 (0.17)	0.73 (0.10)	−1.16 (0.08)	0.37
10	−0.59 (0.06)	−0.03 (0.16)	0.68 (0.26)	0.73
11	−0.55 (0.09)	−0.44 (0.14)	1.14 (0.16)	0.42
12	−0.78 (0.09)	1.17 (0.10)	−0.59 (0.06)	0.17
13	0.87 (0.15)	−0.40 (0.05)	−0.46 (0.25)	0.61
14	0.14 (0.17)	0.30 (0.18)	−0.53 (0.21)	0.86
15	0.36 (0.15)	−0.38 (0.19)	0.08 (0.22)	0.89
16	0.46 (0.23)	−0.01 (0.17)	−0.49 (0.14)	0.85
17	−0.52 (0.23)	0.60 (0.11)	−0.17 (0.18)	0.76
18	0.14 (0.23)	0.58 (0.13)	−0.86 (0.09)	0.64
19	−0.07 (0.33)	0.18 (0.08)	−0.15 (0.06)	0.97
20	−0.67 (0.09)	0.65 (0.20)	−0.07 (0.18)	0.68
21	−0.37 (0.14)	0.47 (0.23)	−0.17 (0.17)	0.85

of the present chapter, they consider another type of manifest variable obtained by censoring a continuous variable. Thus suppose a variable x is observed only if its value is greater than some threshold τ; otherwise we only know that it is in the interval $(-\infty, \tau)$. Such a variable might be described as partly continuous and partly categorical. If we define a new variable

$$x^* = \begin{cases} x & \text{if } x > \tau, \\ \tau & \text{if } x \leq \tau, \end{cases} \tag{7.53}$$

the transformation from x to x^* is called the *tobit* relation. For further details, see Amemiya (1984).

There is a way of dealing with ordered categorical variables within the framework of the sufficiency principle which merits further investigation. We saw in Chapter 5 that the multinomial model involved parameters which can be regarded as category scores (the $\boldsymbol{\alpha}$s of (5.20)). An ordering of the categories implies an ordering of the αs, and thus we could use this ordering information by maximising the likelihood subject to the appropriate inequalities on the αs. This is more general than the cumulative response probability model because it allows us to impose partial orders. For example, if we have ordered responses such as *strongly agree, agree, disagree, strongly disagree* together with *don't know*, we may not know where to place *don't know* in the ranking. We can, therefore, order the four ranked categories and leave the fitting procedure to determine where to place the *don't know* category as in Example 5.12.2.

The maximisation of a likelihood subject to order restraints is a problem in isotonic regression, but general solutions for functions such as the likelihoods arising here do not appear to be available. However, there is one case in which the solution is immediate. If we maximise *without restraints* and find that the resulting parameter estimates satisfy the required order restrictions, then the solutions of the restrained and unrestrained problems are identical. Modest inversions of the ordering, within the limits of sampling variation, might be regarded as consistent with the predicted order. This is clearly a matter where further research is needed.

8

Relationships between latent variables

8.1 Scope

In this chapter we focus on two closely related topics which, in the factor analysis tradition, are usually referred to as *confirmatory factor analysis* and *linear structural relations modelling* (LISREL). Some work has been done in this area for categorical or non-normal continuous manifest variables, but we shall concentrate mainly on the normal model. Nevertheless we shall set the results in the broader framework of the GLLVM partly to provide a springboard for future research and partly to give greater insight into the scope and limitations of the models. We shall make no attempt to provide a comprehensive account of linear structural relations modelling which has been the subject of several books (Bollen 1989; Bollen and Long 1993; Cuttance and Ecob 1987; Kaplan 2009; Lee 2007; Loehlin 1989; Mulaik 2009b; Saris and Stronkhorst 1984) and a number of widely used software packages such as LISREL (Jöreskog and Sörbom 2006), Mplus (Muthén and Muthén 2010), AMOS (Arbuckle 2006) and EQS (Bentler 2008). Instead our aim will be to locate these methods within the framework we have adopted in the earlier part of this book. In this way we shall be able to identify serious issues of identifiability which place limits on the usefulness of the methods.

8.2 Correlated latent variables

In formulating the GLLVM we supposed the latent variables to be standard normal and uncorrelated. The normality, we emphasised, was a matter of convention chosen because of its mathematical convenience. The justification for this lay in the fact

Latent Variable Models and Factor Analysis: A Unified Approach, Third Edition.
David Bartholomew, Martin Knott and Irini Moustaki.

that the form of the prior (latent) distribution was indeterminate unless arbitrary constraints were imposed on the conditional distributions of the manifest variables. More informally, if the latent variable is a construct then we are at liberty to 'construct' it to have any distribution we please.

The assumption that the latent variables be uncorrelated might seem less easy to defend since there seem to be no prior reasons to expect substantively interesting latent variables to be independent. However, independence of the latent variables was simply a convenient starting point since we found that oblique rotations in the factor space generated other models which did have correlated factors. The reason for not introducing correlated factors in the first place was simply that their correlation matrix would not have been identified. Different factor structures having different correlation structures would have yielded exactly the same likelihood. In practice, therefore, the correlation structure was 'estimated' in such a case by choosing that rotation which seemed most meaningful in substantive terms. There could be no empirical ground for preferring one rotation to another.

There are, however, circumstances in which there is prior knowledge, or a hypothesis we wish to explore, about the factor structure, which removes the indeterminacy and so makes it possible to estimate the correlations between the factors as well as their loadings. This arises when there are constraints on the factor loadings. Most commonly these are in the form of loadings assumed to be zero but could be in the form of equalities among subsets of loadings. If the loadings matrix $\mathbf{\Lambda}$ of the GLLVM contains zeros in particular positions then they will not, in general, persist under linear transformations. If the only admissible solutions are those with zeros in these positions then rotation will destroy that structure.

In these circumstances it may be possible to estimate the correlation matrix of the factors in addition to the non-zero loadings. We then replace the standard normal prior distribution with a standard multivariate normal distribution with correlation matrix $\mathbf{\Phi}$. In the case of the normal linear factor model the covariance matrix then becomes

$$\mathbf{\Sigma} = \mathbf{\Lambda}\mathbf{\Phi}\mathbf{\Lambda}' + \mathbf{\Psi}. \tag{8.1}$$

The sufficient condition for the model to be identified is that the number of unknown parameters on the right-hand side of (8.1) be no larger than $\frac{1}{2}p(p-1)+p$, which is the number of sample variances and covariances. If f is the number of free parameters in $\mathbf{\Lambda}$, then on the right-hand side of (8.1) there are

$$f + \frac{1}{2}q(q-1) + p$$

parameters and this must not exceed $\frac{1}{2}p(p+1)$.

Software packages, such as LISREL and Mplus, provide maximum likelihood estimates of the parameters together with standard errors and various other quantities relevant to inference about the model.

8.3 Procrustes methods

A variation on the above approach is first to fit an unrestricted model and then to estimate a rotation matrix $\mathbf{M}$ such that the pattern of the matrix of the rotated loadings is as close as possible to that desired. The correlation matrix for the factors obtained by this method will then be $\mathbf{\Phi} = \mathbf{MM}'$. This procedure is very similar to that used in the search for simple structure when we look for a rotation yielding a loadings matrix with only one non-zero element in each row. The difference is that in confirmatory mode the zeros are specified in advance whereas in the search for simple structure *any* such solution is acceptable. The name 'Procrustes' is taken from the figure in Greek mythology who adjusted the lengths of his victims to fit his bed. The problem is a long-standing one to which contributions have been made by Green (1952), Cliff (1966), Schönemann (1966), Lawley and Maxwell (1971, Sections 6.5 and 6.6) and Brokken (1983). The Procrustes approach gives a better fit at the price of a poorer reproduction of the desired pattern of loadings. Current practice uses the readily available software for estimating the confirmatory factor analysis model with fixed zeros, as in Section 8.2, rather than Procrustes methods.

8.4 Sources of prior knowledge

There are a number of different circumstances in which prior knowledge about the structure of the loading matrix may be available, and these give rise to different forms of analysis.

The first arises when an exploratory analysis shows some loadings to be close to zero. We may then suspect that those loadings should be exactly zero and wish to test that hypothesis using a new set of data. A zero coefficient means that the latent variable to which it is attached exerts no influence on the manifest variable corresponding to that row of the loading matrix. This may be a plausible hypothesis because we often find sets of variables for which a one-factor model provides a good fit. If some or all of these variables were to be embedded in a larger set depending on other factors we would not expect those items to be influenced by any other factor introduced in this fashion.

Another circumstance where we may wish to impose constraints is where there is a dominant general factor. It often happens that the loadings of the first factor turn out to be large and roughly equal. One might then wish to test the hypothesis that they were precisely equal. This situation arises with the logit/normal model of Chapter 4. If, in that case, $\alpha_{i1} = \alpha_1$ $(i = 1, 2, \ldots, p)$ we obtain the random effects Rasch model. The first component is then simply the sum of the binary indicators. This has a simple direct interpretation as the 'number of correct answers' (in the educational test context) and this model permits a stronger form of measurement (see Bartholomew (1996), Section 8.5).

In essence we are seeking empirical justification for using a simpler model by looking at sub-models nested within a larger family. If there are at least two factors involved, the simplification resulting from constraining the loadings matrix may enable us to estimate the correlation between the factors as we saw above.

8.5 Linear structural relations models

Another circumstance in which we may wish to estimate the correlation matrix of the factors is when we are interested in elucidating causal relationships between latent variables. Such analyses for manifest variables have a long history going back to Wright (1921) and have been given a new impetus by the development of so-called graphical models; see, for example, Whittaker (1990) and Cox and Wermuth (1996). It is a natural extension of such models to suppose that the causal relationships exist between the latent variables, rather than the manifest variables. As we shall see in Section 8.6, linear relationships among the latent variables determine the correlation matrix. Conversely, if we can estimate the correlation matrix we may hope to infer what linear structural equations have given rise to it.

We begin with a special, but important, case of the general model. Suppose there are q latent variables $\mathbf{y}' = (y_1, y_2, \ldots, y_q)$. For y_j we have a set of indicators $\mathbf{x}'_j = (x_{j1}, x_{j2}, \ldots, x_{jp_j})$, where p_j is the number of indicators for y_j. If we assume that the indicators and latent variables are related in the manner of a standard normal factor model, we may write

$$\mathbf{x}_j = \boldsymbol{\mu}_j + \boldsymbol{\lambda}_j y_j + \mathbf{e}_j \quad (j = 1, 2, \ldots, q), \tag{8.2}$$

where $\boldsymbol{\lambda}_j$ is a column vector of loadings. These equations may be collected together and written as

$$\mathbf{x} = \boldsymbol{\mu} + \boldsymbol{\Lambda}\mathbf{y} + \mathbf{e}, \tag{8.3}$$

where the vectors $\mathbf{x}_j$, $\boldsymbol{\mu}_j$ and $\mathbf{e}_j$ are stacked vertically to give $\mathbf{x}$, $\boldsymbol{\mu}$ and $\mathbf{e}$ respectively. The matrix $\boldsymbol{\Lambda}$ then has the form

$$\boldsymbol{\Lambda} = \begin{bmatrix} \boldsymbol{\lambda}_1 & \mathbf{0} & \mathbf{0} & \ldots & \mathbf{0} \\ \mathbf{0} & \boldsymbol{\lambda}_2 & \mathbf{0} & \ldots & \mathbf{0} \\ \mathbf{0} & \mathbf{0} & \boldsymbol{\lambda}_3 & \ldots & \mathbf{0} \\ \vdots & \vdots & \vdots & & \vdots \\ \mathbf{0} & \mathbf{0} & \mathbf{0} & \ldots & \boldsymbol{\lambda}_q \end{bmatrix}, \tag{8.4}$$

which corresponds to what, in another context, we have called simple structure.

Provided that the estimation problem is identifiable we can now proceed to estimate the parameters, including $\boldsymbol{\Phi}$, the correlation matrix of the ys. The number of free parameters in $\boldsymbol{\Lambda}$ is $f = \sum_{j=1}^{q} p_j = p$ and so the problem will not be identifiable if

$$2p + \frac{1}{2}q(q-1) > \frac{1}{2}p(p+1)$$

or $q(q-1) > p(p-3)$. If $q = 4$, for example, p must be as least equal to 6 which implies an average of at least $2\frac{1}{2}$ indicators per latent variable.

The foregoing merely tells us something about the intercorrelations of the latent variables. We may wish to go further and explore the causal relationships which gave rise to those correlations. Patterns of such relationships are conveniently expressed as path diagrams, of which the following is a very simple example:

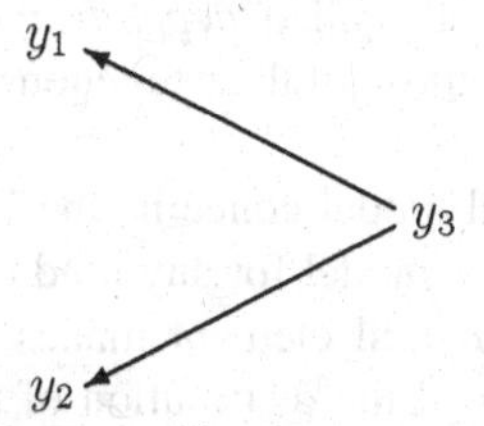

This says that y_1 and y_2 are both influenced by y_3 but are not directly related to one another. There will be, of course, a correlation between y_1 and y_2 resulting from their common dependence on y_3, but there is no causal link. If we are prepared to make an assumption about the relationships in the path diagram we can link their parameters to the correlation matrix of the ys and so have a means of estimating them.

Suppose that we assume that y_1 and y_2 have a simple linear regression on y_3 so that

$$y_1 = a_1 + b_1 y_3 + \epsilon_1 \quad \text{and} \quad y_2 = a_2 + b_2 y_3 + \epsilon_2,$$

with the usual distributional assumptions about the error terms. We are not now free to specify the means and variances of all the ys but only that of y_3. If we suppose $y_3 \sim N(0, 1)$ then

$$\begin{aligned} E(y_1) &= a_1, & E(y_2) &= a_2, \\ \text{var}(y_1) &= b_1^2 + \delta_1, & \text{var}(y_2) &= b_2^2 + \delta_2, \\ \text{corr}(y_1, y_3) &= \frac{b_1}{\sqrt{b_1^2 + \delta_1}}, & \text{corr}(y_2, y_3) &= \frac{b_2}{\sqrt{b_2^2 + \delta_2}}, \\ \text{corr}(y_1, y_2) &= \frac{b_1 b_2}{\sqrt{b_1^2 + \delta_1}\sqrt{b_2^2 + \delta_2}} = \text{corr}(y_1, y_3)\text{corr}(y_2, y_3), \end{aligned} \tag{8.5}$$

where δ_1 and δ_2 are the variances of ϵ_1 and ϵ_2, respectively. If we can estimate the correlation matrix of $\mathbf{y}$ then we can estimate the standardised regression coefficients

$b_i/\sqrt{b_i^2+\delta_i}$ $(i = 1, 2)$ of the structural relationships. Further, we can check whether the last relationship of (8.5) holds and so obtain some idea of the plausibility of the model.

In rudimentary form, this simple example illustrates the idea behind linear structural relationship modelling and its connection with confirmatory factor analysis. Before moving on to give a general treatment we make three observations.

First, a distinction has to be made between y_3 on the one hand and y_1 and y_2 on the other. We are free to choose the distribution of y_3 and, in particular, its mean and variance. There is no such freedom with y_1 and y_2 because, once y_3 is specified, the distributions of y_1 and y_2 are determined by the regression assumption. In general, the free latent variables, whose distributions are arbitrary, are called *independent* or *exogenous*. The remainder are called *dependent* or *endogenous* because their distributions are determined by those of the exogenous variables and the structure of the system.

Second, since the structural model concerns the latent variables only it can, in principle, be made the basis of a model for any kind of indicators, whether normal, categorical or mixed. The structural element makes its presence felt through the matrix $\mathbf{\Phi}$ which is a parameter of the distribution of the ys, not the xs. Essentially there are two models – the *structural model* concerning the ys and the *measurement model* which relates the xs to the ys.

Third, the model can be described in terms of prior and conditional distributions in exactly the same way as the models considered in the earlier chapters.

8.6 The LISREL model

We now generalise the simple example of the previous section to the general linear structural relations model as originally formulated by Jöreskog (see, for example, Jöreskog (1977) and Jöreskog and Sörbom (1977), which also gives references to earlier work). In the following section we shall set the model in a yet more general framework which allows a wider choice of measurement models. In this section the measurement part of the model is the normal linear factor model.

8.6.1 The structural model

We shall adopt a non-standard notation in order to emphasise the links with our treatment of earlier models. This is crucial for the important conclusions we wish to draw later about serious problems of model identification. The latent variables are divided into two groups – the *independent* and the *dependent* variables. The independent variables are denoted by $\mathbf{y}_1' = (y_{11}, y_{12}, \ldots, y_{1q_1})$ and the dependent by $\mathbf{y}_2' = (y_{21}, y_{22}, \ldots, y_{2q_2})$, with $q_1 + q_2 = q$ and $\mathbf{y}' = (\mathbf{y}_1', \mathbf{y}_2')$. The structural part of the model is then

$$\mathbf{B}\mathbf{y}_2 = \mathbf{G}\mathbf{y}_1 + \boldsymbol{\zeta}, \tag{8.6}$$

with $E(\mathbf{y}_2) = \mathbf{0}$, $E(\mathbf{y}_1) = \mathbf{0}$ and $E(\boldsymbol{\zeta}) = \mathbf{0}$. It is further assumed that $\boldsymbol{\zeta}$ is uncorrelated with $\mathbf{y}_1$ and that $\mathbf{B}$ is non-singular. The error term $\boldsymbol{\zeta}$ has a diagonal covariance matrix denoted by $\boldsymbol{\Theta}_\zeta$. The model is said to be *recursive* if the elements of $\mathbf{y}_2$ can be arranged as a sequence (usually temporal) so that each y_2 depends on $\mathbf{y}_1$ and $\boldsymbol{\zeta}$, but only on those other y_{2j} which precede it in the sequence.

8.6.2 The measurement model

This is the standard factor model $\mathbf{x} = \boldsymbol{\mu} + \boldsymbol{\Lambda}\mathbf{y} + \mathbf{e}$, but with a special structure. It is assumed that the xs can be partitioned into two parts $\mathbf{x}' = (\mathbf{x}_1', \mathbf{x}_2')$ of dimensions p_1 and p_2 respectively ($p_1 + p_2 = p$) such that the members of $\mathbf{x}_1$ depend only on those latent variables in $\mathbf{y}_1$, and those in $\mathbf{x}_2$ on those in $\mathbf{y}_2$. In other words, for measurement purposes, the independent and dependent variables have no common indicators. This being so, we can write

$$\boldsymbol{\Lambda} = \begin{bmatrix} \boldsymbol{\Lambda}_{\mathbf{x}_1} & \mathbf{0} \\ \mathbf{0} & \boldsymbol{\Lambda}_{\mathbf{x}_2} \end{bmatrix}. \tag{8.7}$$

We can similarly partition the error term into two parts, $\mathbf{e}' = (\mathbf{e}_1', \mathbf{e}_2')$, with diagonal covariance matrices $\boldsymbol{\Psi}_{e_1}$ and $\boldsymbol{\Psi}_{e_2}$ respectively.

8.6.3 The model as a whole

Under these assumptions the covariance matrix of $\mathbf{x}$ turns out to be

$$\boldsymbol{\Sigma} = \begin{bmatrix} \boldsymbol{\Lambda}_{\mathbf{x}_1}\boldsymbol{\Phi}_{\mathbf{y}_1}\boldsymbol{\Lambda}_{\mathbf{x}_1}' + \boldsymbol{\Psi}_1 & \boldsymbol{\Lambda}_{\mathbf{x}_1}\boldsymbol{\Phi}_{\mathbf{y}_1}\mathbf{G}'\mathbf{B}'^{-1}\boldsymbol{\Lambda}_{\mathbf{x}_2}' \\ \boldsymbol{\Lambda}_{\mathbf{x}_2}\boldsymbol{\Phi}_{\mathbf{y}_1}\mathbf{G}'\mathbf{B}'^{-1}\boldsymbol{\Lambda}_{\mathbf{x}_1}' & \boldsymbol{\Lambda}_{\mathbf{x}_2}(\mathbf{B}^{-1}\mathbf{G}'\boldsymbol{\Phi}_{\mathbf{y}_1}\mathbf{G}\mathbf{B}'^{-1} + \mathbf{B}^{-1}\boldsymbol{\Theta}_\zeta\mathbf{B}'^{-1})\boldsymbol{\Lambda}_{\mathbf{x}_2}' + \boldsymbol{\Psi}_2 \end{bmatrix}, \tag{8.8}$$

where $\boldsymbol{\Phi}_{\mathbf{y}_1}$ is the covariance matrix of $\mathbf{y}_1$. The matrix (8.8) is obtained by first showing that $\boldsymbol{\Phi}$, the covariance matrix of the ys, is given by

$$\boldsymbol{\Phi} = \begin{bmatrix} \boldsymbol{\Phi}_{\mathbf{y}_1} & \boldsymbol{\Phi}_{\mathbf{y}_1}\mathbf{G}'\mathbf{B}'^{-1} \\ \boldsymbol{\Phi}_{\mathbf{y}_1}\mathbf{G}'\mathbf{B}'^{-1} & \mathbf{B}^{-1}\mathbf{G}'\boldsymbol{\Phi}_{\mathbf{y}_1}\mathbf{G}\mathbf{B}'^{-1} + \mathbf{B}^{-1}\boldsymbol{\Theta}_\zeta\mathbf{B}'^{-1} \end{bmatrix} \tag{8.9}$$

and then substituting this into

$$\boldsymbol{\Sigma} = \boldsymbol{\Lambda}\boldsymbol{\Phi}\boldsymbol{\Lambda}' + \boldsymbol{\Psi}. \tag{8.10}$$

The parameters are estimated by minimising some measure of the distance between $\boldsymbol{\Sigma}$ as given by (8.10) and the sample covariance matrix $\mathbf{S}$. Unweighted least squares (ULS), generalised least squares (GLS), generally weighted least squares (WLS), and maximum likelihood (ML) are some of the methods used for estimating structural equation models (see also Bollen (1989), chapters 4 and 9). More specifically, the ULS estimator minimises the function

$$F_{\mathrm{ULS}} = \frac{1}{2}\,\mathrm{trace}[(\mathbf{S} - \boldsymbol{\Sigma})^2].$$

This is a minimisation of the sum of squares of the differences between each element of the sample covariance matrix and the covariance matrix under the specified model. The main advantage of the ULS estimation is that it leads to consistent estimates without requiring any distributional assumptions for the data. However, the ULS method is not invariant to scale transformations and the estimates obtained are different when a covariance or a correlation matrix is used.

The GLS estimator minimises the function

$$F_{\mathrm{GLS}} = \frac{1}{2}\,\mathrm{trace}[\{\mathbf{S}^{-1/2}(\mathbf{S} - \boldsymbol{\Sigma})\mathbf{S}^{-1/2}\}^2].$$

This criterion minimises a sum of squares of weighted differences between each element of the sample covariance matrix and the covariance matrix under the specified model.

The WLS estimator minimises

$$F_{\mathrm{WLS}} = (\mathbf{s} - \boldsymbol{\sigma})'\mathbf{W}^{-1}(\mathbf{s} - \boldsymbol{\sigma}),$$

where $\mathbf{W}^{-1}$ is a $p \times (p+1)/2$ positive definite weight matrix, and $\mathbf{s}$, $\boldsymbol{\sigma}$ are vectors of the elements in the lower triangle, including the diagonal elements, of $\mathbf{S}$, $\boldsymbol{\Sigma}$ respectively. Any positive definite $\mathbf{W}$ will produce consistent estimates which are not necessarily efficient. To obtain estimates with the same optimal properties as ML estimates (asymptotic multivariate normality and asymptotic efficiency), choose $\mathbf{W}$ as a consistent estimator of the covariance matrix of $\mathbf{s}$. It is possible to choose $\mathbf{W}$ so that the WLS estimate is the same as the ULS or GLS estimators.

The ML estimate maximises the likelihood or, in the context of the fit function, minimises

$$F_{\mathrm{ML}} = \ln|\boldsymbol{\Sigma}| + \mathrm{trace}(\mathbf{S}\boldsymbol{\Sigma}^{-1}) - \ln|\mathbf{S}| - p.$$

All the above estimators are consistent, but only the WLS and ML can be asymptotically efficient with asymptotic distributions which are multivariate normal to allow significance testing.

It is usual in practice to work with the standardised variables, and $\boldsymbol{\Sigma}$ then becomes the correlation matrix. The fitting can be done by using LISREL or the other covariance structure analysis programs mentioned earlier.

If the indicator variables are categorical or non-normal the measurement model must be replaced by one of the other exponential family models as appropriate.

No software is currently available for such extensions, but there is no problem, in principle, about providing it. However, in view of the serious questions of identifiability which we shall raise in the following sections it may be better to explore alternative approaches of the kind to be discussed in Section 9.4.

8.7 Adequacy of a structural equation model

One way of testing the adequacy of the specified model is to use the value of the fit function evaluated at the final estimates. More specifically, in large samples $(n-1)F_{\mathrm{GLS}}$ and $(n-1)F_{\mathrm{ML}}$ have an asymptotic chi-squared distribution when the model is correct with degrees of freedom equal to $\frac{1}{2}p(p+1)-t$, where t is the total number of free parameters being estimated and p is the total number of observed variables. For the GLS fit function there is the requirement that the weight matrix be a consistent estimator of the sample covariance matrix. For the ML fit function the requirement for the chi-squared test statistic is multivariate normality of the observed variables when a covariance matrix is being analysed. The chi-squared test statistics are very sensitive to the sample size and to deviations from normality.

Because of the limitation of the chi-squared statistic mentioned above, alternative standardised forms of the chi-squared statistic have been proposed in the literature which are less sensitive to the sample size and to the assumption that the model holds exactly in the population. Those statistics are known as fit indices. The best-known fit indices are discussed below. Fit indices are based on the sample and estimated (reproduced) covariance matrices. There is a bewildering array of such indices. We enumerate below those that are more robust.

1. The root mean square error of approximation (RMSEA) takes into account the fact that the model might hold approximately rather than exactly in the population. Browne and Cudeck (1993) suggested that values smaller than 0.05 indicate a close fit. A value greater than 0.1 indicates a poor fit. A p-value and a confidence interval can be computed.
2. The comparative fit index (CFI) takes values between 0 and 1. CFI values close to 1 indicate a very good fit. The CFI computes the relative improvement in the overall fit of the fitted model compared to a null model; this is usually the independence model (where the observed variables are assumed to be uncorrelated).
3. A similar measure to the CFI is the relative fit index (RFI). Values close to 1 indicate a very good fit.
4. The standardised root mean square residual (SRMR) is a standardised summary of the average covariance residuals. Covariance residuals are discrepancies between the sample covariance matrix and the reproduced covariance matrix estimated under the model, assuming that the model is correct. The smaller the SRMR is, the better the fit. A value less than 0.08 indicates a good fit.

5. For the Tucker–Lewis or non-normed fit index (NNFI) a value of 1 indicates a very good fit, while a value greater than 1 might indicate overfitting. Values smaller than 0.9 indicate a poor fit.

6. For Bollen's incremental fit index (IFI) a value of 1 again indicates a very good fit, while a value greater than 1 might indicate overfitting. Values smaller than 0.9 indicate a poor fit. The IFI is reported to be less variable than the NNFI.

The CFI, NNFI and IFI are relative fit indices. They compare the chi-squared statistic of the fitted model with the chi-squared of a baseline model, usually the independence model (the degrees of freedom of each model are taken into account). Indices such as the RMSEA are calculated using the chi-squared statistics and the degrees of freedom of the fitted model.

8.8 Structural relationships in a general setting

It is interesting and instructive to look at the general question of relationships between latent variables from the perspective taken in Chapter 2. We shall, therefore, retrace our steps and consider the topics of the earlier part of this chapter from that point of view.

The fundamental idea is that if the dependencies among a set of manifest variables **x** can be explained by their common dependence on a set of latent variables **y**, then their joint distribution admits the representation

$$f(\mathbf{x}) = \int h(\mathbf{y}) \prod_{i=1}^{p} g_i(x_i \mid \mathbf{y}) d\mathbf{y}. \tag{8.11}$$

The impossibility of estimating $h(\mathbf{y})$, for example, then followed from the fact that $f(\mathbf{x})$ remains unchanged by transformations of **y**. However, by restricting the conditional distributions $\{g_i(x_i \mid \mathbf{y})\}$ to be members of an exponential family we were able to find functions of **x** which were sufficient for **y** independently of any assumptions about the prior distribution of **y**. We shall return to the relevance of this fact later. First we look at how the questions about relationships between latent variables appear when translated into the general framework.

Simple structure means that each x is dependent on no more than one y. We can then partition the **x** vector according to which latent variable the members of the groups thus formed belong to. Suppose that there are q_j xs which depend on y_j and that the xs are so ordered that those depending on y_1 come first, those depending on y_2 come second and so on. The product term in (8.11) then decomposes into

$$\prod_{j=1}^{q} \prod_{i=Q_{j-1}+1}^{Q_j} g_i(x_i \mid y_j), \tag{8.12}$$

where $Q_j = q_0 + q_1 + \cdots + q_j$ and $Q_0 \equiv 0$. If the ys are independent we then have $f(\mathbf{x}) = f(\mathbf{x}_1) f(\mathbf{x}_2) \cdots f(\mathbf{x}_q)$, which reduces the problem to q separate one-factor models, as it should.

If $\mathbf{y}$ is assumed to have a standard normal multivariate distribution with correlation matrix $\mathbf{\Phi}$, then $\mathbf{\Phi}$ can be estimated by maximum likelihood. The assumption of simple structure somewhat simplifies the likelihood but, in particular cases, there may be considerable further simplifications as we shall see.

8.9 Generalisations of the LISREL model

The core of the LISREL model is the relationship between the dependent (endogenous) and the independent (exogenous) latent variables. This specifies the distribution of $\mathbf{y}_2$ given $\mathbf{y}_1$ which, in its most general form, may be written $h(\mathbf{y}_2 \mid \mathbf{y}_1)$. We thus replace $h(\mathbf{y})$ in (8.11) by $h(\mathbf{y}_1)\, h(\mathbf{y}_2 \mid \mathbf{y}_1)$.

The measurement part of the model supposes that the indicators divide into two groups – those that depend on $\mathbf{y}_1$ and those that depend on $\mathbf{y}_2$. This means that the product in (8.11) factorises into two parts:

$$\prod_{i=1}^{p} g_i(x_i \mid \mathbf{y}) = \prod_{i=1}^{p_1} g_i(x_i \mid \mathbf{y}_1) \prod_{i=p_1+1}^{p} g_i(x_i \mid \mathbf{y}_2). \tag{8.13}$$

This implies that $f(\mathbf{x})$ admits the representation

$$f(\mathbf{x}) = \int h(\mathbf{y}_1) h(\mathbf{y}_2 \mid \mathbf{y}_1) \prod_{i=1}^{p_1} g_i(x_i \mid \mathbf{y}_1) \prod_{i=p_1+1}^{p} g_i(x_i \mid \mathbf{y}_2)\, d\mathbf{y}. \tag{8.14}$$

If all the distributions $\{g_i\}$ are members of the exponential family, the first product within the integral will give rise to q_1 statistics sufficient for $\mathbf{y}_1$ and the second product to q_2 which are sufficient for $\mathbf{y}_2$. The parameters which appear in the products in the integrand relate to the measurement part of the model. Our earlier work suggests that these latter estimates will not depend critically on the choice of prior. However, this fact has an unfortunate consequence for our present purposes. We have already seen that the inherent arbitrariness of the prior means that we can say very little about its form. If, as we showed in the case of the latent class and latent trait models, the difference between a two-point and a normal prior is scarcely distinguishable, what hope is there for establishing the form of the relationships between a set of latent variables? The structural part of the present model is contained in the conditional distribution $h(\mathbf{y}_2 \mid \mathbf{y}_1)$, and especially in $E(\mathbf{y}_2 \mid \mathbf{y}_1)$. Transformations of $\mathbf{y}_1$ and $\mathbf{y}_2$ which leave $f(\mathbf{x})$ unchanged could do considerable violence to the supposed form of the relationships between the ys. It is important to emphasise that the lack of identifiability we are talking about here is not concerned with parameter values, in the usual sense, but with different models which have the same empirical consequences.

This simple analysis shows the LISREL model of Section 8.6 in a different light. It demonstrates that, even if a linear model fits the data, there may be other models which fit them equally well. We now illustrate this point by means of examples.

8.10 Examples of models which are indistinguishable

We shall construct two models, in which all the latent variables are categorical, whose covariance structure is identical to a given LISREL model. In order to do this we shall represent the categorical variables by indicator variables taking the values 0 and 1. In the measurement part of the model we shall suppose that the manifest variables depend linearly on these indicators. The covariances of the manifest variables will then be functions of the covariances of the indicators. Two models will then be indistinguishable if their latent indicators have the same covariance matrix. We do not, therefore, need to specify the details of the measurement sub-model; all that is needed is the assumption that the manifest variables are continuous with linear regressions on the indicators of the latent variables. The alternative models are thus specified by the conditional distribution of $\mathbf{y}_2$ on $\mathbf{y}_1$. Both examples are adapted from Bartholomew (1993).

Model I is a standard LISREL model with two independent latent variables, y_{11} and y_{12}, and one dependent latent variable, y_2. The structural sub-model is

$$y_2 = G_1 y_{11} + G_2 y_{12} + \zeta, \tag{8.15}$$

where ζ has mean zero, variance Θ, and is uncorrelated with y_{11} and y_{12}; y_{11} and y_{12} are standard normal with correlation coefficient Ψ_{12}.

Model II has one independent latent variable, y_1, and one dependent variable, y_2. The dependent variable, y_2, is binary with outcomes coded 0 and 1; y_1 has three categories which are coded by a bivariate indicator $\mathbf{u} = (u_1, u_2)$ taking the values (1, 0), (0, 1) and (0, 0). The joint distribution of y_2 and $\mathbf{u}$, specified by the marginal distribution and the conditional distribution of y_2 given $\mathbf{u}$ (the structural model), is set out below:

$y_2 \setminus \mathbf{u}$	(1, 0)	(0, 1)	(0, 0)	
1	$p_1\pi_1$	$p_2\pi_2$	$p_3\pi_3$	$\bar{\pi}$
0	$p_1(1-\pi_1)$	$p_2(1-\pi_2)$	$p_3(1-\pi_3)$	$1-\bar{\pi}$
	p_1	p_2	p_3	1

From this table we can find the covariance matrix of (y_2, u_1, u_2). To facilitate comparison with model I, we shall first standardise u_1 and u_2 since the LISREL formulation assumes the independent latent variables to be standardised.

Model III is the same as model II except that the trichotomous independent latent variable is replaced by two binary variables, y_{11} and y_{12}. The joint distribution is again specified by the marginal distribution of the independent variables and the conditional distribution of the dependent variable, as follows:

$$P\{y_{11} = i, y_{12} = j\} = p_{ij} \quad (i, j = 1, 2)$$

and

$$P\{y_2 = 1 \mid y_{11} = i,\ y_{12} = j\} = \pi_{ij} \quad (i, j = 1, 2).$$

The joint distribution is then as follows:

$y_2 \setminus (y_{11}, y_{12})$	(1, 1)	(1, 0)	(0, 1)	(0, 0)	
1	$\pi_{11}p_{11}$	$\pi_{10}p_{10}$	$\pi_{01}p_{01}$	$\pi_{00}p_{00}$	$\bar{\pi}$
0	$(1-\pi_{11})p_{11}$	$(1-\pi_{10})p_{10}$	$(1-\pi_{01})p_{01}$	$(1-\pi_{00})p_{00}$	$1-\bar{\pi}$
	p_{11}	p_{10}	p_{01}	p_{00}	1

Again we shall standardise the independent variables before computing the covariance matrix.

The three covariance matrices which result are as follows: for model I,

$$\begin{bmatrix} 1 & \Phi_{12} & G_2 + G_1\Phi_{12} \\ \Phi_{12} & 1 & G_1 + G_2\Phi_{12} \\ G_2 + G_1\Phi_{12} & G_1 + G_2\Phi_{12} & G_1^2 + G_2^2 + G_1G_2\Phi_{12} + \Theta \end{bmatrix};$$

for model II,

$$\begin{bmatrix} 1 & -\sqrt{\dfrac{p_1p_2}{(1-p_1)(1-p_2)}} & (\pi_2 - \bar{\pi})\sqrt{\dfrac{p_2}{1-p_2}} \\ -\sqrt{\dfrac{p_1p_2}{(1-p_1)(1-p_2)}} & 1 & (\pi_1 - \bar{\pi})\sqrt{\dfrac{p_1}{1-p_1}} \\ (\pi_2 - \bar{\pi})\sqrt{\dfrac{p_2}{1-p_2}} & (\pi_1 - \bar{\pi})\sqrt{\dfrac{p_1}{1-p_1}} & \bar{\pi}(1-\bar{\pi}) \end{bmatrix};$$

and for model III,

$$\begin{bmatrix} 1 & \dfrac{p_{11} - p_{.1}p_{1.}}{\sqrt{p_{.1}(1-p_{.1})p_{1.}(1-p_{1.})}} & (\bar{\pi}_{.1} - \bar{\pi})\sqrt{\dfrac{p_{.1}}{1-p_{.1}}} \\ \dfrac{p_{11} - p_{.1}p_{1.}}{\sqrt{p_{.1}(1-p_{.1})p_{1.}(1-p_{1.})}} & 1 & (\bar{\pi}_{1.} - \bar{\pi})\sqrt{\dfrac{p_{1.}}{1-p_{1.}}} \\ (\bar{\pi}_{.1} - \bar{\pi})\sqrt{\dfrac{p_{.1}}{1-p_{.1}}} & (\bar{\pi}_{1.} - \bar{\pi})\sqrt{\dfrac{p_{1.}}{1-p_{1.}}} & \bar{\pi}(1-\bar{\pi}) \end{bmatrix},$$

where $p_{1.} = p_{10} + p_{11}$, $p_{.1} = p_{01} + p_{11}$, $\bar{\pi}_{1.} = (p_{11}\pi_{11} + p_{10}\pi_{10})/p_{1.}$ and similarly for $\bar{\pi}_{.1}$.

The matrix for model I involves four unknowns, that for model II has five and that for model III has seven. One might, therefore, expect to be able to find many examples of models II and III having the same covariance matrix as model I. However, there are constraints imposed by the fact that the defining probabilities of models II and III must lie in the unit interval and, in model II, the covariance of u_1 and u_2 must be negative. Nevertheless it is still easy to find models of all three types which yield identical matrices. For example if, for model I,

$$G_1 = 0.189, \quad G_2 = 0.094, \quad \Phi_{12} = -0.500, \quad \Theta = 0.204,$$

for model II,

$$p_1 = p_2 = p_3 = 0.333, \quad \pi_1 = 0.6, \quad \pi_2 = 0.4, \quad \pi_3 = 0.2,$$

and for model III,

$$p_{11} = p_{00} = 0.125, \quad p_{10} = p_{01} = 0.375,$$

$$\pi_{11} = 0.700, \quad \pi_{10} = 0.489, \quad \pi_{01} = 0.300, \quad \pi_{00} = 0.132,$$

we find in each case,

$$\text{cov}(\mathbf{y}) = \begin{bmatrix} 0.240 & 0.142 & 0 \\ 0.142 & 1.000 & -0.500 \\ 0 & -0.500 & 1.000 \end{bmatrix}.$$

These examples are a simple extension of the idea used in Section 6.14 showing how latent profile models and linear factor models could have identical covariance structures.

As a result of a LISREL analysis we *can* therefore say whether or not a model with a specified linear latent structure is consistent with the data, but we *cannot* go on to infer that it is the *only* latent structure model having that property. Unless we have further non-empirical information favouring the linear model, no substantive claims for it can be justified. Just as the latent variables themselves are constructs, so also is the model connecting them.

The ready availability of good software for LISREL models has made it very easy to carry out analyses, but this has not been matched by a corresponding concern with the interpretation which the results will bear. However, there have been a number of moves to provide alternative methods which allow a rather more critical approach to the data. These are reviewed in Section 9.5.

8.11 Implications for analysis

This entire chapter has been concerned with estimating the covariance matrix of a set of latent variables, or of parameters which are functions of these covariances. Almost all of the discussion has been concerned with the normal factor model and its generalisations, although we have shown that the extension to all models derived from the exponential family is almost immediate. In the case of confirmatory factor analysis the theory required is a straightforward development of that already given for the normal factor model for which good software is available.

When we move on to LISREL models which specify a linear structure for the latent variables, serious problems emerge. These arise because the prior distribution of the latent variables, which incorporates the structural part of the model, is poorly determined by the data. In addition to the possible lack of identifiability of the linear model itself, we demonstrated that distinct models involving different numbers of categorical latent variables could give rise to exactly the same covariance structure. For these reasons, any inference made from a structural model rests heavily on the substantive assumptions made about the nature of the latent variables and their relationships. The essence of the structural part of these assumptions is often expressed in terms of a path diagram of postulated causal linkages. These may place quite severe constraints on the form of the conditional distribution $h(\mathbf{y}_2 \mid \mathbf{y}_1)$ of (8.14) and hence reduce the degree of arbitrariness. Nevertheless it remains unclear how far the key structural relationships would survive arbitrary transformations of the latent space. The global character of the analysis, in which the measurement and structural models are conflated, makes it difficult to diagnose the source of any misfit.

A partial way out of this difficulty is offered by the proposal to separate the two stages of the analysis. First we scale the indicator variables for each latent variable, using appropriate measurement models, and then check their validity before proceeding to the second stage. We use the score thus derived as the raw material for the structural analysis. The fact that our general approach yields scores (components) which are sufficient for the latent variables means that no information about the latter is lost in the process.

There are two problems with this course of action. One is that the correlations among the components always underestimate the correlations among the latent variables. However, this underestimation is slight if a reasonable large number of indicators are used for each latent variable. In any case, adjustments can be made to correct the underestimation. The estimation of regression coefficients can, as Croon and Bolck (1997) have shown, be made to a good approximation by using the Bartlett scores which are scaled components. This form of analysis needs further investigation, but appears to us to offer a better way forward.

The second difficulty is that we cannot even be sure that our individual measurement models are genuinely scaling a continuous latent variable. We saw in Chapter 6 how difficult it was to discriminate between a continuous normal prior and one with probability masses concentrated at a few points.

This last point suggests abandoning the structural part of the model altogether. After all, as Sobel (1994) pertinently points out, it is not clear how a mere construct can *cause* anything. Once the measurement sub-models have yielded their sufficient statistics, we can treat these statistics as variables in their own right. They are usually linear functions of the indicators and, as such, it is often possible to name what they are measuring in exactly the same manner as we do with principal components. It is then reasonable to make these quantities the basis of a structural path analysis without any reference to an underlying latent variable model. Such a model is, after all, no more than a construct and it is of no substantive concern whether or not regressions and correlations between components correctly estimate its purely hypothetical parameters. When we add to the argument the fact that working with components enables us to explore the form of the structural relations, the case against the traditional LISREL analysis is formidable.

It might appear that our treatment of structural relationships has been inconsistent with our use of posterior analysis in earlier chapters. There we argued that, after the xs have been observed, inferences about $\mathbf{y}$ should be made conditional on $\mathbf{x}$. Thus we were led to construct measures of $\mathbf{y}$ from the posterior distribution $h(\mathbf{y} \mid \mathbf{x})$ – in particular, its expectation $E(\mathbf{y} \mid \mathbf{x})$. If this is characterised as a Bayesian approach then what we have been using in this chapter is a frequentist approach. We have been treating components as functions of $\mathbf{x}$ and determining their properties by reference to the unconditional distribution of $\mathbf{x}$.

A seemingly more consistent approach to, for example, the regression of y_2 on y_1 discussed in Section 9.5 would have been to consider $E(y_2 \mid y_1, \mathbf{x})$. This is the regression of y_2 on y_1, after $\mathbf{x}$ has been observed. However, this is not what is required in the present instance. We used $E(y \mid \mathbf{x})$ to scale an *individual* for whom the indicator $\mathbf{x}$ has been recorded. In the problems of this chapter we are not interested in what can be said about the relationship between y_1 and y_2 for any particular individual. We want to know about the relationship for the whole population of individuals, and so the distribution unconditioned by $\mathbf{x}$ is relevant. That is, we are interested in the structure of the prior distribution $h(\mathbf{y})$, as we noted in Section 8.9.

We conclude with the observation that, in the model-based approach to statistical methodology, everything is conditional on the model being true. When we come to models for relationships between latent variables we have reached a point where so much has to be assumed that one might justly conclude that the limits of scientific usefulness have been reached, if not exceeded.

9

Related techniques for investigating dependency

9.1 Introduction

One of the main purposes of a latent variable model is to explain the dependencies among a set of manifest variables. It does this by showing that if the latent variables are held fixed, then the manifest variables are conditionally independent. There are other methods which have the same aim, without (explicitly) introducing latent variables. In this chapter we review a number of them and compare them withe corresponding latent variable models. We begin with principal components analysis.

9.2 Principal components analysis

9.2.1 A distributional treatment

In this section we shall introduce several ways of looking at principal components analysis (PCA). We begin with one which facilitates the comparison with factor analysis, with which it is sometimes confused. In fact, PCA is sometimes regarded as one method of factor analysis and it is so treated in, for example, the SPSS software. PCA was introduced by Hotelling (1933) who appears to have regarded it as an alternative to factor analysis, but it was quickly absorbed into the armoury of factor analysts among whom it was sometimes known as 'Hotelling's method'. One of the purposes of this section is to show its relationship to factor analysis, as we have developed it in this book, and to identify the circumstances under which the two methods are very similar. Latent variable models achieve their aim in two stages: first, they introduce independent, or at worst, highly structured random variables; secondly, they attempt to interpret them individually. Independent random variables

Latent Variable Models and Factor Analysis: A Unified Approach, Third Edition.
David Bartholomew, Martin Knott and Irini Moustaki.

are easier to interpret than dependent variables, and it is there that the object of the analysis lies. Throughout this section, unless otherwise stated, we shall assume manifest variables to be standardised with zero mean and unit standard deviation. According to the normal linear factor model, the joint distribution of the manifest variables is multivariate normal. This will be our starting point for the development of PCA.

A set of correlated normal variables can be linearly transformed into a set of independent normal variables in a routine way as follows. If the manifest variables are listed in a vector $\mathbf{x}$ of dimension p then we may choose

$$\mathbf{y} = \mathbf{A}\mathbf{x}. \tag{9.1}$$

We require the ys to be uncorrelated, which means that cov($\mathbf{y}$) must be diagonal – that is,

$$\text{cov}(\mathbf{y}) = \text{E}(\mathbf{y}\mathbf{y}') = \text{E}(\mathbf{A}\mathbf{x}\mathbf{x}'\mathbf{A}') \tag{9.2}$$

is diagonal. If, as before, we denote the covariance matrix of $\mathbf{x}$ by $\boldsymbol{\Sigma}$, then $\mathbf{A}$ must be such that $\mathbf{A}\boldsymbol{\Sigma}\mathbf{A}'$ is diagonal. Since $\boldsymbol{\Sigma}$ is a symmetric non-negative matrix it may be expressed in the form

$$\boldsymbol{\Sigma} = \mathbf{M}'\boldsymbol{\Delta}\mathbf{M}, \tag{9.3}$$

where $\mathbf{M}$ is an orthogonal matrix of eigenvectors and $\boldsymbol{\Delta}$ is a diagonal matrix of eigenvalues which are all non-negative. If, therefore, we choose $\mathbf{M} = \mathbf{A}$, $\boldsymbol{\Sigma}$ is diagonalised. The covariance matrix of y will therefore be

$$\text{cov}(\mathbf{y}) = \boldsymbol{\Delta}. \tag{9.4}$$

We may list the variables in $\mathbf{y}$ so that the eigenvalues are in decreasing order of magnitude. The ys are called the *principal components*, so by making this transformation we have arrived at a new set of independent normal variables whose relative importance is reflected in their associated variances given, for y_i, by δ_i $(i = 1, 2 \dots p)$.

The principal components are given by

$$\mathbf{y} = \mathbf{M}\mathbf{x}. \tag{9.5}$$

To make the comparison with factor analysis we need to express $\mathbf{x}$ in terms of $\mathbf{y}$, which is easily done. Because $\mathbf{M}$ is an orthogonal matrix its inverse is the same as its transpose. Thus

$$\mathbf{x} = \mathbf{M}'\mathbf{y}.$$

This is still not in a form suitable for our purposes because in the factor model the factors, corresponding to the principal components, all have unit variance. Thus, to facilitate the comparison we need to make a further transformation, scaling the *y*s so that each has unit variance. This is achieved by pre-multiplying $\mathbf{y}$ by $\mathbf{\Delta}^{-1/2}$. So if $\mathbf{z} = \mathbf{\Delta}^{-1/2}\mathbf{y}$, we shall have

$$\mathbf{x} = \mathbf{M}'\mathbf{\Delta}^{1/2}\mathbf{z} = \mathbf{Bz}.$$

Suppose we now arbitrarily partition $\mathbf{z}$ into two parts such that $\mathbf{z}_q$ consists of the first q elements of $\mathbf{z}$ and $\mathbf{z}_{p-q}$ the remainder. Similarly, we partition $\mathbf{B}$ so that $\mathbf{B}_q$ consists of the first q columns of $\mathbf{B}$ and $\mathbf{B}_{p-q}$ the remainder. We may then write

$$\mathbf{x} = \mathbf{B}_q\mathbf{z}_q + \mathbf{e}^*, \quad \text{where } \mathbf{e}^* = \mathbf{B}_{p-q}\mathbf{z}_{p-q}. \tag{9.6}$$

The extent to which (9.6) agrees with (3.3) depends on the comparison of $\mathbf{e}$ of that equation and $\mathbf{e}^*$ of (9.6). It is clear that $\mathbf{e}^*$ is independent of $\mathbf{B}_q\mathbf{z}_q$ because there are no *z*s in common. However, the elements of $\mathbf{e}^*$ are not, in general, independent because they are all functions of $\mathbf{z}_{p-q}$. If the first few principal components account for a high proportion of the total variation, the error terms will be small and there will be little difference between principal components and the corresponding factor model. This fact justifies the common use of PCA to determine the value of q in factor analysis. The number of eigenvalues judged to be 'large' will correspond approximately with the number of factors needed.

There is another way of demonstrating the similarity of factor analysis and PCA which arises out of older approaches to factor analysis which were not model-based. This is on the same lines as the methods of fitting without normality assumptions which we used in Chapter 3. The idea there was to make the actual and theoretical values of the correlation matrix as close as possible. According to the factor model we are looking for a matrix $\mathbf{\Lambda}$ such that

$$\mathbf{R} - \mathbf{\Psi} = \mathbf{\Lambda}\mathbf{\Lambda}', \tag{9.7}$$

where $\mathbf{R}$ is the correlation matrix of the *x*s and $\mathbf{\Psi}$ is the diagonal matrix of residual variances. If $\mathbf{\Psi}$ is known, the equation is satisfied exactly because $\mathbf{R} - \mathbf{\Psi}$ is symmetrical, semi-definite and so can be expressed exactly in the required form. In the case of $\mathbf{\Psi} = \mathbf{0}$ we are back to principal components. As we have already seen, methods of factor analysis often estimate the loadings for a fixed value of $\mathbf{\Psi}$ and, from that point on, the fitting can be regarded as carrying out a PCA on the matrix $\mathbf{R} - \mathbf{\Psi}$. PCA can then be regarded as one of a family of methods which chooses the communalities (the complements of the ψs) to be unity.

To complete our discussion of the relationship between PCA and factor analysis we show that, in certain circumstances, PCA provides the solution of the estimation problem for a particular member of the normal linear factor model. This member, which is known as the *isotropic normal model*, has $\psi_i = \psi$ for $i = 1, \ldots, p$. This particular model appears to have been first considered in this connection by Lawley (1953) and Anderson and Rubin (1956). A further account is given in Basilevsky

(1994, pp. 361–362), and the whole problem has been considered in a broader context by Tipping and Bishop (1999).

As in our earlier treatment of the normal linear factor model with p variables and q factors, we are considering the model having covariance structure

$$\mathbf{\Sigma} = \mathbf{\Lambda}\mathbf{\Lambda}' + \psi\mathbf{I}. \tag{9.8}$$

We first establish a relationship between the eigenvalues and vectors of $\mathbf{\Sigma}$ and $\mathbf{\Lambda}\mathbf{\Lambda}'$. The eigenvalues and vectors of $\mathbf{\Sigma}$ are given by

$$(\mathbf{\Sigma} - \delta\mathbf{I})\mathbf{u} = 0. \tag{9.9}$$

Substituting for $\mathbf{\Sigma}$, this equation is equivalent to

$$(\mathbf{\Lambda}\mathbf{\Lambda}' + \psi\mathbf{I} - \delta\mathbf{I})\mathbf{u} = [\mathbf{\Lambda}\mathbf{\Lambda}' - (\delta - \psi)\mathbf{I}]\mathbf{u}. \tag{9.10}$$

It is clear from (9.9) and (9.10) that $\mathbf{\Sigma}$ and $\mathbf{\Lambda}\mathbf{\Lambda}'$ have the same eigenvectors and that the eigenvalues of $\mathbf{\Lambda}\mathbf{\Lambda}'$ are less than those of $\mathbf{\Sigma}$ by the amount ψ. However, because $\mathbf{\Sigma}$ and $\mathbf{\Lambda}\mathbf{\Lambda}'$ are both positive definite matrices, their eigenvalues must be non-negative. This requirement is not met if, for any eigenvalue, $\delta < \psi$. Hence the equivalence required by (9.10) implies that $\delta_i = \psi$ for all those eigenvalues of $\mathbf{\Sigma}$ which are less than or equal to ψ. This means that the $p-q$ smallest eigenvalues of $\mathbf{\Sigma}$ must be equal.

It follows that we may write

$$\mathbf{\Lambda} = \mathbf{U}\begin{pmatrix} (\delta_1 - \psi)^{1/2} & 0 & 0 & . & 0 \\ 0 & (\delta_2 - \psi)^{1/2} & 0 & . & 0 \\ . & . & . & . & . \\ . & . & . & . & . \\ 0 & 0 & . & . & (\delta_q - \psi)^{1/2} \end{pmatrix}, \tag{9.11}$$

where $\mathbf{U}$ is the $p \times q$ matrix of eigenvectors associated with the q largest eigenvalues of $\mathbf{\Sigma}$ and $\delta_{q+1} = \delta_{q+2} = \cdots = \delta_p = \psi$.

In practice, we shall not know $\mathbf{\Sigma}$. The sample covariance provides an estimate, of course, but it is highly unlikely that its $p-q$ eigenvalues will be exactly equal. However, the maximum likelihood estimator of this common value, which is equal to ψ, can easily be determined, as shown, for example, in Tipping and Bishop (1999). It turns out that

$$\hat{\psi} = \frac{\sum_{i=q+1}^{p} \hat{\delta}_i}{p-q}, \tag{9.12}$$

where $\hat{\delta}_i$ is the ith eigenvalue of the observed covariance matrix.

We are thus able to estimate the parameters of the linear normal isotropic model from the eigendecomposition of the covariance matrix and without any iteration. If we happened to encounter a sample $p \times p$ covariance matrix whose $p-q$ smallest eigenvalues were roughly equal, then the principal components analysis of that matrix would be essentially the same as that for the corresponding isotropic normal factor model with q factors.

It is worth mentioning that the foregoing analysis treats q, the number of factors, as fixed. The larger q becomes, the greater will be the likelihood until we reach the limiting case when $q = p$. At that point the factor model becomes degenerate and it and PCA coincide.

Our main interest here has been in showing the close link between PCA and factor analysis. The emphasis of Tipping and Bishop (1999) was rather different. They wished to show that PCA could be regarded as a model-based technique and that methods already available for factor analysis could be used to obtain principal components. One might pause to consider why this should be desirable since determining principal components is, if anything, rather simpler than the iterative methods usually needed for factor analysis. The reason is that methods are available for model-based techniques which are useful in applications which lie beyond the range of the problems considered here. For example, there are ways of dealing with missing values which presuppose a probability model, and the authors' interest in image compression and visualisation involving complex projection methods requires multiple PCA models. It is when moving in such directions that the benefits of a model-based formulation are realised.

9.2.2 A sample-based treatment

The foregoing treatment was designed to show the close link between PCA and factor analysis at a distributional level. This is not the usual approach, but it may be linked to the following sample-based treatment as follows. In this case we start with a sample of n multivariate observations as set out in the matrix below. However, the prime object now is not to make inferences about the population from which the sample been drawn. If it was we could, of course, estimate the correlation matrix and from it the coefficients of the components, and then make inferences about their values, but our main concern now is with the structure of the data.

We suppose that a random sample of n observations has been obtained on p variables. These may be set out in an $n \times p$ data matrix,

$$\mathbf{X} = \begin{pmatrix} x_{11} & x_{12} & . & . & x_{1p} \\ x_{21} & x_{22} & . & . & x_{2p} \\ . & . & . & . & . \\ . & . & . & . & . \\ x_{n1} & x_{n2} & . & . & x_{np}. \end{pmatrix}. \tag{9.13}$$

The rows may be regarded as the coordinates of n points in p-dimensional space, and from that perspective PCA is concerned with the dimensionality of the space occupied by the points. More exactly, we ask what is the smallest dimensional space into which the points can be adequately fitted.

The procedure is first to find a linear combination of the *x*s which accounts for as much of the total variation of the *x*s as possible. This is the first principal component. The second is a linear combination of the *x*s which is uncorrelated with the first component and accounts for as much of the remaining variance as possible. The extraction proceeds in this fashion, forming at each stage a linear combination which is uncorrelated with all its predecessors and accounts for as much of the remaining variance as possible.

By basing PCA on correlations we have, implicitly, measured all variables in units of their standard deviations. It would be perfectly possible to carry out the analysis with different units of measurement for each variable basing the analysis on the covariance matrix. However, it is important to recognise that the resulting principal components will depend on the particular units used. The technique is, therefore, only likely to be practically useful when the units are arbitrary and so can reasonably be scaled by the standard deviations.

Example 9.2.1 We use Example 3.17.2 to make a comparison between solutions produced by factor analysis and principal component analysis. A PCA was carried out on the correlation matrix given in Table 3.8 and produced the eigenvalues shown in Table 9.1. Only the first two eigenvalues are greater than 1 and together they explain 80.5% of the total variance of the manifest variables. This is also shown in the scree plot in Figure 3.2. When the factor analysis model was fitted we also adopted a two-factor solution as being the most interpretable for the correlation among the eight variables. The first two component loadings and the factor loadings from the two-factor analysis model already given in Table 3.10 are presented in Table 9.2. The loadings of the first component are all large and positive, measuring a general state of physique. The second component is a contrast between the first four items (lankiness) and the last four (stockiness). Since the aim is to compare the PCA and factor analysis solutions, we orthogonally rotate both solutions to achieve a match. The rotated solutions are very close to each other, indicating that PCA and factor analysis can produce in some cases very close loadings.

Table 9.1 Physical data: variance explained by each principal component.

Component	Variance	%	Cumulative %
1	4.67	58.41	58.41
2	1.77	22.14	80.55
3	0.48	6.01	86.56
4	0.42	5.27	91.83
5	0.23	2.92	94.74
6	0.19	2.33	97.08
7	0.14	1.72	98.79
8	0.10	1.21	100.00

Table 9.2 Physical data: unrotated and matched PCA and factor analysis (FA) solutions.

	Unrotated solution				Matched solution			
	PCA		FA		PCA		FA	
Variable	$\hat{\alpha}_1$	$\hat{\alpha}_2$	$\hat{\lambda}_1$	$\hat{\lambda}_2$	$\hat{\alpha}_1^*$	$\hat{\alpha}_2^*$	$\hat{\lambda}_1^*$	$\hat{\lambda}_2^*$
1	0.859	−0.372	0.885	0.219	−0.864	−0.359	−0.856	−0.315
2	0.842	−0.441	0.939	0.108	−0.848	−0.429	−0.838	−0.437
3	0.813	−0.459	0.907	0.109	−0.820	−0.447	−0.812	−0.418
4	0.840	−0.395	0.876	0.184	−0.846	−0.383	−0.829	−0.339
5	0.758	0.525	0.304	0.905	−0.750	0.536	−0.759	0.579
6	0.674	0.533	0.254	0.756	−0.666	0.543	−0.634	0.484
7	0.617	0.580	0.192	0.739	−0.608	0.589	−0.573	0.504
8	0.671	0.418	0.325	0.598	−0.665	0.428	−0.604	0.313

9.2.3 Unordered categorical data

There is yet another way of looking at PCA which helps us to see its link with the latent variable models for categorical data discussed in Chapter 5. This starts by considering how the idea of PCA might be carried over when the xs are categorical. This approach starts from what is known as the singular value decomposition of the data matrix $\mathbf{X}$ of standardised observations which we have taken as being $n \times p$ and having rank p – this may be expressed in the form

$$\mathbf{X} = \mathbf{U}\boldsymbol{\Delta}^{1/2}\mathbf{V}', \tag{9.14}$$

where $\boldsymbol{\Delta}^{1/2}$ is a positive definite diagonal matrix whose elements are known as the singular values of the $n \times p$ matrix $\mathbf{X}$; $\mathbf{U}$ and the $p \times p$ matrix $\mathbf{V}$ have orthogonal columns. It follows that

$$(1/n)\mathbf{X}'\mathbf{X} = (1/n)\mathbf{V}\boldsymbol{\Delta}\mathbf{V}'. \tag{9.15}$$

From this expression it is clear that the singular values are the square roots of the eigenvalues of the correlation matrix of the xs which, in turn, are the variances of the principal components. The matrix $\mathbf{V}$ is the matrix of eigenvectors of the covariance matrix, and we have seen that these provide the coefficients in the principal components.

If the xs are categorical, we can still write the data matrix in a form similar to that in (9.13), but the columns have to be treated differently as in Chapter 5. The value x_{ij} is replaced by an indicator *vector* with one element for each category. In this vector a 1 indicates the category into which the individual falls and the other elements are 0.

The relationship between PCA and the singular value decomposition worked because the xs were standardised so that the matrix $(1/n)\mathbf{X}'\mathbf{X}$ was the correlation matrix. If the matrix is constructed from categorical data we first standardise $\mathbf{X}$ for scale by dividing each column by the corresponding diagonal element in $(1/n)\mathbf{X}'\mathbf{X}$, *then* standardise each column of $\mathbf{X}$ by *its* mean. Now $\mathbf{X}'\mathbf{X}$ becomes a standardised 'Burt' matrix of all the two-way contingency tables and provides one way of carrying out multiple correspondence analysis; see Greenacre (1984, p. 140).

It thus appears that the latent variable model of Chapter 5 stands in a similar relationship to multiple correspondence as does normal linear factor analysis to PCA. However, the connection goes deeper. In Sections 5.6 and 5.7 we derived approximations to the maximum likelihood estimators for the polytomous logit model, and these turned out to be the same as the multiple correspondence scores – as we noted for the case of binary data at the end of Section 5.7. We know of no work on the closeness of this approximation. If it turned out that the approximation was very good, there would be no need for the much more difficult maximum likelihood estimation for the polytomous ordered model. In any event the link at a conceptual level is remarkable.

9.2.4 Ordered categorical data

Multiple correspondence analysis makes no use of any ordering of the categories. After the event, of course, the scores allocated to the categories would indicate an ordering and those allocated to the principal dimension might reveal an unsuspected ordering. However, when the ordering is available *a priori* one would hope to make explicit use of that information in the analysis exactly as we did in Section 5.8.4. By assuming, for example, that there was a normal distribution underlying the categories one could estimate the correlation matrix using, say, the mid-points of each group. From that stage it would be possible to carry out a principal components analysis and so to make some inference about the dimensionality of the data. However, to construct the actual principal components the best we could do would be to take these mid-points, substituting them into the formulae for the components. This could give no more than a coarse approximation, and one might well prefer to use the full latent variable model.

An alternative approach to finding principal components for categorical variables is described in Linting *et al.* (2007). This is actually described as a method of nonlinear PCA but it repeatedly assigns scores to categories and then performs a standard PCA, treating the scores as if they were values of a variable. These scores may be chosen in variety of ways according to the assumptions one wishes to make about the variables which are supposed to underlie the categories. In essence, however, the method involves a principal components analysis of the assigned scores.

9.3 An alternative to the normal factor model

A key idea underlying all of the methods in this book, and reiterated at the beginning of this chapter, is that the interdependence of a set of variables may be explained by supposing them to depend on latent variables. This was demonstrated by showing that, if the latent variables were held fixed, the interdependence of the manifest variables vanished. It is then tempting to regard the latent variables as real, in some sense, and to suppose that the causal relationships with the manifest variables are also real. This is not a legitimate conclusion because there may be other models which may have exactly the same observable consequences. We now describe one such model which is almost as old as factor analysis itself but is still little known.

The origin of factor analysis is usually dated to a seminal paper by Spearman (1904). He proposed and championed what was described as the 'two-factor' model which supposed that an individual's score on a test item was made up of two parts – a common part which affected all individuals and a specific part peculiar to that individual. He described the common part as general ability, usually denoted, following Spearman, by g. Expressed in modern terms, this becomes what we would now call the normal linear one-factor model. Spearman's specific factor has now become the error term of the standard model. Thurstone and others subsequently generalised this and it is now subsumed into the normal linear factor model. There is little doubt that Spearman regarded g as measuring some real characteristic of the brain; initially he referred to this as 'mental energy', but later he was less specific about what it might be. Nevertheless it continues to lie at the foundation of much work in the intelligence field (see, for example, Jensen (1998)).

A decade later Godfrey Thomson challenged Spearman's interpretation, (Thomson 1916) and the debate between them continued, inconclusively, for 30 years. Thomson's approach was to propose another model which predicted exactly the same correlation matrix but implied no general factor. He thereby intended to show that if at least two models predicted the same correlation matrix then neither could be interpreted as being the 'truth'. Thomson was handicapped by the fact that he relied, initially, on sampling experiments conducted with dice and, latterly, on an incomplete mathematical argument. The valid point which he was making seems to have been missed by subsequent workers who, without exception, followed Spearman. Thomson never claimed that Spearman's model was wrong, merely that it was not unique.

Initially, Thomson spoke of his *sampling theory*, but we shall use the term he introduced subsequently, referring to it as the *bonds model*. The essence of the model is as follows:

1. The relevant part of the brain involved in answering test items consists of N *bonds*. (Thomson did not know what a bond might be, though he often spoke of neural arcs.)

2. When an item is attempted a sample of bonds is selected at random (the harder the item the more bonds will be needed). A bond which is selected may be described as *active* or *switched on*.

3. Each bond contributes the same (expected) amount to the score.

Since Thomson's death, the bonds model seems to have been ignored until a version, in modern dress, was given in Bartholomew (2007). It may be written

$$x_i = a_{i1}e_{i1} + a_{i2}e_{i2} + \cdots + a_{iN}e_{iN}, \tag{9.16}$$

where $e_{ij} \sim N(\mu, \sigma_i^2)$, $\Pr(a_{ij} = 1) = \pi_i$ and $\Pr(a_{ij} = 0) = 1 - \pi_i$ $(i = 1, \ldots, p)$. The *e*s and *a*s are mutually independent. In the same paper it was shown that the correlation matrix of the *x*s was of exactly the same form as produced by Spearman's *g*-model.

This result can be generalised by supposing that each test score is arrived at by summing the outcome of *k* independent 'passes'. In practice each pass is supposed to focus on a different aspect of the item – its verbal or spatial aspects, for example. The final score is then the sum of the scores for the individual passes:

$$x_i = x_{i1} + x_{i2} + \cdots x_{ik} \quad (i = 1, 2, \ldots p). \tag{9.17}$$

In this formula a second subscript is added to the *x*s to denote the pass. If the passes are independent it is easy to determine the correlation structure. The first step is to calculate the relevant expectations conditional on the bonds which are selected and then take expectations over selections. This shows that the covariances are identical to those for the normal linear factor model. This was done in Bartholomew *et al.* (2009), for the case of *k* passes. If π_{ij} is the probability that any bond is selected for the *i*th item on the *j*th pass, the correlation coefficient of x_i and x_j is

$$\begin{aligned}
\operatorname{corr}(x_i, x_j) &= \frac{\sum_{r=1}^{k} \pi_{ri}\pi_{rj}}{\sqrt{\sum_{r=1}^{k} \pi_{ri} \sum_{r=1}^{k} \pi_{rj}}} \\
&= \sum_{r=1}^{k} \left(\frac{\pi_{ri}}{\sqrt{\sum_{r=1}^{k} \pi_{ri}}} \frac{\pi_{rj}}{\sqrt{\sum_{r=1}^{k} \pi_{rj}}} \right) \\
&= \sum_{r=1}^{k} \lambda_{ri}\lambda_{rj} \quad (i, j = 1, 2, \ldots, p).
\end{aligned} \tag{9.18}$$

This has exactly the same form as the correlation for the linear factor model originating with Spearman, as we intended to show. Using this formula, we can easily move from the πs of the bonds model to the λs of the factor model. However, if we move in the opposite direction, it may happen that some of the πs turn out to be negative. In such cases there is no bonds model equivalent to the factor model, but it may still be possible to achieve positivity by rotating the factor solution.

9.4 Replacing latent variables by linear functions of the manifest variables

In Chapter 8 we moved away from the determination of a set of latent variables, which could be regarded as explaining the correlations among a set of manifest variables, to the study of the relationships among the latent variables themselves. The principal tool, which is widely used for this purpose, is structural equation modelling which we considered briefly in Chapter 8. In particular, we were interested in exploring causal relationships when there was a temporal ordering of the latent variables. In this section we explore other ways of achieving much the same end without explicitly introducing latent variables.

Earlier in this chapter we discussed PCA, which is a way of determining linear combinations of manifest variables which can be given a meaning in their own right. It is natural to wonder whether it might be better to investigate causal relationships between such linear combinations rather than the unobservable latent variables which they purport to indicate. There is a ready-made technique in *path analysis* which aims to map the pattern of causal relationships between a set of individual manifest variables. In a similar manner we might treat the aforementioned linear combinations in the same way. Some such attempts have been made, and it is instructive to consider them here.

One of the earliest attempts is known as soft modelling or partial least squares (PLS). This was proposed by Wold (1975, 1980, 1982). A recent, extended, treatment is given in Esposito Vinzi *et al.* (2010). This approach makes no use of a probability model, though it is very similar to LISREL in the way it conceptualises the problem. What are called 'blocks' of manifest variables are selected from which a linear estimate of each underlying latent variable is to be 'estimated'. These estimates are then used to explore the linear relationships between the latent variables which they represent. The procedure is iterative. At one stage weights are estimated for the manifest variables within each block and, at the next, regression analysis is used on the measures thus formed. The method does not appear to have undergone further theoretical development, but has commended itself as a practical tool. A useful summary is given by Schneeweiss (1993), whose main purpose is to show that under certain circumstances the PLS method and the LISREL model give very similar results. Since the PLS method does not depend on a model there is no way of testing its adequacy and, like LISREL methods, it provides a global analysis in which the measures and their relationships are estimated simultaneously and interdependently. It does not, therefore meet the requirement of greater transparency.

A number of writers have expressed unease about LISREL analysis and have sought some means of making the implications more transparent. An obvious route is to separate the measurement and the structural modelling aspects. One could then examine the adequacy of the measurement model before going on to look at the relationships between the latent variables. Strategies of this kind appear to have been first proposed by Burt (1973, 1976) and taken up by Hunter and Gerbing (1982), Anderson and Gerbing (1988, 1992) and Lance *et al.* (1988). The idea here is to go one step further than Wold by replacing the latent variables by linear functions suggested by a probability model in which the variables are assumed to have normal distributions. The question is then whether an analysis carried out using these functions instead of the latent variables themselves is satisfactory. Since the analysis depends on the correlation coefficients, some insight can be gained by comparing the correlation coefficients between the latent variables with those of their manifest substitutes.

9.5 Estimation of correlations and regressions between latent variables

We can illustrate the position by considering just two latent variables, y_1 and y_2. Let us suppose that we have two random variables, y_1 and y_2, such that

$$E(y_1) = 0, \quad \mathrm{var}(y_1) = 1, \quad E(y_2 \mid y_1) = by_1, \quad \mathrm{var}(y_2 \mid y_1) = \sigma^2. \tag{9.19}$$

Then it follows that the correlation of y_1 and y_2 is

$$\rho(y_1, y_2) = b/\sqrt{b^2 + \sigma^2}. \tag{9.20}$$

The proposal we are considering is to replace y_1 and y_2 by their sufficient statistics (components) X_1 and X_2, say. The question then to be answered is whether the regression coefficient of X_2 on X_1 and the correlation coefficient of X_1 and X_2 are sufficiently close to the corresponding coefficients for y_1 and y_2.

Let us suppose that we have a group of indicators $\mathbf{x}_1' = (x_{11}, x_{12}, \ldots, x_{1p_1})$ for y_1 and $\mathbf{x}_2' = (x_{21}, x_{22}, \ldots, x_{2p_2})$ for y_2, with $p_1 + p_2 = p$. The corresponding sufficient statistics (components) are, from (3.4), given by

$$\mathbf{X} = \begin{pmatrix} X_1 \\ X_2 \end{pmatrix} = \begin{pmatrix} \boldsymbol{\lambda}_1' & \mathbf{0} \\ \mathbf{0} & \boldsymbol{\lambda}_2' \end{pmatrix} \boldsymbol{\Psi}^{-1}\mathbf{x}, \tag{9.21}$$

where $\mathbf{x}' = (\mathbf{x}_1', \mathbf{x}_2')$. The covariance matrix of $\mathbf{X}$ is

$$\mathrm{cov}(\mathbf{X}) = \boldsymbol{\Gamma\Phi\Gamma} + \boldsymbol{\Gamma}, \tag{9.22}$$

where $\boldsymbol{\Phi}$ is the *covariance* matrix of $\mathbf{y}$ and

$$\boldsymbol{\Gamma} = \boldsymbol{\Lambda}'\boldsymbol{\Psi}^{-1}\boldsymbol{\Lambda} = \begin{pmatrix} \boldsymbol{\lambda}_1' & \mathbf{0} \\ \mathbf{0} & \boldsymbol{\lambda}_2' \end{pmatrix} \begin{pmatrix} \boldsymbol{\Psi}_1^{-1} & \mathbf{0} \\ \mathbf{0} & \boldsymbol{\Psi}_2^{-1} \end{pmatrix} \begin{pmatrix} \boldsymbol{\lambda}_1 & \mathbf{0} \\ \mathbf{0} & \boldsymbol{\lambda}_2 \end{pmatrix},$$

where we have partitioned $\boldsymbol{\Psi}$ to match the partitioning of $\boldsymbol{\Lambda}$ and $\mathbf{x}$. We then obtain

$$\boldsymbol{\Gamma} = \begin{pmatrix} \Gamma_1 & 0 \\ 0 & \Gamma_2 \end{pmatrix},$$

where

$$\Gamma_i = \boldsymbol{\lambda}_i'\boldsymbol{\Psi}_i^{-1}\boldsymbol{\lambda}_i = \sum_{j=1}^{p_i} \frac{\lambda_{ij}^2}{\Psi_j} \quad (i = 1, 2).$$

It then follows that

$$\mathrm{cov}(X_1, X_2) = \Gamma_1\Gamma_2\Phi_{12}, \quad \mathrm{var}(X_i) = \Gamma_i^2\Phi_{ii} + \Gamma_i \quad (i = 1, 2),$$

and hence that

$$\begin{aligned}\rho(X_1, X_2) &= \frac{\Gamma_1\Gamma_2\Phi_{12}}{\sqrt{\Gamma_1^2 + \Gamma_1}\sqrt{\Gamma_2^2(b^2 + \sigma^2) + \Gamma_2}} \\ &= \frac{\Gamma_1\Gamma_2\rho(y_1, y_2)\sqrt{b^2 + \sigma^2}}{\sqrt{\Gamma_1^2 + \Gamma_1}\sqrt{\Gamma_2^2(b^2 + \sigma^2) + \Gamma_2}} \\ &= \sqrt{\frac{\Gamma_1}{\Gamma_1 + 1}}\sqrt{\frac{\Gamma_2(b^2 + \sigma^2)}{\Gamma_2(b^2 + \sigma^2) + 1}}\rho(y_1, y_2). \end{aligned} \tag{9.23}$$

Since Γ_1 and Γ_2 are both positive,

$$\rho(X_1, X_2) < \rho(y_1, y_2).$$

The correlation between the components thus underestimates the true correlation between the latent variables. This is a generalisation of a term known among psychologists as *attenuation*. The latter refers to the fact that, if two variables are both measured with error, the true correlation will be underestimated by their estimated correlation. That situation is, essentially, a special case of the result derived here with $p_1 = p_2 = 1$.

The regression coefficient of X_2 on X_1 is

$$\frac{\mathrm{cov}(X_1, X_2)}{\mathrm{var}(X_1)} = \frac{\Gamma_2\Phi_{12}}{1 + \Gamma_1} = \frac{\Gamma_2 b}{1 + \Gamma_1}. \tag{9.24}$$

In general, therefore, the regression coefficients will not be the same.

Two questions now arise. How serious are the discrepancies revealed by this analysis? Can they be corrected? The difference between the two correlation coefficients depends on Γ_1 and Γ_2. If both are large, the factors $\sqrt{\Gamma_1/(\Gamma_1 + 1)}$ and $\sqrt{\Gamma_2(b^2 + \sigma^2)/(\Gamma_2(b^2 + \sigma^2) + \Gamma_2)}$ will be close to one and the discrepancy will be negligible. This will happen if p_1 and p_2 are large (i.e. there is a large number of indicators), if the λs are large and, especially, if any of the ψ_i are small. One can easily correct for attenuation by scaling $\rho(X_1, X_2)$ by the estimated values of $\sqrt{\Gamma_1/(\Gamma_1 + 1)}$ and $\sqrt{\Gamma_2(b^2 + \sigma^2)/(\Gamma_2(b^2 + \sigma^2) + \Gamma_2)}$.

The regression coefficient is scale-dependent, and it is thus possible that the estimation of $b_{y_2 \mid y_1}$ can be improved by applying a scaling factor to the components. This point was investigated by Croon and Bolck (1997) who suggested that each component should be multiplied by a scalar c such that

$$E(c_i X_i \mid y_i) = y_i \quad (i = 1, 2).$$

This requires that

$$E(c_i \boldsymbol{\lambda}_i' \boldsymbol{\Psi}_i^{-1} x_i \mid y_i) = c_i \boldsymbol{\lambda}_i' \boldsymbol{\Psi}_i^{-1} \boldsymbol{\lambda}_i y_i = y_i$$

or that $c_i = 1/\Gamma_i$. If we choose this scaling then the regression coefficient of the second scaled component on the first is

$$b_{c_2X_2 \mid c_1X_1} = \frac{c_2}{c_1} \frac{\Gamma_2}{\Gamma_1 + 1} b = \frac{\Gamma_1}{\Gamma_1 + 1} b.$$

Under the conditions outlined above the regression coefficient of the scaled components will be almost equal to, but slightly less than, the true regression coefficient. Components scaled in this way are the *Bartlett scores* mentioned in Section 3.15.

This analysis suggests that there is not likely to be a serious discrepancy between LISREL and the more *ad hoc* method outlined here if there is a large number of indicators for each latent variable and if the $\boldsymbol{\psi}$s are small – in other words, if the latent variables are well determined by the associated manifest variables. In quantitative terms one must obviously look at the estimated parameter values. However, the ubiquity of easily used software for fitting linear structural equation models appears to have inhibited further exploration of this alternative.

9.6 Q-Methodology

This is the name given to the analysis in which the rows and columns of the data matrix are first transposed. The key to understanding what Q-analysis all about is to focus on the meaning of the correlation coefficients themselves. In order to make the point more clearly we will use terminology from educational testing. Thus in the matrix of (9.13) the columns would represent test items and the rows individuals or subjects; the entries would be test scores. The ordinary correlation coefficient between the test scores for, say, a pair of individuals is a measure of the similarity of those two individuals. Similarly, if we take the correlation of two test items across all individuals we get a measure of how close those test items are with respect to that set of individuals. R-analysis is then concerned with the relationship between variables and Q-analysis with the relationships between persons.

In principle, one can carry out either PCA or factor analysis starting from the P matrix or the Q matrix. Thomson (1951, Part V), traces the origin of this idea as far back as 1912, but Stephenson was undoubtedly the chief architect and advocate of the method. His first contribution was in Stephenson (1935).

There was a good deal of controversy in the early days, particularly between Stephenson on the one hand and Burt (1973) on the other. Burt regarded R- and Q-methodology as equivalent ways of extracting the basic factorial information from the correlation matrix, whereas Stephenson saw them as quite different. As we shall shortly see, Stephenson was essentially correct and Thomson (1951), although more circumspect, clearly understood the weakness of Burt's position. We shall not pursue the history of this controversy but place the problem in the context of modern factor analysis.

Consider the matrix $\mathbf{X}'\mathbf{X}$, where $\mathbf{X}$ is the data matrix defined in (9.13). This is symmetric and of dimension $p \times p$. It will thus have p non-negative eigenvalues and associated eigenvectors satisfying

$$\mathbf{X}'\mathbf{X}\mathbf{u} = \lambda\mathbf{u}. \tag{9.25}$$

Next, suppose we pre-multiply each side of this equation by $\mathbf{X}$. We shall then have

$$(\mathbf{X}\mathbf{X}')(\mathbf{X}\mathbf{u}) = \lambda(\mathbf{X}\mathbf{u}). \tag{9.26}$$

This shows that $\mathbf{X}'\mathbf{X}$ and its transpose, $\mathbf{X}\mathbf{X}'$, have the same positive eigenvalues, and eigenvectors equal to $\mathbf{u}$ and $\mathbf{X}\mathbf{u}$ respectively. There is thus a duality in the eigensolutions of the two matrices which means that if we have the solution of one we can easily find that of the other.

If we regard the decomposition represented by (9.25) as the R-analysis of the matrix $\mathbf{X}'\mathbf{X}$ and that by (9.26) as the Q-analysis then it is clear that the two are equivalent. The question is whether this duality carries over to the covariance or correlation matrices computed from $\mathbf{X}$ and $\mathbf{X}'$. Burt assumed that it did, at least approximately, but Thomson saw difficulties and produced numerical examples to make his point (Thomson 1951, Chapter XVII).

The equivalence which Burt looked for exists only under very special circumstances. These occur when $\mathbf{X}$ is *doubly centred*. This means that the row sums of $\mathbf{X}$ and the column sums are zero. In this case $(1/n)\mathbf{X}'\mathbf{X}$ is the covariance matrix of $\mathbf{X}$ and $(1/n)\mathbf{X}\mathbf{X}'$ is the covariance matrix of $\mathbf{X}'$. In this special case a PCA in R-mode is equivalent to a PCA in Q-mode. One might then expect that the corresponding factor analyses would be very similar and hence that any factors uncovered in the two cases would be essentially the same. It is easy to make the column sums, say, of $\mathbf{X}$ zero by subtracting the column mean from each element, but attempting to do the same simultaneously for the rows tends to destroy the associations which it is the purpose of the analysis to uncover.

Although we have shown that the R- and Q-versions of the analysis are not equivalent in general, there is nothing to prevent us from forming the correlation matrix of the transposed matrix and asking how the results of a factor or principal components analysis might be interpreted. This is exactly what has become known as Q-factor analysis.

It must be recognised straight away that there can be no probability model behind Q-analysis. Anything we do will be purely descriptive. There is, in fact, an obvious and better way to analyse a matrix of similarities. This is provided by *non-metric multidimensional scaling*. Such an analysis would locate each variable in a space of small dimension in which distances apart reflected the similarities as measured, for example by the 'between-persons' correlation coefficients. This method is illustrated in Bartholomew *et al.* (2008). However, in the 1920s when Q-analysis was first mooted no such method was available and the natural thing seemed to factor-analyse the transposed matrix and classify the variables into groups by means of rotations to something like 'simple structure'. Factor analysis was, in effect, being used as a form of cluster analysis because it was the only technique then available.

Subsequently, Q-methodology became detached from both psychology and statistics and exists as a separate entity with its own journal, conferences and so on. The journal in which relevant publications occur is called *Operant Subjectivity*m and Stephenson's ideas are now propagated by the International Society for the Scientific Study Of Subjectivity. It is, perhaps, appropriate to observe that although the original link with factor analysis has become tenuous, if not tendentious, the existence of this field of study bears testimony to fertile ground which factor analysis has provided for the generation of new methods of statistical analysis.

9.7 Concluding reflections of the role of latent variables in statistical modelling

We conclude the chapter and the book with some general remarks about latent variable models. Some of the foregoing material may have led the reader to wonder whether it is really necessary to introduce latent variables. For if we can achieve much the same objectives without introducing latent variables at all, a case has to be made for what may seem the unnecessary complication of introducing additional variables into the model. Arguments can be put on either side but they must, inevitably, involve substantive questions of meaningfulness and usefulness. Viewed purely theoretically, latent variable models, like any other probability models, are mathematical statements about probability distributions, Theoretical deductions can be made from them about other distributions in a routine fashion. It seems to us that any arguments, which we do not intend to pursue in detail here, must reckon with the following points.

First, much of science, especially social science, deals with quantities which are essentially latent. We use the word 'quantities' because they occur, typically, in statements which imply a magnitude, if only in the sense of an ordering. Thus, psychologists speak about abilities as something varying from one individual to another and about which one can at least make statements of relative magnitude. The same is true of attitudes which can exist in varying degrees. Sociologists speak of one political party being more right wing than another, even though the degree of conservatism possessed by a party is not something which can be directly measured. Instead they will have to depend on various indicators which are believed to be related to conservatism. There is, therefore, a close correspondence between the way that social scientists think and the latent variable models constructed to give expression to that thinking. It is no accident that factor analysis was invented by Charles Spearman, a psychologist, and developed almost exclusively by psychologists. Similarly, although structural equation modelling was first introduced in *Biometrika* by Karl Jöreskog, a statistician, most of the subsequent development has been in the hands of social scientists. In a sense, the process of modelling mirrors very closely the way we think about relationships, which makes no fundamental distinction between latent and manifest variables.

Secondly, the essential idea behind our analysis is that we stand between the data, which are already available, and the latent variables, which can never be observed. It is thus appropriate that the analysis should be based on distributions which look at the latent variables conditional on the manifest variables – the data. This procedure neatly circumvents one of the problems which statisticians have rarely noticed but which has exercised a good many psychologists. This arises from the fact that the manifest variables are seldom samples drawn randomly from any population of variables. The inferences derived in this book are all made conditional on those particular variables which have been used. They therefore do not depend on how the variables were obtained. In practice the questions used in a sample survey will be those which the surveyors have though relevant or practical. They will not have been selected from some population of possible questions – at least not in any formal manner. Psychologists have attempted to formalise this by introducing what they call the *domain*, and Williams (1978), for example, has attempted to carry out a formal analysis, using this concept, to place the notion of a factor score on a formal footing. By making the inferences conditional on the data we may hope to obtain relevant information about the latent variables which is valid whatever manifest variables happen to be to hand. The dimensionality of the manifest sample space is then irrelevant.

If, for example, we were to use PCA instead of factor analysis, the first principal component might commend itself as a suitable proxy for something like general intelligence. But we would get a different linear combination of manifest variables if we had used a different set of manifest variables. It is difficult to see how one could be preferred to another. It is worth remembering that the choice of a linear function for the components was, itself, an arbitrary choice motivated more by convenience than optimality in any sense. In the general approach to latent variable modelling which we have adopted, linearity arises naturally from the very general model we have proposed.

Thirdly, the reality of latent variables has no bearing upon the validity of the methods described in this book. In a purely formal sense, latent variables are treated just like manifest variables. In effect, they may be regarded as variables on which all observations are missing. From a practical point of view, however, their role is radically different. They are introduced because the scientific model we are trying to determine requires their presence. In other words, their meaning derives from the scientific interpretation we put upon them. This does not mean that they are real in the sense that something like the price of potatoes in a particular wholesale market is real – that is, it is independently verifiable by different observers. It is worth adding that we might have a similar discussion about whether the parameters which appear in statistical models are real. They are unobservable just like latent variables. Indeed, in a fully Bayesian treatment, both parameters and latent variables would be treated as random variables.

No statistical model is real in that sense, nor is it intended to be. Rather, it aims to capture part of reality to a sufficient degree to give genuine insight and to provide a tool for changing things. Latent variables models do just that.

Software appendix

Many of the standard packages such as SPSS and S-Plus, but also more specialised software such as LISREL, GLLAMM in STATA, and CEFA, can be used for some of the analyses which we have performed but occasionally we have used and reported the results of using packages such as IRTPRO, Latent GOLD, Mplus and WinBUGS. We provide here a short description of the version and characteristics of each of these.

- IRTPRO (Item Response Theory for Patient-Reported Outcomes) is distributed by Scientific Software International, Inc. (available from August 2011). The program fits various unidimensional and multidimensional confirmatory and exploratory factor analysis models for binary, nominal, ordinal, continuous and mixed observed variables. The estimation is based on the E-M algorithm with fixed or adaptive quadrature points. Other estimation methods such as the Metropolis–Hastings Robbins–Monro algorithm (Cai 2010) are available. IRTPRO 2 was used in Examples 4.13.1–4.13.3 and 5.12.1–5.12.4.
- Latent GOLD 4.5 is a latent class and finite mixture program. Latent GOLD can fit latent class and latent trait models for any type of observed variables as well as models with mixed latent variables (hybrid models). A log-linear model representation is also used. The program is distributed by Statistical Innovations. Latent GOLD was used in Examples 6.9.1, 6.9.2, 6.13.3 and 7.8.3.
- Mplus 5 is a general structural equation modelling program that can analyse any combination of observed variables. It runs exploratory and confirmatory factor analysis and performs all the analysis available in a structural equation modelling framework both for cross-sectional and longitudinal data. Mplus was used in Examples 3.17.1 and 3.17.2.
- WinBUGS 1.4 is a freely available program which can be used for fitting complex statistical models using Markov chain Monte Carlo (MCMC) methods ('BUGS' stands for Bayesian inference Using Gibbs Sampling). The program produces estimates, standard deviations as well as monitoring and convergence diagnostics plots and statistics. WinBUGS 1.4 was used in Example 4.13.2. The program can be downloaded from http://www.mrc-bsu.cam.ac.uk/bugs/

Finally, R packages for fitting various item response theory models have been developed and can be found at http://cran.r-project.org/web/views/Psychometrics.html.

Latent Variable Models and Factor Analysis: A Unified Approach, Third Edition.
David Bartholomew, Martin Knott and Irini Moustaki.

References

Aitkin M, Anderson D and Hinde J 1981 Statistical modelling of data on teaching styles. *Journal of the Royal Statistical Society A* **144**, 419–461.

Akaike H 1983 Information measures and model selection. *Bulletin of the International Statistical Institute* **50**, 277–290.

Akaike H 1987 Factor analysis and AIC. *Psychometrika* **52**, 317–332.

Albert JH 1992 Bayesian estimation of normal ogive item response curves using Gibbs sampling. *Journal of Educational Statistics* **17**, 251–269.

Albert JH and Chib S 1993 Bayesian analysis of binary and polychotomous response data. *Journal of the American Statistical Association* **88**, 669–679.

Amemiya T 1984 Tobit models: a survey. *Journal of Econometrics* **24**, 3–61.

Amemiya Y and Anderson TW 1985 Asymptotic chi-square tests for a large class of factor analysis models. Technical Report 13, Stanford University.

Andersen EB 1973 Conditional inference and multiple choice questionnaires. *British Journal of Mathematical and Statistical Psychology* **26**, 31–44.

Andersen EB 1977 Sufficient statistics and latent trait models. *Psychometrika* **42**(1), 69–81.

Andersen EB 1980a Comparing latent distributions. *Psychometrika* **45**(1), 121–134.

Andersen EB 1980b *Discrete Statistical Models with Social Science Applications*. North-Holland, Amsterdam.

Andersen EB and Madsen M 1977 Estimating the parameters of the latent population distribution. *Psychometrika* **42**(3), 357–374.

Anderson JC and Gerbing DW 1984 Statistical inference in factor analysis. *Psychometrika* **49**, 155–173.

Anderson JC and Gerbing DW 1988 Structural equation modelling in practice: a review and recommended two-step approach. *Psychological Bulletin* **103**, 411–423.

Anderson JC and Gerbing DW 1992 Assumptions and comparative strength of the two-step approach: comment on Furnell & Yi. *Sociological Methods and Research* **20**, 321–333.

Anderson TW 1959 Some scaling models and estimation procedures in the latent class model In *Probability and Statistics* (ed. Grenander U) John Wiley & Sons, Inc. New York pp. 9–38.

Anderson TW 1984 *An Introduction to Multivariate Statistical Analysis* 2nd edn. John Wiley & Sons, Inc., New York.

Anderson TW and Rubin H 1956 Statistical inference in factor analysis. *Third Berkeley Symposium of Mathematical Statistics and Probability* (ed. Neyman J) University of California Press, Berkeley vol. 5, pp. 111–150.

Latent Variable Models and Factor Analysis: A Unified Approach, Third Edition.
David Bartholomew, Martin Knott and Irini Moustaki.

Arbuckle JL 2006 *AMOS 7.0 – Programming: Reference Guide*. Amos Development Corporation.

Archer CO and Jennrich RI 1973 Standard errors for rotated factor loadings. *Psychometrika* **38**, 581–592.

Arminger G and Küsters U 1988 Latent trait models with indicators of mixed measurement level In *Latent Trait and Latent Class Models* (ed. Langeheine R and Rost J) Plenum Press New York.

Arminger G and Muthén B 1998 A Bayesian approach to nonlinear latent variables models using Gibbs sampler and the Metropolis-Hastings algorithm. *Psychometrika* **63**, 271–300.

Baker FB 1998 An investigation of the item parameter recovery characteristics of a Gibbs sampling procedure. *Applied Psychological Measurement* **22**, 153–169.

Barankin EW and Maitra EP 1963 Generalisation of the Fisher-Darmois-Koopman-Pitman theorem on sufficient statistics. *Sankhyā A* **25**, 217–244.

Barnes S and Kaase M 1979 *Political Action: Mass Participation in Five Western Democracies*. Sage Publications, Beverly Hills, CA.

Bartholomew DJ 1980 Factor analysis for categorical data. *Journal of the Royal Statistical Society B* **42**(3), 293–321.

Bartholomew DJ 1981 Posterior analysis of the factor model. *British Journal of Mathematical and Statistical Psychology* **43**, 93–99.

Bartholomew DJ 1984 The foundations of factor analysis. *Biometrika* **71**, 221–232.

Bartholomew DJ 1985 Foundations of factor analysis: Some practical implications. *British Journal of Mathematical and Statistical Psychology* **38**, 1–10.

Bartholomew DJ 1987 *Latent Variable Models and Factor Analysis* vol. 40 of *Griffin's Statistical Monographs and Courses*. Charles Griffin, London.

Bartholomew DJ 1993 Estimating relationships between latent variables. *Sankhyā* **55**(3), 409–419.

Bartholomew DJ 1995a Social measurement: A statistical perspective In *Applied Statistics – Recent Developments* (ed. Frohn J, Gather U, Stute W and Thöni H) Vandenhoeck & Ruprecht Göttingen.

Bartholomew DJ 1995b Spearman and the origin and development of factor analysis. *British Journal of Mathematical and Statistical Psychology* **48**, 211–220.

Bartholomew DJ 1996 *The Statistical Approach to Social Measurement*. Academic Press, San Diego, CA.

Bartholomew DJ 2007 Three faces of factor analysis In *Factor Analysis at 100: Historical Developments and Future Directions* (ed. Cudeck R and MacCallum R) Lawrence Erlbaum Associates Mahwah, NJ pp. 9–21.

Bartholomew DJ and Knott M 1993 Scale construction by maximising reliability In *Statistical Modelling and Latent Variables* (ed. Haagen K, Bartholomew DJ and Deistler M) North Holland Amsterdam pp. 17–26.

Bartholomew DJ and Knott M 1999 *Latent Variable Models and Factor Analysis* vol. 7 of *Kendall Library of Statistics* 2nd edn. Arnold, London.

Bartholomew DJ and Leung SO 2002 A goodness of fit test for sparse 2^p contingency tables. *British Journal of Mathematical and Statistical Psychology* **55**, 1–15.

Bartholomew DJ and McDonald RP 1986 The foundations of factor analysis: a further comment. *British Journal of Mathematical and Statistical Psychology* **39**, 228–229.

Bartholomew DJ and Schuessler KF 1991 Reliability of attitude scores based on a latent trait model. *Sociological Methodology* **21**, 97–123.

Bartholomew DJ and Tzamourani P 1999 The goodness-of-fit of latent trait models in attitude measurement. *Sociological Methods and Research* **27**, 525–546.

Bartholomew DJ, Bassin EL and Schuessler KF 1993 Properties of a latent trait reliability coefficient. *Sociological Methods & Research* **22**(2), 163–192.

Bartholomew DJ, de Menezes L and Tzamourani P 1997 Latent trait and latent class models applied to survey data In *Application of Latent Trait and Latent Class Models in the Social Sciences* (ed. Rost J and Langeheine R) Waxmann Münster.

Bartholomew DJ, Deary IJ and Lawn M 2009 A new lease of life for Thomson's bonds model of intelligence. *Psychological Review* **116**, 567–579.

Bartholomew DJ, Steele F, Moustaki I and Galbraith J 2008 *Analysis of Multivariate Social Science Data* 2nd edn. CRC Press, Boca Raton, FL.

Bartlett MS 1937 The statistical conception of mental factors. *British Journal of Psychology* **28**, 97–104.

Bartlett MS 1950 Tests of significance in factor analysis. *British Journal of Psychology (Statistical Section)* **3**, 77–85.

Bartlett MS 1953 Factor analysis in psychology as a statistician sees it *Uppsala Symposium of Psychology and Factor Analysis* Almqvist & Wicksell Uppsala pp. 23–34.

Basilevsky A 1994 *Statistical Factor Analysis and Related Methods. Theory and Applications* Wiley Series in Probability and Mathematical Statistics. John Wiley & Sons, Inc., New York.

Béguin AA and Glas CAW 2001 MCMC estimation and some model-fit analysis of multidimensional IRT models. *Psychometrika* **66**(4), 541–562.

Bentler PM 1968 Alpha-maximized factor analysis (alphamax): its relation to alpha and canonical factor analysis. *Psychometrika* **33**, 335–345.

Bentler PM 2008 *EQS 6: Structural Equations Program Manual.* Multivariate Software, Encino, CA.

Bergsma W, Croon M and Hagenaars JA 2009 *Marginal Models*. Springer.

Birnbaum A 1968 Some latent trait models and their use in inferring an examinee's ability In *Statistical Theories of Mental Test Scores* (ed. Lord FM and Novick MR) Reading, MA pp. 425–435.

Birnbaum A 1969 Statistical theory for logistic mental test models with a prior distribution of ability. *Journal of Mathematical Psychology* **6**, 258–276.

Bock RD 1972 Estimating item parameters and latent ability when responses are scored in two or more nominal categories. *Psychometrika* **37**(1), 29–51.

Bock RD and Aitkin M 1981 Marginal maximum likelihood estimation of item parameters: application of an EM algorithm. *Psychometrika* **46**(4), 443–459.

Bock RD and Moustaki I 2007 Item response theory in a general framework In *Psychometrics* (ed. Rao CR and Sinharay S) Elsevier Boston.

Bollen KA 1989 *Structural Equations with Latent Variables.* John Wiley & Sons, Inc., New York.

Bollen KA 1995 Structural equation models that are nonlinear in latent variables: A least squares estimator In *Sociological Methodology* (ed. Marsden P) Blackwell Cambridge, MA.

Bollen KA and Long JS 1993 *Testing Structural Equation Models*. Sage Publications, Newbury Park, CA.

Bollen KA and Paxton P 1998 Two-stage least squares estimation of interaction effects In *Interaction and Nonlinear Effects in Structural Equation Models* (ed. Schumacker R and Marcoulides G) Lawrence Erlbaum Associates Mahwah, NJ pp. 125–151.

Boomsma A 1985 Nonconvergence, improper solutions, and starting values in LISREL maximum likelihood estimation. *Psychometrika* **50**(2), 229–242.

Bozdogan H and Ramirez DE 1986 Model selection approach to the factor model problem, parameter parsimony, and choosing the number of factors. Research report, University of Virginia, Department of Mathematics.

Brokken FB 1983 Orthogonal Procrustes rotation maximising congruence. *Psychometrika* **48**, 343–349.

Brooks SP and Roberts GO 1998 Convergence assessment techniques for Markov chain Monte Carlo. *Statistics and Computing* **8**, 319–335.

Browne MW 1982 Covariance structures In *Topics in Applied Multivariate Analysis* (ed. Hawkins DM) Cambridge University Press Cambridge pp. 72–141.

Browne MW 1984 Asymptotically distribution-free methods for the analysis of covariance structures. *British Journal of Mathematical and Statistical Psychology* **37**, 62–83.

Browne MW and Cudeck R 1993 Alternative ways of assessing model fit In *Testing Structural Equation Models* (ed. Bollen KL and Long JS) Sage Newbury Park, CA.

Browne MW, Cudeck R, Tateneni K and Mels G 1998 CEFA: Comprehensive exploratory factor analysis. Technical report, Ohio State University. Software and documentation available at http://faculty.psy.ohio-state.edu/browne/software.php.

Burt RS 1973 Confirmatory factor analytic structures and the theory of the construction process. *Sociological Methods and Research* **2**, 131–190.

Burt RS 1976 Interpretational confounding of unobserved variables in structural equation models. *Sociological Methods and Research* **5**, 3–52.

Busemeyer J and Jones L 1983 Analysis of multiplicative combination rules when the causal variables are measured with error. *Psychological Bulletin* **93**, 549–562.

Cai L 2010 High-dimensional exploratory item factor analysis by a Metropolis-Hastings Robbins-Monro algorithm. *Psychometrika* **75**, 33–57.

Cai L, du Toit SHC and Thissen D 2011 *IRTPRO: Flexible, multidimensional, multiple categorical IRT modeling*. Scientific Software International, Chicago.

Cai L, Maydeu-Olivares A, Coffman DL and Thissen D 2006 Limited information goodness of fit testing of item response theory models for sparse 2^p tables. *British Journal of Mathematical and Statistical Psychology* **59**, 173–194.

Cattell RB 1978 *The Scientific Use of Factor Analysis in Behavioral and Life Sciences*. Plenum Press, New York.

Chang HH and Stout W 1993 The asymptotic posterior normality of the latent trait in an IRT model. *Psychometrika* **58**(1), 37–52.

Chatfield C and Collins C 1980 *An Introduction to Multivariate Analysis*. Chapman & Hall, London.

Chatterjee S 1984 Variance estimation in factor analysis: an application of the bootstrap. *British Journal of Mathematical and Statistical Psychology* **37**, 252–262.

Chen WH and Thissen D 1997 Local dependence indexes for item pairs using item response theory. *Journal of Educational and Behavioral Statistics* **22**, 265–289.

Christoffersson A 1975 Factor analysis of dichotomized variables. *Psychometrika* **40**(1), 5–32.

Cliff N 1966 Orthogonal rotation to congruence. *Psychometrika* **31**, 33–42.

Clogg CC 1979 Some latent structure models for the analysis of Likert-type data. *Social Science Research* **8**, 287–301.

Collins LM, Fidler PL, Wugalter SE and Long JD 1993 Goodness-of-fit testing for latent class models. *Multivariate Behavioral Research* **28**(33), 375–389.

Comrey AL 1962 The minimum residual method of factor analysis. *Psychology Reports* **11**, 15–18.

Comrey AL and Ahumada A 1964 An improved procedure and program for minimum residual factor analysis. *Psychology Reports* **15**, 91–96.

Cowles MK and Carlin BP 1996 Markov chain Monte Carlo convergence diagnostics: a comparative review. *Journal of the American Statistical Association* **91**, 883–904.

Cox DR and Wermuth N 1996 *Multivariate Dependencies* Monographs on Statistics and Applied Probability 2nd edn. Chapman & Hall, London.

Cronbach LJ 1951 Coefficient alpha and the internal structure of tests. *Psychometrika* **48**, 99–111.

Croon M 1990 Latent class analysis with ordered latent classes. *British Journal of Mathematical and Statistical Psychology* **43**, 171–192.

Croon M and Bolck A 1997 On the use of factor scores in structural equations models. Technical Report 97.10.102/7, Work and Organization Research Centre, Tilburg University. WORC Paper.

Cudeck R and MacCallum RC (eds) 2007 *Factor Analysis at 100: Historical Developments and Future Directions.* Lawrence Erlbaum Associates, Mahwah, NJ.

Cudeck R and O'Dell LL 1994 Applications of standard error estimates in unrestricted factor analysis: significance tests for factor loadings and correlations. *Psychological Bulletin* **115**, 475–487.

Cuttance P and Ecob R (eds) 1987 *Structural Modelling by Example.* Cambridge University Press, Cambridge.

de Boeck P 2008 Random item IRT models. *Psychometrika* **73**, 533–559.

de Menezes LM 1999 On fitting latent class models for binary data: the estimation of standard errors. *British Journal of Mathematical and Statistical Psychology* **52**, 149–168.

de Menezes LM and Bartholomew DJ 1996 New developments in latent structure analysis applied to social attitudes. *Journal of the Royal Statistical Society A* **159**, 213–224.

Dempster AP, Laird NM and Rubin DB 1977 Maximum likelihood from incomplete data via the EM algorithm. *Journal of the Royal Statistical Society B* **39**, 1–38.

Dolby GR 1976 Structural relations and factor analysis. *Mathematical Scientist Supplement No. 1* pp. 25–29. Abstracts from the First Conference of the CSIRO Division of Mathematics and Statistics.

Dunson DB 2000 Bayesian latent variable models for clustered mixed outcomes. *Journal of the Royal Statistical Society B* **62**, 355–366.

Dunson DB 2003 Dynamic latent trait models for multidimensional longitudinal data. *Journal of the American Statistical Association* **98**, 555–563.

Esposito Vinzi V, Chin W, Henseler J and Wang H 2010 *Handbook of Partial Least Squares: Concepts, Methods and Applications in Marketing and Related Areas* Handbook of Computational Statistics. Springer, New York.

Etezadi-Amoli J and McDonald RP 1983 A second generation nonlinear factor analysis. *Psychometrika* **48**(3), 315–342.

Everitt BS 1984 *An Introduction to Latent Variable Models*. Chapman & Hall, London.

Everitt BS and Hand DJ 1981 *Finite Mixture Distributions*. Chapman & Hall, London.

Eysenck SBG, Eysenck HJ and Barrett PT 1985 A revised version of the Psychoticism scale. *Personality and Individual Differences* **6**, 21–29.

Fachel GMG 1986 *The C-type Distribution as an Underlying Model for Categorical Data and its Use in Factor Analysis* PhD thesis University of London.

Feller W 1966 *An Introduction to Probability Theory and Its Applications* vol. 2. John Wiley and Sons, Inc., New York.

Fielding A 1977 Latent structure models In *The Analysis of Survey Data: Exploring Data Structures* (ed. O'Muircheartaigh CA and Payne C) vol. 1. John Wiley & Sons, Ltd. Chichester pp. 125–157.

Formann AK 1992 Linear logistic latent class analysis for polytomous data. *Journal of the American Statistical Association* **87**, 476–486.

Fox JP and Glas CAW 2001 Bayesian estimation of a multilevel IRT model using Gibbs sampling. *Psychometrika* **66**(2), 271–288.

Francis I 1974 Factor analysis: fact or fabrication. *Mathematical Chronicle* **3**, 9–44. Invited address delivered at the Seventh New Zealand Mathematics Colloquium held at Christchurch.

Franzoi SL and Shield SA 1984 The Body Esteem Scale: Multidimensional structure and sex differences in a college population. *Journal of Personality Assessment* **48**, 173–178.

Galton F 1888 Co-relations and their measurement. *Proceedings of the Royal Society* **45**, 135–145.

Gelfand AE and Smith AFM 1990 Sampling-based approaches to calculating marginal densities. *Journal of the American Statistical Association* **85**, 398–409.

Gelman A and Rubin DB 1992 Inference from iterative simulation using multiple sequences. *Statistical Science* **7**, 457–511.

Geman S and Geman D 1984 Stochastic relaxation, Gibbs distributions, and the Bayesian restoration of images. *IEEE Transactions on Pattern Analysis and Machine Intelligence* **6**, 721–741.

Geweke JF 1992 Evaluating the accuracy of sampling-based approaches to calculating posterior moments In *Bayesian Statistics 4* (ed. Bernardo JM, Berger JO, Dawid AP and Smith AFM) Clarendon Press Oxford pp. 169–188.

Geweke JF and Singleton KJ 1980 Interpreting the likelihood ratio statistic in factor models when sample size is small. *Journal of the American Statistical Association* **75**, 133–137.

Glas CAW and Pimental JL 2008 Modeling nonignorable missing data in speeded tests. *Educational and Psychological Measurement* **68**, 907–922.

Goldstein H 1980 Dimensionality, bias, independence and measurement scale problems in latent trait test score models. *British Journal of Mathematical and Statistical Psychology* **33**, 234–246.

Goodman LA 1974 Explanatory latent structure analysis using both identifiable and unidentifiable models. *Biometrika* **61**, 215–231.

Goodman LA 1978 *Analyzing Qualitative/Categorical Data.* Abt Books, Cambridge, MA. Edited by J. Magisdon.

Green BF 1952 The orthogonal approximation of an oblique structure in factor analysis. *Psychometrika* **17**, 429–440.

Greenacre MJ 1984 *Theory and Applications of Correspondence Analysis.* Academic Press, London.

Guttman L 1953 Image theory for the structure of quantitative variables. *Psychometrika* **18**, 277–296.

Haberman SJ 1979 *Analysis of Qualitative Data, Volume 2, New Developments.* Academic Press, New York.

Hakistian AR, Rogers WT and Cattell RB 1982 The behaviour of number-of-factors rules with simulated data. *Multivariate Behavioural Research* **17**, 193–219.

Hambleton RK, Swaminathan H and Rogers HJ 1991 *Fundamentals of Item Response Theory.* Sage Publications, Newbury Park, CA.

Harman HH 1976 *Modern Factor Analysis* 3rd edn. University of Chicago Press, Chicago.

Harman HH and Jones WH 1966 Factor analysis by minimising residuals (Minres). *Psychometrika* **31**, 351–368.

Hastings WK 1970 Monte Carlo sampling methods using Markov chains and their applications. *Biometrika* **57**, 97–109.

Heckman JJ and Singer B (eds) 1985 *Longitudinal Analysis of Labor Market Data.* Cambridge University Press, Cambridge.

Heidelberger P and Welch P 1983 Simulation run length control in the presence of an initial transient. *Operations Research* **31**, 1109–1144.

Heinen T 1996 *Latent Class and Discrete Latent Trait Models Similarities and Differences* Advanced Quantitative Techniques in the Social Sciences. Sage Publications, Thousand Oaks, CA.

Heywood HB 1931 On finite sequences of real numbers. *Proceedings of the Royal Society A* **134**, 486–510.

Hills M 1977 Review of 'An Introduction to Multivariate Techniques for Social and Behavioural Sciences' by S. Bennett and D. Bowers. *Applied Statistics* **26**(3), 339–340.

Holland PW 1981 When are item response models consistent with observed data? *Psychometrika* **46**, 79–92.

Holland PW 1990a The Dutch identity: a new tool for the study of item response models. *Psychometrika* **55**, 5–18.

Holland PW 1990b On the sampling theory foundations of item response theory models. *Psychometrika* **55**, 577–601.

Holland PW and Rosenbaum PR 1985 Conditional association and unidimensionality in monotone latent variable models. Technical report, Educational Testing Service, Princeton, NJ.

Hotelling H 1933 Analysis of a complex of statistical variables into principal components. *Journal of Educational Psychology* **24**, 417–441.

Hotelling H 1957 The relation of the newer multivariate methods to factor analysis. *British Journal of Statistical Psychology* **10**, 69–79.

Huber P, Ronchetti E and Victoria-Feser MP 2004 Estimation of generalized linear latent variable models. *Journal of the Royal Statistical Society B* **66**, 893–908.

Hunter JE and Gerbing DW 1982 Unidimensional measurement, second order factor analysis, and causal models. *Research in Organizational Behavior* **4**, 267–320.

Jennrich RI 1973 Standard errors for obliquely rotated factor loadings. *Psychometrika* **38**, 593–604.

Jennrich RI 2006 Rotation to simple loadings using component loss functions: The oblique case. *Psychometrika* **71**, 173–191.

Jennrich RI and Thayer DT 1973 A note on Lawley's formulas for standard errors in maximum likelihood factor analysis. *Psychometrika* **38**, 571–580.

Jensen AR 1998 *The g Factor: The Science of Mental Ability*. Praeger, Westport, CT.

Johnson RA and Wichern DW 1982 *Applied Multivariate Analysis*. Prentice Hall, Englewood Cliffs, NJ.

Johnson VE and Albert JH 1999 *Ordinal Data Modeling*. Springer-Verlag, New York.

Jöreskog KG 1967 Some contributions to maximum likelihood factor analysis. *Psychometrika* **32**, 443–482.

Jöreskog KG 1969 Efficient estimation in image factor analysis. *Psychometrika* **34**, 51–75.

Jöreskog KG 1970 A general method for analysis of covariance structures. *Biometrika* **57**, 239–251.

Jöreskog KG 1977 Factor analysis by least squares and maximum likelihood methods In *Statistical Methods for Digital Computers* (ed. Enslein K, Ralston W and Wilf HS) vol. 3 John Wiley & Sons, Inc. New York.

Jöreskog KG 1990 New developments in LISREL: analysis of ordinal variables using polychoric correlations and weighted least squares. *Quality and Quantity* **24**, 387–404.

Jöreskog KG 1994 On the estimation of polychoric correlations and their asymptotic covariance matrix. *Psychometrika* **59**, 381–389.

Jöreskog KG and Goldberger AS 1972 Factor analysis by generalized least squares. *Psychometrika* **37**, 243–259.

Jöreskog KG and Moustaki I 2001 Factor analysis of ordinal variables: a comparison of three approaches. *Multivariate Behavioral Research* **36**, 347–387.

Jöreskog KG and Sörbom D 1977 Statistical models and methods for analysis of longitudinal data In *Latent Variables and Socio-economic Models* (ed. Aigner DJ and Goldberger AS) North-Holland Amsterdam.

Jöreskog KG and Sörbom D 2006 *LISREL 8.8 for Windows*. Scientific Software International, Lincolnwood, IL. Computer software.

Jöreskog KG and Yang F 1996 Nonlinear structural equation models: The Kenny-Judd model with interaction effects In *Advanced Structural Equation Modeling: Issues and Techniques* (ed. Marcoulides G and Schumacker R) Lawrence Erlbaum Associates Mahwah, NJ pp. 57–88.

Junker BW 1993 Conditional association, essential independence and monotone unidimensional item response models. *Annals of Statistics* **21**, 1359–1378.

Kaiser HF 1958 The varimax criterion for analytic rotation in factor analysis. *Psychometrika* **23**, 187–200.

Kaiser HF and Caffrey J 1965 Alpha factor analysis. *Psychometrika* **30**(1), 1–14.

Kano Y 1983 Consistency of estimators in factor analysis. *Journal of the Japanese Statistical Society* **13**, 137–174.

Kano Y 1986a Conditions on consistency of estimators in covariance structure model. *Journal of the Japanese Statistical Society* **16**, 75–80.

Kano Y 1986b Consistency conditions on the least squares estimators in a single common factor analysis model. *Annals of the Institute of Statistical Mathematics, A* **39**, 57–68.

Kaplan D 2009 *Structural Equation Modeling: Foundations and Extensions* 2nd edn. Sage, London.

Kenny DA and Judd CM 1984 Estimating the non-linear and interactive effects of latent variables. *Psychological Bulletin* **96**, 201–210.

Kim SH 2001 An evaluation of a Markov chain Monte Carlo method for the Rasch model. *Applied Psyhological Measurement* **25**(2), 163–176.

Knott M 2005 A measure of independence for a multivariate normal distribution and some connections with factor analysis. *Journal of Multivariate Analysis* **96**(2), 374–383.

Knott M and Albanese MT 1992 Large values for maximum likelihood estimates of parameters in latent variable models Unpublished.

Knott M and Albanese MT 1993 Conditional distributions of a latent variable and scoring for binary data. *Revista Brasileira de Probabilidade e Estatística* **6**, 171–188.

Knott M and Bartholomew DJ 1993 Scoring binary responses for maximum reliability. *Psychometrika* **58**, 331–338.

Krane WR and McDonald RP 1978 Scale invariance and the factor analysis of correlation matrices. *British Journal of Mathematical and Statistical Psychology* **31**, 218–228.

Krebs D and Schuessler KF 1987 *Soziale Empfindungen: ein interkultureller Skalenvergleich bei Deutschen und Amerikanern (Social Attitudes: an intercultural Scale Comparison for Germans and Americans)* Monographien : Sozialwissenschaftliche Methoden. Campus Verlag, Frankfurt/Main.

Lance CE, Cornwell JM and Mulaik SA 1988 Limited information parameter estimates for latent or mixed manifest and latent variable models. *Multivariate Behavioral Research* **23**, 171–187.

Langeheine R and Rost J 1988 *Latent Trait and Latent Class Models*. Plenum Press, New York.

Langeheine R, Pannekoek J and van de Pol F 1996 Bootstrapping goodness-of-fit measures in categorical data analysis. *Sociological Methods and Research* **24**(4), 492–516.

Lawley DN 1953 *A Modified Method of Estimation in Factor Analysis and Some Large Sample Results* number 3 in *Nordisk Psychologi's Monograph Series*, Copenhagen: Ejnar Mundsgaards.

Lawley DN 1967 Some new results in maximum likelihood factor analysis. *Proceedings of the Royal Society of Edinburgh A* **67**, 256–264.

Lawley DN and Maxwell AE 1971 *Factor Analysis as a Statistical Method* 2nd edn. Butterworth, London.

Lazarsfeld PF and Henry NW 1968 *Latent Structure Analysis*. Houghton-Mifflin, New York.

Lee SY 1981 A Bayesian approach to confirmatory factor analysis. *Psychometrika* **46**, 153–160.

Lee SY 2007 *Structural Equation Modelling: A Bayesian Approach*. John Wiley & Sons, Inc., New York.

Lee SY and Song XY 2003 Bayesian analysis of structural equation models with dichotomous variables. *Statistics in Medicine* **22**, 3073–3088.

Lee SY and Song XY 2004a Bayesian model comparison of nonlinear latent variable models with missing continuous and ordinal categorical data. *British Journal of Mathematical and Statistical Psychology* **57**, 131–150.

Lee SY and Song XY 2004b Maximum likelihood analysis of a generalized latent variable model with hierarchical mixed data. *Biometrics* **60**, 624–636.

Lee SY and Zhu H 2002 Maximum likelihood estimation of nonlinear structural equation models. *Psychometrika* **67**, 189–210.

Lee SY, Poon WY and Bentler PM 1990a Full maximum likelihood analysis of structural equation models with polytomous variables. *Statistics and Probability letters* **9**, 91–97.

Lee SY, Poon WY and Bentler PM 1990b A three-stage estimation procedure for structural equation models with polytomous variables. *Psychometrika* **55**, 45–51.

Lee SY, Poon WY and Bentler PM 1992 Structural equation models with continuous and polytomous variables. *Psychometrika* **57**, 89–105.

Lee SY, Poon WY and Bentler PM 1995 A two-stage estimation of structural equation models with continuous and polytomous variables. *British Journal of Mathematical and Statistical Psychology* **48**, 339–358.

Lee Y and Nelder JA 2009 Likelihood inference for models with unobservables: another view. *Statistical Science* **24**(3), 255–269.

Leimu H 1983 Työntekijäin työasema ja työpaikkaliikkuvuus erikokoisissa teollisuusyrityksissä. (Blue collar work position and workplace mobility in small and large industrial firms. I. Theoretical points of departure and comparison of workers in small and large industrial firms.). Technical Report C40, University of Turku, Turku.

Leung SO 1992 Estimation and application of latent variable models in categorical data analysis. *British Journal of Mathematical and Statistical Psychology* **45**, 311–328.

Linting M, Meulman JJ, Groenen PJF and Van der Kooij JJ 2007 Nonlinear principal components analysis: Introduction and application. *Psychological Methods* **12**, 336–358.

Loehlin JC 1989 *Latent Variable Models: An Introduction to Factor, Path, and Structural Analysis* 2nd edn. Lawrence Erlbaum Associates, Hillsdale, NJ.

Lopez HF and West M 2004 Bayesian model assessment in factor analysis. *Statistica Sinica* **14**, 41–67.

Macready GB and Dayton CM 1977 The use of probabilistic models in the assessment of mastery. *Journal of Educational Statistics* **2**(2), 99–120.

Magnus JR and Neudecker H 1988 *Matrix Calculus with Applications in Statistics and Econometrics* Wiley Series in Probability and Mathematical Statistics. John Wiley & Sons, Inc., New York.

Maraun MD 1996 Metaphor taken as math: Indeterminacy in the factor analysis model. *Multivariate Behavioral Research* **31**, 517–538.

Martin JK and McDonald RP 1975 Bayesian estimation in unrestricted factor analysis; a treatment for Heywood cases. *Psychometrika* **40**, 505–517.

Martinson EO and Hadman MA 1975 Calculation of the polychoric estimate of correlation in contingency tables. *Applied Statistics* **24**, 272–278.

Masters GM 1985 A comparison of latent trait and latent class analysis of Likert-type data. *Psychometrika* **50**, 69–82.

Masters GN 1982 A Rasch models for partial credit scoring. *Psychometrika* **47**, 149–174.

Masters GN and Wright BD 1997 The partial credit model In *Handbook of Modern Item Response Theory* (ed. van der Linden W and Hambleton RK) Springer New York pp. 351–367.

Maydeu-Olivares A and Joe H 2005 Limited- and full-information estimation and goodness-of-fit testing in 2^n contingency tables: A unified framework. *Journal of the American Statistical Association* **100**(471), 1009–1020.

Maydeu-Olivares A and Joe H 2006 Limited information goodness-of-fit testing in multidimensional contingency tables. *Psychometrika* **71**, 713–732.

McDonald RP 1962 A note on the derivation of the general latent class model. *Psychometrika* **27**(2), 203–206.

McDonald RP 1967a Factor interaction in nonlinear factor analysis. *British Journal of Mathematical and Statistical Psychology* **32**(20), 205–215.

McDonald RP 1967b *Non-linear factor analysis* number 15 in *Psychometric Monographs*. Psychometric Society.

McDonald RP 1967c Numerical methods for polynomial models in nonlinear factor analysis. *Psychometrika* **32**, 77–112.

McDonald RP 1983 Alternative weights and invariant parameters in optimal scaling. *Psychometrika* **48**(3), 377–391.

McDonald RP 1985 *Factor Analysis and Related Methods*. Lawrence Erlbaum Associates, Hillsdale, NJ.

McFadden D 1974 Conditional logit analysis of qualitative choice behaviour In *Frontiers in Econometrics* (ed. Zarembka P) Academic Press New York pp. 105–142.

McFadden D 1982 Qualitative response models In *Advances in Econometrics* (ed. Hildenbrand W) Cambridge University Press Cambridge pp. 1–37.

McHugh RB 1956 Efficient estimation and local identification in latent class analysis. *Psychometrika* **21**(4), 331–347.

McHugh RB 1958 Note on 'efficient estimation and local identification in latent class analysis'. *Psychometrika* **23**(3), 273–274.

McLachlan GJ and Krishnan J 1997 *The EM Algorithm and Extensions*. John Wiley & Sons, Inc., New York.

Meng XL and Schilling S 1996 Fitting full-information item factor models and an empirical investigation of bridge sampling. *Journal of the American Statistical Association* **91**(435), 1254–1267.

Metropolis N, Rosenbluth AW, Rosenbluth MN, Teller AH and Teller E 1953 Equations of state calculations by fast computing machine. *Journal of Chemical Physics* **21**, 1087–1091.

Mislevy RJ and Verhelst N 1990 Modeling item responses when different subjects employ different solution strategies. *Psychometrika* **55**, 195–215.

Molenaar PCW and von Eye A 1994 On the arbitrary nature of latent variables In *Latent Variables Analysis* (ed. von Eye A and Clogg CC) Sage London pp. 226–242.

Moustaki I 1996 A latent trait and a latent class model for mixed observed variables. *British Journal of Mathematical and Statistical Psychology* **49**, 313–334.

Moustaki I 2000 A latent variable model for ordinal variables. *Applied Psychological Measurement* **24**, 211–223.

Moustaki I and Knott M 1997 Generalized latent trait models. Statistics Research Report 36, London School of Economics and Political Science, Statistics Department.

Moustaki I and Knott M 2000a Generalized latent trait models. *Psychometrika* **65**, 391–411.

Moustaki I and Knott M 2000b Weighting for item non-response in attitude scales by using latent variable models with covariates. *Journal of the Royal Statistical Society, Series A* **163**, 445–459.

Moustaki I and Knott M 2005 Computational aspects of the E-M and Bayesian estimation in latent variable models In *New Developments in Categorical Data Analysis for the Social and Behavioral Sciences* (ed. van der Ark L, Croon M and Sijtsma K) Lawrence Erlbaum Associates, Mahwah, NJ pp. 103–124.

Moustaki I and Papageorgiou I 2004 Latent class models for mixed outcomes with applications in archaeometry. *Computational Statistics & Data Analysis* **48**, 659–675.

Mulaik SA 1972 *The Foundations of Factor Analysis*. McGraw-Hill, New York.

Mulaik SA 1986 Factor analysis and Psychometrika: major developments. *Psychometrika* **51**(1), 23–33.

Mulaik SA 2009a *Foundations of Factor Analysis* 2nd edn. CRC Press, Boca Raton, FL.

Mulaik SA 2009b *Linear Causal Modeling with Structural Equations*. CRC Press, Boca Raton, FL.

Muraki E 1990 Fitting a polytomous item response model to Likert-type data. *Applied Psychological Measurement* **14**, 59–71.

Muraki E 1992 A generalized partial credit model: application of an EM algorithm. *Applied Psychological Measurement* **16**, 159–176.

Muraki E and Carlson E 1995 Full-information factor analysis for polytomous item responses. *Applied Psychological Measurement* **19**, 73–90.

Muthén B 1978 Contributions to factor analysis of dichotomous variables. *Psychometrika* **43**(4), 551–560.

Muthén B 1984 A general structural model with dichotomous, ordered categorical and continuous latent variable indicators. *Psychometrika* **49**(1), 115–132.

Muthén B 2008 Latent variable hybrids In *Advances in Latent Variable Mixture Models* (ed. Hancock G and Samuelsen K) Information Age Charlotte, NC pp. 1–24.

Muthén B and Asparouhov T 2006 Item response mixture modeling: Application to tobacco dependence criteria. *Addictive Behaviors* **31**, 1050–1066.

Muthén B and Satorra A 1995 Technical aspects of Muthén's LISCOMP approach to estimation of latent variable relations with a comprehensive measurement model. *Psychometrika* **60**, 489–503.

Muthén L and Muthén B 2010 *Mplus User's Guide* 6th edn. Muthén and Muthén, Los Angeles.

Ogasawara H 1998 Standard errors for rotation matrices with an application to the promax solution. *British Journal of Mathematical and Statistical Psychology* **51**, 163–178.

Okamoto M and Ihara M 1983 A new algorithm for least-squares solution in factor analysis. *Psychometrika* **48**, 597–605.

Olsson U 1979 On the robustness of factor analysis against crude classification of the observations. *Multivariate Behavioral Research* **14**, 485–500.

O'Muircheartaigh C and Moustaki I 1999 Symmetric pattern models: a latent variable approach to item non-response in attitude scales. *Journal of the Royal Statistical Society, Series A* **162**, 177–194.

Owen RJ 1975 A Bayesian sequential procedure for quantal response in the context of adaptive testing. *Journal of the American Statistical Association* **70**, 351–356.

Palomo J, Dunson DB and Bollen K 2007 Bayesian structural equation modeling In *Handbook of Latent Variable and Related Models* (ed. Lee SY) Handbook of Computing and Statistics with Applications Elsevier Amsterdam.

Patz RJ 1996 *Markov Chain Monte Carlo Methods for Item Response Theory Models with Applications for the National Assessment of Educational Progress* PhD thesis Carnegie Mellon University.

Patz RJ and Junker BW 1999a Applications and extensions of MCMC in IRT: multiple item types, missing data, and rated responses. *Journal of Educational and Behavioral Statistics* **24**, 342–366.

Patz RJ and Junker BW 1999b A straightforward aproach to Markov chain Monte Carlo methods for item response models. *Journal of Educational and Behavioral Statistics* **24**, 146–178.

Pearson K 1901 On lines and planes of closest fit to a system of points in space. *Philosophical Magazine (6)* **2**, 557–572.

Pennell R 1972 Routinely computable confidence intervals for factor loadings using the 'jack-knife'. *British Journal of Mathematical and Statistical Psychology* **25**, 107–114.

Pickering RM and Forbes JF 1984 A classification of Scottish infants using latent class analysis. *Statistics in Medicine* **3**, 249–259.

Rabe-Hesketh S, Skrondal A and Pickles A 2002 Reliable estimation of generalized linear mixed models using adaptive quadrature. *Stata Journal* **2**(1), 1–21.

Raftery AL and Lewis S 1992a Comment: One long run with diagnostics: Implementation strategies for Markov chain Monte Carlo. *Statistical Science* **7**, 493–497.

Raftery AL and Lewis SM 1992b How many iterations in the Gibbs sampler? In *Bayesian Statistics 4* (ed. Bernardo JM, Berger, J. O.and Dawid AP and Smith AFM) Clarendon Press Oxford pp. 763–773.

Rao CR 1955 Estimation and tests of significance in factor analysis. *Psychometrika* **20**, 93–111.

Rao CR 1973 *Linear Statistical Inference and Its Applications* 2nd edn. John Wiley & Sons, Inc., New York.

Rasch G 1960 *Probabilistic Models for Some Intelligence and Attainment Tests*. Pædagogiske Institut, Copenhagen.

Rasch G 1961 On general laws and the meaning of measurement in psychology In *Proceedings of the Fourth Berkeley Symposium on Mathematical Statistics and Probability* (ed. Neyman J), vol. 4, pp. 321–334. University of California Press, Berkeley.

Reiser M 1996 Analysis of residuals for the multinomial item response model. *Psychometrika* **61**, 509–528.

Reiser M and VandenBerg M 1994 Validity of the chi-square test in dichotomous variable factor analysis when expected frequencies are small. *British Journal of Mathematical and Statistical Psychology* **47**, 85–107.

Rizopoulos D and Moustaki I 2008 Generalized latent variable models with non-linear effects. *British Journal of Mathematical and Statistical Psychology* **24**, 415–438.

Rosenbaum PR 1984 Testing the conditional independence and monotonicity assumptions of item response theory. *Psychometrika* **49**(3), 425–435.

Rost J 1990 Rasch models in latent classes: An integration of two approaches to item analysis. *Applied Psychological Measurement* **14**(3), 271–282.

Rost J and Langeheine R (eds) 1997 *Applications of Latent Trait and Latent Class Models in the Social Sciences*. Waxmann, Münster.

Rubin DB and Thayer DT 1982 EM algorithms for ML factor analysis. *Psychometrika* **47**, 69–76.

Samejima F 1969 Estimation of latent ability using a response pattern of graded scores. *Psychometrika, Monograph Supplement No. 17*.

Sammel RD, Ryan LM and Legler JM 1997 Latent variable models for mixed discrete and continuous outcomes. *Journal of the Royal Statistical Society, B* **59**, 667–678.

Saris WE and Stronkhorst LH 1984 *Causal Modelling in Non-experimental Research: An Introduction to the LISREL Approach*. Sociometric Research Foundation, Amsterdam.

Schilling S and Bock RD 2005 High-dimensional maximum marginal likelihood item factor analysis by adaptive quadrature. *Psychometrika* **70**, 533–555.

Schneeweiss H 1993 Consistency at large in models with latent variables In *Statistical Modelling and Latent Variables* (ed. Haagen K, Bartholomew DJ and Deistler M) North Holland Amsterdam pp. 299–320.

Schönemann PH 1966 A generalized solution of the orthogonal Procrustes problem. *Psychometrika* **31**, 1–10.

Schönemann PH 1981 Power as a function of communality in factor analysis. *Bulletin of the Psychometric Society* **17**(1), 57–60.

Schuessler KF 1982 *Measuring Social Life Feelings* Jossey-Bass Social and Behavioral Science Series. Jossey-Bass, San Francisco.

Schumacker R and Marcoulides G (eds) 1998 *Interaction and Nonlinear Effects in Structural Equation Modeling*. Lawrence Erlbaum Associates, Mahwah, NJ.

Schwarz G 1978 Estimating the dimension of a model. *Annals of Statistics* **6**, 461–464.

Sclove S 1987 Application of model-selection criteria to some problems of multivariate analysis. *Psychometrika* **52**, 333–343.

Seber GAF 1984 *Multivariate Observations*. John Wiley & Sons, Inc., New York.

Shi JQ and Lee SY 1997a A Bayesian estimator of factor score in confirmatory factor model with polytomous, censored or truncated data. *Psychometrika* **62**, 29–50.

Shi JQ and Lee SY 1997b Estimation of factor scores with polytomous data by the E-M algorithm. *British Journal of Mathematical and Statistical Psychology* **50**, 215–226.

Shi JQ and Lee SY 1998 Bayesian sampling-based approach for factor analysis model with continuous and polytomous data. *British Journal of Mathematical and Statistical Psychology* **51**, 233–252.

Skrondal A and Rabe-Hesketh S 2004 *Generalized Latent Variable Modeling: Multilevel, Longitudinal, and Structural Equation Models*. Chapman & Hall/CRC, Boca Raton, FL.

Smith GA and Stanley G 1983 Clocking g: Relating intelligence and measures of timed performance. *Intelligence* **7**, 353–368.

Sobel ME 1994 Causal inference in latent variable models In *Latent Variables Analysis* (ed. van Eye A and Clogg CC) Sage Publications Thousand Oaks, CA pp. 3–35.

Song XY and Lee SY 2004 Bayesian analysis of two-level nonlinear structural equation models with continuous and polytomous data. *British Journal of Mathematical and Statistical Psychology* **57**, 29–52.

Song XY and Lee SY 2006 Bayesian analysis of structural equation models with nonlinear covariates and latent variables. *Multivariate Behavioral Research* **41**, 337–365.

Spearman C 1904 General intelligence, objectively determined and measured. *American Journal of Psychology* **15**, 201–293.

Spiegelhalter D, Thomas A, Best N and Gilks W 1996 BUGS: Bayesian inference using Gibbs sampling. Technical report, MRC Biostatistics Unit, Institute of Public Health, Cambridge.

Stephenson W 1935 Correlating persons instead of tests. *Character and personality* **4**, 17–24.

Stout W 1987 A nonparametric approach for assessing latent trait unidimensionality. *Psychometrika* **52**, 589–617.

Stout W 1990 A new item response theory modeling approach with applications to unidimensionality assessment and ability estimation. *Psychometrika* **55**, 293–325.

Tabachnick BG and Fidell LS 1996 *Using Multivariate Statistics* 3rd edn. Harper & Row, New York.

Takane Y and de Leeuw J 1987 On the relationship between item response theory and factor analysis of discretized variables. *Psychometrika* **52**, 393–408.

Tanaka JS and Huba GJ 1985 A fit index for covariance structure models under arbitrary GLS. *British Journal of Mathematical and Statistical Psychology* **38**, 197–201.

Tanner MA and Wong WH 1987 The calculation of posterior distributions by data augmentation. *Journal of the American Statistical Association* **82**, 528–540.

Thissen D and Steinberg L 1984 A response model for multiple choice items. *Psychometrika* **49**, 501–519.

Thissen D and Steinberg L 1986 A taxonomy of item response models. *Psychometrika* **51**, 567–577.

Thomson GH 1916 A hierarchy without a general factor. *British Journal of Psychology* **8**, 271–281.

Thomson GH 1951 *The Factorial Analysis of Human Ability* 5th edn. University of London Press, London.

Thurstone LL 1947 *Multiple Factor Analysis*. University of Chicago Press, Chicago.

Tipping ME and Bishop CM 1999 Probabilistic principal component analysis. *Journal of the Royal Statistical Society B* **21**(3), 611–622.

Titterington DM, Smith AFM and Makov UE 1985 *Statistical Analysis of Finite Mixture Distributions*. John Wiley & Sons, Ltd, Chichester.

Tucker LR, Koopman RF and Linn RL 1969 Evaluation of factor analytic research procedures by means of simulated correlation matrices. *Psychometrika* **34**, 421–459.

Tuma NB and Hannan MT 1984 *Social Dynamics, Models and Methods*. Academic Press, San Diego, CA.

Tzamourani P and Knott M 2002 Fully semiparametric estimation of the two-parameter latent trait model for binary data In *Latent Variable and Latent Structure Models* (ed. Marcoulides GA and Moustaki I) Lawrence Erlbaum Associates Mahwah NJ pp. 63–84.

Upton GJ 1980 Contribution to the discussion of Bartholomew (1980). *Journal of the Royal Statistical Society, B* **42**, 316.

van der Linden W and Hambleton RK (eds) 1997 *Handbook of Modern Item Response Theory*. Springer, New York.

van Driel OP 1978 On various causes of improper solutions in maximum likelihood factor analysis. *Psychometrika* **43**, 225–243.

Vermunt JK and Magidson J 2000 *Latent GOLD 4.5 User's Guide*. Statistical Innovations, Belmont, MA.

Wall M and Amemiya Y 2000 Estimation of polynomial structural equation models. *Journal of the American Statistical Association* **95**, 929–940.

Wall M and Amemiya Y 2001 Generalized appended product indicator procedure for nonlinear structural equation analysis. *Journal of Educational and Behavioural Statistics* **26**, 1–29.

Wedel M and DeSarbo WS 1995 A mixture likelihood approach for generalized linear models. *Journal of Classification* **12**, 21–55.

White H 1982 Maximum likelihood estimation of misspecified models. *Econometrics* **50**, 1–25.

Whittaker J 1990 *Graphical Models in Applied Multivariate Analysis*. John Wiley & Sons, Inc., New York.

Whittle P 1953 On principal components and least square methods of factor analysis. *Skandinavisk Aktuarietidskrift* **35**, 223–239.

Williams JS 1978 A definition for the common factor analysis model and the elimination of problems of factor score indeterminacy. *Psychometrika* **43**, 293–306.

Wold H 1975 Soft modelling by latent variables: the non-linear iterative partial least squares (NIPALS) approach In *Perspectives in Probability and Statistics: Papers in honour of M. S. Bartlett* (ed. Gani J) Academic Press London pp. 117–142.

Wold H 1980 Model construction and evaluation when theoretical knowledge is scarce: theory and application of partial least squares In *Evaluation of Econometric Models* (ed. Kmenta J and Ramsey JB) Academic Press London pp. 47–74.

Wold H 1982 Soft modelling: the basic design and some extensions In *Systems under Indirect Observation: Causality, Structure, Prediction, Part II* (ed. Jöreskog K and Wold H) North Holland Amsterdam.

Wollack JA, Bolt DM, Cohen AS and Lee YS 2002 Recovery of item parameters in the nominal response model: A comparison of marginal maximum likelihood estimation and Markov chain Monte Carlo estimation. *Applied Psychological Measurement* **26**(3), 339–352.

Wright DB, Gaskell G and O'Muircheartaigh C 1994 Flashbulb memory assumptions: Using national surveys to explore cognitive phenomena. Technical Report 11, London School of Economics Methodology Institute.

Wright S 1921 Correlation and causation. *Journal of Agricultural Research* **20**, 557–585.

Wu CF 1983 On the convergence properties of the EM algorithm. *Annals of Statistics* **11**, 95–103.

Yang Jonsson F 1997 *Simulation Studies of the Kenny-Judd Model* PhD thesis Uppsala University.

Zegers FE and ten Berge JMF 1983 A fast and simple computational method of minimum residual factor analysis. *Multivariate Behavioral Research* **18**, 331–340.

Zhu H and Lee SY 1999 Statistical analysis of nonlinear factor analysis models. *British Journal of Mathematical and Statistical Psychology* **52**, 225–242.

Zucchini W and MacDonald IL 2009 *Hidden Markov Models for Time Series: An Introduction Using R* Monographs on Statistics and Applied Probability. Chapman & Hall/CRC, Boca Raton, FL.

Author index

Ahumada, A., 62
Aitkin, M., 90, 92, 164
Akaike, H., 58, 201
Albanese, M.T., 42, 100, 109
Albert, J.H., 31, 103
Amemiya, Y., 26, 63
Amemiya, T., 210
Andersen, E.B., 16, 17, 23, 44
Anderson, T.W., 12, 15, 17, 60, 62, 63, 65, 67, 164, 231, 239
Arbuckle, J.L., 213
Archer, C.O., 56
Arminger, G., 26, 144, 192, 209
Asparouhov, T., 186

Baker, F.B., 31
Barankin, E.W., 22
Barnes, S., 154
Barrett, P.T., 114, 115
Bartholomew, D.J., ii, 12, 13, 16, 26, 42, 46, 72, 74, 98, 99, 101, 112, 113, 168, 170, 215, 224, 237, 238, 243
Bartlett, M.S., 26, 58, 73
Basilevsky, A., 15, 44, 63, 232
Bassin, E.L., 112
Beguin, A.A., 31
Bentler, P.M., 16, 71, 142, 143, 213
Bergsma, W., 178
Best, N., 108, 112
Birnbaum, A., 16, 23, 31
Bishop, C.M., 232, 233
Bock, R.D., 44, 90, 92, 93, 123
Bolck, A., 227, 241
Bollen, K.A., 16, 26, 31, 213, 220
Bolt, D.M., 31
Boomsma, A., 67
Bozdogan, H., 59
Brokken, F.B., 215
Brooks, S.P., 108
Browne, M.W., 56, 59, 71, 221
Burt, R.S., 239, 242
Busemeyer, J., 27

Caffrey, J., 45, 70
Cai, L., 99, 110, 116, 149, 155, 160
Carlin, B.P., 108
Carlson, E., 136
Cattell, R.B., 15, 64
Chamberlain, G., 18
Chang, H.H., 43, 100
Chatfield, C., 16
Chatterjee, S., 56
Chen, W.H., 115, 155
Chib, S., 31
Chin, W., 239
Christoffersson, A., 96, 97
Cliff, N., 215
Clogg, C.C., 170, 176
Coffman, D.L., 99
Cohen, A.S., 31
Collins, L.M., 16, 98
Comrey, A.L., 62
Cornwell, J.M., 239
Cowles, M.K., 108

Latent Variable Models and Factor Analysis: A Unified Approach, Third Edition.
David Bartholomew, Martin Knott and Irini Moustaki.

Cox, D.R., 216
Cronbach, L.J., 45
Croon, M., 178, 185, 227, 241
Cudeck, R., 15, 55, 56, 71, 221
Cuttance, P., 213

Dayton, C.M., 158, 167
de Leeuw, J., 89
de Menezes, L., 56, 112, 165, 168, 170
Deary, I.J., 238
Dempster, A.P., 53, 90, 162
DeSarbo, W.S., 180
Dolby, G.R., 72
du Toit, S.H.C., 110, 116, 149, 155, 160
Dunson, D.B., 31

Ecob, R., 213
Esposito Vinzi, V., 239
Etezadi-Amoli, J., 26
Everitt, B.S., 16, 157, 181
Eysenck, H.J., 114, 115
Eysenck, S.B.G., 114, 115

Fachel, G.M.G., 64, 67
Fidell, L.S., 70
Fidler, P.L., 98
Fielding, A., 12
Forbes, J.F., 57, 158, 170, 174
Formann, A.K., 157
Fox, J.P., 31
Francis, I., 56, 59
Franzoi, S.L., 155, 177

Galbraith, J., ii, 99, 243
Galton, F., 13
Gaskell, G., 204
Gelfand, A.E., 31, 102
Gelman, A., 108, 109
Geman, D., 102
Geman, S., 102
Gerbing, D.W., 67, 239
Geweke, J.F., 58, 108
Gilks, W., 108, 112
Glas, C.A.W., 31, 110
Goldberger, A.S., 60
Goldstein, H., 40
Goodman, L.A., 157, 162, 170, 180
Green, B.F., 215
Greenacre, M.J., 236
Groenen, P.J.F., 236
Guttman, L., 63

Haberman, S.J., 157, 162
Hadman, M.A., 141
Hagenaars, J.A., 178
Hakistian, A.R., 64
Hambleton, R.K., 16, 136
Hand, D.J., 181
Hannan, M.T., 18
Harman, H.H., 15, 62, 79
Hastings, W.K., 105
Heckman, J.J., 18
Heidelberger, P., 108
Heinen, T., 16, 40, 157, 162
Henry, N.W., 16, 157, 182
Henseler, J., 239
Heywood, H.B., 67
Hills, M., 16
Hinde, J., 164
Hoem, J.M., 18
Holland, P.W., 41, 42, 100
Hotelling, H., 15, 16, 229
Huba, G.J., 59
Huber, P., 93
Hunter, J.E., 239

Ihara, M., 62

Jöreskog, K.G., 16, 52, 53, 60, 63, 141–144, 154, 192, 213, 218
Jennrich, R.I., 56, 81
Jensen, A.R., 237
Joe, H., 99, 148
Johnson, R.A., 31, 56
Jones, W.H., 27, 62
Judd, C.M., 26
Junker, B.W., 31, 100, 106, 107

Kaase, M., 154
Kaiser, H.F., 45, 69, 70

Kano, Y., 65
Kaplan, D., 213
Kenny, D.A., 26
Kim, S.H., 31
Knott, M., 31, 39, 42, 62, 74, 100, 109, 110, 192, 204
Koopman, R.F., 56
Krane, W.R., 66
Krebs, D., 84, 112
Krishnan, J., 53
Kuha, J., 42
Küsters, U., 144, 192

Laird, N.M., 53, 90, 162
Lance, C.E., 239
Langeheine, R., 16, 98, 157
Lawley, D.N., 15, 50, 51, 55, 56, 59, 215, 231
Lawn, M., 238
Lazarsfeld, P.F., 16, 157, 182
Lee, Y., 17, 26, 31, 68, 94, 142, 143, 148
Lee, S.-Y., 213
Legler, J.M., 93, 192
Leimu, H., 152
Leung, S.O., 99, 152
Lewis, S., 108
Linn, R.L., 56
Linting, M., 236
Loehlin, J.C., 213
Long, J., 98
Lopez, J., 31, 109

MacCullum, R., 15
MacDonald, I.L., 18
Macready, G.B., 158, 167
Madsen, M., 17
Magidson, J., 160, 164, 177
Magnus, J.R., 50, 51, 69
Maitra, E.P., 22
Makov, U.E., 17
Maraun, M.D., 12, 41, 72
Marcoulides, G., 26
Martin, J.K., 68
Martinson, E.O., 141
Masters, G.M., 140, 170, 176
Maxwell, A.E., 15, 50, 51, 56, 59, 215
Maydeu-Olivares, A., 99, 148
McDonald, R.P., 15, 26, 66, 68, 70
McFadden, D., 144
McHugh, R.B., 162
McLachlan, G.J., 53
Mels, G., 56, 71
Meng, X.L., 94
Metropolis, N., 105
Meulman, J.J., 236
Mislevy, R.J., 186
Molenaar, P.C.W., 184
Moran, P.A., 89
Moustaki, I., ii, 11, 26, 27, 31, 44, 99, 110, 136, 143, 144, 154, 192, 204, 205, 207, 243
Mulaik, S.A., 14, 15, 213, 239
Muraki, E., 136, 140
Muthén, B., 16, 26, 56, 70, 71, 75, 81, 97, 110, 142, 186, 191, 213
Muthén, L., 16, 56, 70, 71, 75, 81, 142, 213

Nelder, J.A., 17
Neudecker, H., 50, 51, 69

O'Dell, L.L., 55, 56
O'Muircheartaigh, C., 110, 204
Ogasawara, H., 56
Okamoto, M., 62
Olsson, U., 141
Owen, R.J., 31

Palomo, J., 31
Pannekoek, J., 98
Papageorgiou, I., 11, 207
Patz, R.J., 31, 106, 107
Paxton, P., 26
Pearson, K., 16
Pennell, R., 56
Pickering, R.M., 57, 158, 170, 174
Pimental, J.L., 110
Poon, W.-Y., 142, 143

Rabe-Hesketh, S., 12, 93
Raftery, A.L., 108
Ramirez, D.E., 59
Rao, C.R., 51, 63, 70
Rasch, G., 140
Reiser, M., 99
Rizopoulos, D., 26, 27
Roberts, G.O., 108
Rogers, W.T., 16, 64
Ronchetti, E., 93
Rosenbaum, P.R., 100
Rosenbluth, A.W., 105
Rosenbluth, M.N., 105
Rost, J., 16, 157, 186
Rubin, D.B., 17, 53, 65, 90, 108, 109, 162, 231
Ryan, L., 93, 192

Sörbom, D., 142, 143, 154, 192, 218
Samejima, F., 136
Sammel, R.D., 93, 192
Saris, W.E., 213
Satorra, A., 142
Savage, L.J., 43
Schönemann, P.H., 58, 215
Schilling, S., 93, 94
Schneeweiss, H., 43, 46, 239
Schuessler, K.F., 46, 84, 101, 112
Schumacker, R., 26
Schwarz, G., 59, 201
Sclove, S., 202
Scott Long, J., 213
Seber, G.A.F., 16, 56, 57
Shi, J.-Q., 31, 68, 143, 148
Shield, S.A., 155, 177
Singer, B., 18
Singleton, K.J., 58
Skrondal, A., 12, 93
Smith, G.A., 17, 31, 75, 102
Smith, A.F.M., 102
Sobel, M.E., 3, 227
Song, X.-Y., 26, 31, 94
Spearman, C., 12, 237
Spiegelhalter, D., 108, 112
Stanley, G., 75
Steele, F., ii, 99, 243
Steinberg, L., 122, 123
Stephenson, W., 242
Stout, W., 43, 100
Stronkhorst, L.H., 213
Swaminathan, H., 16

Tabachnick, B.G., 70
Takane, Y., 89
Tanaka, J.S., 59
Tanner, M.A., 31
Tateneni, K., 56, 71
Teller, A.H., 105
Teller, E., 105
ten Berge, J.M.F., 62
Thayer, D.T., 53, 56
Thissen, D., 99, 110, 115, 116, 122, 123, 149, 155, 160
Thomas, A., 108, 112
Thomson, G.H., 13, 15, 73, 237, 242, 243
Thurstone, L.L., 14, 15, 237
Tipping, M.E., 232, 233
Titterington, D.M., 17
Tucker, L.R., 56
Tuma, N.B., 18
Tzamourani, P., 39, 98, 112, 113

Upton, G.J., 42

van de Pol, F., 98
Van der Kooij, J.J., 236
van der Linden, W., 136
van Driel, O.P., 67
Vandenberg, M., 99
Verhelst, N., 186
Vermunt, J.K., 160, 164, 177
Victoria-Feser, M.P., 93
von Eye, A., 184

Wall, M., 26
Wang, H., 239
Wedel, M., 180
Welch, P., 108
Wermuth, N., 216

West, M., 31, 109
White, H., 27
Whittaker, J., 216
Whittle, P., 17
Wichern, D.W., 56
Williams, J.S., 46, 245
Wold, H., 239
Wollack, J.A., 31
Wong, W.H., 31
Wright, B.D., 140
Wright, D.B., 204
Wright, S., 216
Wu, C .F., 162
Wugalter, S.E., 98

Yang, F., 26

Zegers, F.E., 62
Zhu, H., 26, 94
Zucchini, W., 18

Subject index

ability testing, 112
acceptance/rejection sampling, 31
accident proneness, 17
adaptive quadrature, 93
Akaike's information criterion (AIC), 58, 79, 201, 207
alpha factor analysis, 70
AMOS, 213
assumption (or axiom) of conditional (or local) independence, 7
asymptotic standard errors, 29, 38, 141, 148, 165, 166, 182, 205
asymptotic theory of maximum likelihood, 55
asymptotic variance–covariance matrix, 94, 99, 142, 165
attenuation, 241
attitude scaling, 109
average posterior variance, 101

Bartlett scores, 73, 227, 242
Bayes estimates, 31
Bayes' theorem, 7, 41, 163
Bayesian approach, 32, 68
Bayesian estimation, 26, 30, 31, 102
Bayesian inference, 30, 43, 102
Bayesian information criterion (BIC), 59, 201
Bayesian minimal sufficient statistic, 192
Bernoulli distribution, 20, 193, 196
Bernoulli random variable, 4, 21, 24, 196
binary data
  examples, 167
  latent trait models, 83
  special case, 134
binary response, 33, 83, 86, 87, 95, 100
binomial distribution, 193
body satisfaction, 154, 177
bonds model, 13, 237, 238
bootstrap, 38, 56, 57, 112, 201
bootstrap samples, 98
BUGS, 108, 112, 247
Burt matrix, 236
business confidence, 2

canonical correlation, 70
categorical responses, 144
categorical variables, 236
category scores, 124, 127, 141, 149, 150, 152, 154
Cauchy inequality, 73
causal relationships, 216, 239
CEFA, 56, 71, 247
chi-squared test, 97, 110, 221
choice model, 146
cluster analysis, 244
coefficient alpha, 45
common factor, 13
communality, 49, 75, 77
comparative fit index (CFI), 221
complementary log link function, 40

Latent Variable Models and Factor Analysis: A Unified Approach, Third Edition.
David Bartholomew, Martin Knott and Irini Moustaki.

component score, 112, 113, 152, 153
components, 21, 26, 37, 41, 42, 49, 86, 121, 151, 166, 167, 193–195, 203, 227, 228, 240, 241
conditional density, 6, 7
conditional distributions, 8, 19, 22, 27
conditional expectation, 203
conditional probability, 120, 123, 137, 140
confidence intervals, 56
confirmatory factor analysis, 50, 213, 218
confirmatory mode, 215
conservatism, 2
consistency, 64
continuous variable, 87
convergence, 92, 94
convergence diagnostics, 108
correlated latent variables, 213
correlation coefficient, 141, 149, 152
correlation matrix, 13, 14, 50, 59, 61, 65, 66, 69, 70, 79, 85, 93
correspondence analysis, 124, 127, 149, 151
covariance matrix, 28, 29, 35, 42, 49, 58, 59, 65, 66, 97, 99, 133, 214, 219, 224, 227, 230, 232, 234, 235, 240, 243
covariance structure, 224
Cramér–von Mises statistic, 108
Cronbach's alpha, 45
cross-validation, 57
cumulant generating function, 42
cumulative probability, 171, 208
cumulative response function, 137

data augmentation, 103
density function, 146
dependent latent variables, 218, 219, 223–225
diagonal elements, 128
diagonal matrix of variances, 9
diagonality condition, 36
difficulty parameter, 43
dimension reduction, 1, 22
dimensionality, 1
discrete (multivariate) prior distribution, 24
discrimination parameter, 43, 86
distribution function, 86, 89, 90
distribution theory, 8
distribution-free, 27, 28
divergence of estimation algorithm, 109

E-M algorithm, 53, 90, 92, 93, 112, 119, 125, 138, 148, 162, 164, 174, 182, 189, 195, 199, 200
econometric models, 18
educational testing, 16, 83, 84, 86, 109
eigenvalues, 130, 132
eigenvectors, 130, 132
employment data, 153
endogenous latent variables, 218, 223
EQS, 16, 213
equivalence, 87, 88, 127, 145
exogenous latent variables, 218, 223
exponential
  parameter, 27
exponential family, 20, 23, 27, 29, 32, 41, 48, 85, 101, 121, 136, 185, 191, 192, 195, 196, 199, 220, 222, 223, 227
  one-parameter, 19, 29
exponential type, 22
extreme-value distribution, 145–147
Eysenck's Personality Questionnaire, 114

factor
  general, 14
  loadings, 86
  specific, 14
factor analysis, 1, 9, 10, 12
factor loadings, 31, 35, 47, 49, 55, 56, 69, 70, 76, 117
factor model, 10, 13, 16, 24
factor scores, 41, 42, 44, 46, 71, 72, 74
factor space, 69, 71
factors, 2
finite mixture model, 17

fit indices, 221
frequency distribution, 84, 112

gamma distribution, 18, 20, 105, 194
Gauss–Hermite quadrature, 92, 93, 126, 139, 187
general intelligence, 2
general linear latent variable model (GLLVM), 19–21, 23–25, 27, 33–35, 37, 38, 44, 86, 100, 121, 161, 191–193, 195, 198, 213
  fitting by maximum likelihood, 29
  fitting the models, 27
  interpretation, 35
  properties, 20
  sampling error, 38
  special cases, 21, 22, 24
  vector-valued, 121
generalisability, 70
generalisability coefficient, 46
generalised least squares, 60, 65, 90, 93, 94, 96, 219
generalised linear model, 20
generalised partial credit model, 140
generally weighted least squares, 220
Gibbs sampler algorithm, 33, 94, 143
Gibbs sampling, 31, 102, 103, 106, 112
GLLAMM, 247
goodness of fit, 58, 63
goodness-of-fit test, 5, 6, 27, 29, 65, 67, 69, 78, 79, 84, 97–99, 113, 117, 148, 149, 152, 154, 167, 176, 201
graded response model, 136
graphical models, 216

Heywood cases, 31, 50, 51, 53, 67, 69, 75, 77, 109
hidden factor, 4
hidden variables, 18
Hillsborough football disaster, 204
Hotelling's method, 229
hybrid models, 11, 186

identifiability, 178, 180, 223, 227
image factor analysis, 63
incremental fit index (IFI), 222
independence chain, 105
independent latent variables, 218, 219, 223–225
index of generalisability, 45
indistinguishable models, 224
inference, 32, 33, 40, 41
information matrix, 27, 94
interaction, 26
intercept, 85
intercultural version, 112
IRTPRO, 110, 116, 149, 155, 160, 247
isotonic regression, 211
isotropic normal model, 231
item characteristic curve, 85
item response curve, 43, 161
item response function, 85, 100, 161
iterative weighted least squares analysis, 126

jackknife, 56
Jacobian, 180
joint distribution, 84, 100, 102, 222, 224, 225
joint probability, 162, 188
joint probability function, 171, 172

Kaiser–Guttman criterion, 64

label switching, 31, 32, 109
Laplace approximations, 93
latent class analysis, 12
latent class model, 17, 22, 24, 41, 83, 98, 109, 157, 223
  allocation of individuals, 174
  for binary data as a latent trait model, 159
  ordered, 171
  unordered variables, 170
  with binary manifest variables, 158
Latent GOLD, 160, 164, 177, 247
latent profile models, 24, 181, 184, 226
latent structure analysis, 16, 157
latent trait models, 16, 24, 98, 149, 157, 159–161, 166, 176, 186, 200, 223

latent variable methods
broader theoretical view, 6
classification, 11, 24
closely related fields, 17
historical context, 12
statistical problem, 1
latent variables, 2, 200
intercorrelations, 217
relationships between, 213
Law School Admission Test, 84, 109–111, 160, 165
least squares
posterior analysis, 72
least squares methods, 59
choice of q, 63
life satisfaction data, 176, 177
likelihood function, 27, 29
approximation, 95, 96
limited-information estimation method, 154
limited-information goodness-of-fit test, 99, 113
limited-information maximum likelihood estimation, 142
linear factor models, 226
linear normal isotropic model, 232
linear structural relations modelling (LISREL), 16, 154, 192, 213, 214, 239, 242, 247
link function, 21, 25
LISREL model, 218, 224, 226, 227
generalisations, 223
measurement model, 219
structural model, 218
loading matrix, 35–37
log-likelihood, 129, 130, 138, 162, 172, 187
bivariate, 143
univariate, 143
log-likelihood ratio statistic, 67, 97
logistic distribution, 141, 147
logistic response functions, 85
logit model, 119, 121
polytomous data, 123, 124
logit response function model, 145
logit/normal model, 84, 87, 90, 100–102, 109, 123, 136, 144, 159, 161, 187, 215
manifest variables, 3, 11
of mixed type, 191
marginal frequencies, 128, 129, 148
marginal probabilities, 84, 88, 98
Markov chain, 18, 30, 33, 102, 105, 108
Markov chain Monte Carlo (MCMC), 30–32, 102, 108
mastery model, 167
maximum likelihood
composite, 143
maximum likelihood estimation, 22, 27, 29, 39, 44, 48, 50, 52, 53, 59, 61, 64, 65, 77, 81, 90, 92, 94, 96, 126, 140, 142, 151, 162, 181, 184, 187, 195, 214, 220, 223
binary data, 172
marginal, 31
polytomous logit model, 125, 236
with ordinal variables, 138
maximum likelihood estimators, 17, 126
approximate, 95, 127
sampling properties, 94
maximum likelihood method, 141
measurement model, 218, 223, 227, 239
metrical variables, 6, 11
Metropolis–Hastings algorithm (M-H), 31, 33, 102, 103, 105, 106, 112
minimum trace, 62
minres method, 62
missing values, 110
mixed latent class models, 202
moment generating function, 42
Monte Carlo E-M (MCEM), 93
Monte Carlo integration, 32, 94
Monte Carlo methods, 93
Mplus, 16, 56, 70, 71, 75, 77, 78, 81, 192, 213, 214, 247
multi-group factor analysis, 44
multinomial distribution, 193

multinomial probability, 129
multiple correspondence analysis, 136, 149, 236
multiple correspondence scores, 236
multivariate data analysis, 1
multivariate normal distribution, 8, 9

natural parameter, 23
Newton–Raphson, 92
nominal model, 148
non-linear models, 25–27
non-linear principal components analysis, 236
non-metric multidimensional scaling, 243
non-normed fit index (NNFI), 222
normal distributions, 6, 10
normal linear factor model (NLFM), 24, 38, 47, 87, 119, 134
  approximate methods for estimating Ψ, 62
  constraints, 50
  distributional properties, 48
  fitting without normality assumptions, 59
normal random variables, 48, 196
normalising constant, 32, 103, 105, 106
nuisance parameter, 194–196

oblique rotations, 35, 56, 69, 70, 77, 214
  geomin, 81
  OBLIMIN, 70, 77
one-dimensional latent space, 26
one-factor latent trait model, 208
one-factor logit/normal model, 117
one-factor model, 26, 46, 103, 150
one-to-one transformation, 9, 37
operant subjectivity, 244
ordered categorical data, 236
ordered categorical variables, 208
ordered latent classes, 185
ordering of categories, 136
orthogonal rotation, 34, 38, 56, 69, 132
orthogonal transformation, 10, 124, 132

pairwise residuals, 113
parametric bootstrap method, 98, 165
partial credit model, 140
partial least squares (PLS), 239
path analysis, 239
pattern loading matrix, 70
Perron–Frobenius theorem, 130
phi-coefficient matrix, 95, 96, 136
physical data, 234
physical variables, 79
PLS, 239
Poisson distribution, 17, 194
Poisson variable, 29
political action survey, 154
polychoric correlations, 141, 142, 191, 209
polyserial correlations, 191, 209
polytomous data
  analysis using logit model, 149
  latent trait models, 119
  more than one factor, 130
  one factor, 127
  unordered, 170, 174
polytomous logit model
  maximum likelihood estimation, 125
positive response, 86, 122
posterior analysis, 20, 41, 100, 113, 147, 203, 228
  least squares, 72
  reliability, 74
  the normal case, 71
  with binary manifest variables, 166
  with metrical manifest variables, 180
  with unordered manifest variables, 171
posterior covariance matrix, 71
posterior density, 141
posterior density function, 120
posterior distribution, 5, 9, 20–22, 30–33, 39, 42, 43, 100, 102
posterior mean, 100, 101
posterior probability, 41

posterior variance, 71, 101
principal axis method, 63
principal components, 11, 12, 21, 96, 136, 228, 230
principal components analysis (PCA), 16, 56, 61, 64, 229, 233, 235, 236, 239, 242
principal factor method, 56, 61
principle of precise measurement, 43
prior distribution, 9, 19, 22, 32, 33, 39, 43, 44, 85, 88, 100, 108, 159
prior information, 31
probit link, 154
probit/normal model, 31, 86, 93, 96, 102, 103, 123, 136, 154
Procrustes methods, 215
proportional odds model, 137, 140, 154
psychometric inference, 44
Psychometrika, 14

Q-analysis, 15, 46, 242, 243
Q-methodology, 242
quality of life, 2
quartimax, 69

R-methodology, 242
random effects Rasch model, 44, 86
random score, 40
random variables, 6
random walk chain, 105
Rasch model, 136, 140, 215
reference category, 121
regression coefficients, 217, 227
regression scores, 74
reification, 2
rejection probability, 105
rejection sampling, 103, 106
relative fit index, 221
reliability, 74, 101
reproduced covariance matrix, 221
residuals, 201
response
  binomial, 29
  gamma, 28
response function, 21, 34, 46, 85, 87, 88, 109, 120, 122, 161
response function model, 136, 141, 143, 148, 149
response patterns, 83, 84, 92, 98, 99
  pooling, 148
reversible jump MCMC, 31
root mean square error of approximation (RMSEA), 221
rotated factors, 56
rotated solution, 95, 184
rotation, 9, 31, 33, 35, 69, 124, 214
  see also oblique and orthogonal
rotational indeterminacy, 109

S-Plus, 247
saddle point, 164
sample correlation matrix, 64, 66, 69
sample covariance matrix, 96, 198, 219–221
sample size, 56
sample standard deviations, 65
sampling behaviour, 55
sampling error of parameter estimates, 38
sampling properties
  of maximum likelihood estimators, 201
sampling variables, 43, 46
sampling variation of estimators, 55
sandwich estimator, 27
scale-invariant estimators, 66
scaled components, 242
score function, 53–55
Scottish infants, 174
scree plot, 79
scree test, 64
shape parameters, 39
simple structure, 36, 37
simulated states, 33
single binary latent variable, 4
single latent variable, 122, 138
singular value decomposition, 37
small industry data, 152

Smith and Stanley data, 75
social life feelings, 84, 112
social surveys, 1
specific factor, 13, 237
specific variances, 47
SPSS, 229, 247
staff assessment data, 149
standard deviations, 56
standard errors, 27, 55, 56, 94, 110, 112, 119, 150, 164, 214
standard form, 66
standard linear factor model, 140
standard normal distribution, 9
standardised χ^2 local dependence (LD) index, 115, 155
standardised loadings, 69, 75
standardised root mean square residual (SRMR), 221
STATA, 247
stationarity conditions, 51
stationary distribution, 33
statistical inference, 44
stochastic subject rationale, 41
structural equation modelling, 239
structural model, 40, 218, 227
structural path analysis, 228
structural relationships, 218, 222
structure loading, 70
sufficiency, 87, 121
sufficiency principle, 22, 119, 141, 208–210
sufficient statistic, 25, 26, 37, 48, 54, 100, 159, 228, 240
  q-dimensional, 22
  minimal, 21–23

Taylor expansion, 127, 129, 131
terracotta data, 207
test–retest correlation, 101, 113
tetrachoric correlations, 93, 97, 191
tetrad differences, 13
thresholds, 140
tobit relation, 210
transition kernel, 33, 105
transition probabilities, 30
two-factor model, 26, 27, 79, 154
two-factor solution, 155
Type I distribution, 146

underlying factor model, 142
underlying latent variable model, 228
underlying variable, 96, 191
underlying variable approach
  and response function, 88
  modelling categorical data, 87
underlying variable model, 89, 90, 119, 136, 140–143, 145, 147, 148, 206, 208
  alternative, 144
  binary, 136
unidimensionality, 100
univariate distributions, 84
unobserved heterogeneity, 18
unordered categorical data, 235
unweighted least squares, 219

varimax, 69

weighted least squares, 142
welfare measurement, 207
WinBUGS, 247